Physics of Ion–Ion and Electron–Ion Collisions

NATO ADVANCED STUDY INSTITUTES SERIES

A series of edited volumes comprising multifaceted studies of contemporary scientific issues by some of the best scientific minds in the world, assembled in cooperation with NATO Scientific Affairs Division.

Series B: Physics

Recent Volumes in this Series

Volume 76 — Metal Hydrides
edited by Gust Bambakidis

Volume 77 — Nonlinear Phenomena at Phase Transitions and Instabilities
edited by T. Riste

Volume 78 — Excitations in Disordered Systems
edited by M. F. Thorpe

Volume 79 — Artificial Particle Beams in Space Plasma Studies
edited by Bjørn Grandal

Volume 80 — Quantum Electrodynamics of Strong Fields
edited by W. Greiner

Volume 81 — Electron Correlations in Solids, Molecules, and Atoms
edited by Jozef T. Devreese and Fons Brosens

Volume 82 — Structural Elements in Particle Physics and Statistical Mechanics
edited by J. Honerkamp, K. Pohlmeyer, and H. Römer

Volume 83 — Physics of Ion-Ion and Electron-Ion Collisions
edited by F. Brouillard and J. W. McGowan

Volume 84 — Physical Processes in Laser-Materials Interactions
edited by M. Bertolotti

Volume 85 — Fundamental Interactions: *Cargèse 1981*
edited by Maurice Lévy, Jean-Louis Basdevant, David Speiser, Jacques Weyers, Maurice Jacob, and Raymond Gastmans

This series is published by an international board of publishers in conjunction with NATO Scientific Affairs Division

A Life Sciences B Physics	Plenum Publishing Corporation London and New York
C Mathematical and Physical Sciences	D. Reidel Publishing Company Dordrecht, The Netherlands and Hingham, Massachusetts, USA
D Behavioral and Social Sciences E Applied Sciences	Martinus Nijhoff Publishers The Hague, The Netherlands

Physics of Ion–Ion and Electron–Ion Collisions

Edited by

F. Brouillard
Institute of Physics
Catholic University of Louvain
Louvain–La–Neuve, Belgium

and

J. W. McGowan
University of Western Ontario
London, Ontario, Canada

PLENUM PRESS • NEW YORK AND LONDON
Published in cooperation with NATO Scientific Affairs Division

Library of Congress Cataloging in Publication Data

Main entry under title:

Physics of ion-ion and electron-ion collisions.

(NATO advanced study institutes series. Series B, Physics; v. 83)
"Proceedings of a NATO Advanced Summer Institute on the Physics of Ion-Ion and
Electron-Ion Collisions, held Sept. 13–26, 1981, in Baddeck, Nova Scotia, Canada"—
T.p. verso.
Bibliography: p.
Includes index.
1. Electron-ion collisions-Congresses. 2. Ion-ion collisions—Congresses. I. Brouil-
lard, F., 1937–   . II. McGowan, J. W. III. NATO Advanced Summer Institute on
the Physics of Ion-Ion Collisions (1981: Baddeck, N.S.) IV. Title. V. Series.
QC794.6.C6P49 1982   539.7′54   82-13309
ISBN 0-306-41105-9

Proceedings of a NATO Advanced Summer Institute on the Physics of Ion-Ion and
Electron-Ion Collisions, held September 13–26, 1981,
in Baddeck, Nova Scotia, Canada

©1983 Plenum Press, New York
A Division of Plenum Publishing Corporation
233 Spring Street, New York, N.Y. 10013

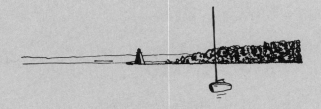

We're outward bound from BADDECK pier
Skipper LUKE came aboard with a crate of beer

    Stormy weather boys, Stormy weather boys
    when the wind blows the boat will go.

And at the helm is YONG-KI-KIM
The little cod fishes aren't afraid of him

We're a cautious crew, for the boat's on hire
And the oatmeal porridge has caught on fire

Then up jumped a mermaid covered in slime
So we took her down the foc'sle and we had a good time

At last we're home, both safe and sound
Our skipper hasn't run the boat aground

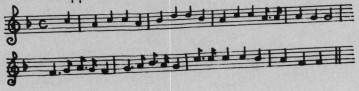

- Johannes Hasted composuit

High above the acid rainfall, Where the sun must sometimes shine
There the atoms in their glory, Ionize and recombine

Oh my darlings, Oh my darlings, Oh my darlings ions mine
Thou art lost and gone forever if but once you recombine

In a cavity, in a discharge, Back in nineteen forty-nine,
FRED BIONDI waved his wand-y And the ions recombine

In the labs of PEART and DOLDER, By the banks of coaly TYNE,
There's inclined beams, intertwined beams, There are beams of either sign.

At the plasma physics meetings, Mid the crates of beer and wine
NORMAN BARDSLEY, TOM O'MALLEY, Dissociate and recombine.

Tokamaks and Stellagators, Eating Taxes, yours and mine,
And all that they are heating Is Tungsten twenty-nine.

Recombination

1981

Air-Clementine (Trad)

CAPE

Baddeck

Sydney

BRETON

ISLAND

- Johannes Hasted composuit

PREFACE

Some of the earliest civilizations regarded the universe as
organized around four principles, the four "elements" earth, water,
air, and fire. Fire, which was the source of light and as such
possessed an immaterial quality related to the spiritual world, was
clearly the most impressive of these elements, although its quantitative importance could not have been properly discerned. Modern science has changed the names, but macroscopic matter is still
divided into four states. The solid, liquid, and gaseous states
are ordinary states, but the fourth state of matter, the plasma
state, has retained a somewhat extraordinary character. It is now
recognized that most of the matter of the universe is in the ionized
state, but on the earth, the plasma state is still the exception.
Hence the importance and also the difficulty of investigations
dealing with ionized matter, which have been greatly furthered by
the development of thermonuclear fusion research.

The study of matter in the ionized state comprises a large
diversity of problems belonging to many different branches of physics. A number of them relate to the microscopic properties of
plasmas and concern the structure and the collisional behavior of
atomic constituents. Although they are clearly of basic importance,
their relevance to thermonuclear research was at first overlooked,
at a time when most of the effort was concentrated on designing
fusion devices and understanding macroscopic phenomena, mostly of
an electromagnetic nature. At present, however, increased maturity
and a more systematic approach to the fusion problem have revealed
that better knowledge of the underlying atomic collisional processes
is an urgent necessity.

Very specific to the collisional processes in an ionized gas are, of course, the collisions between charged particles, i.e., electron-ion and ion-ion collisions. But electron-ion and ion-ion collisions are also specifically more difficult to investigate than collisions with neutral species. In theoretical studies the scarcity of data regarding the excited states of ions is a major handicap. In experimental studies one has to face the difficulty of working with comparatively low densities of colliding species and vanishingly small observable signals. Nevertheless, significant progress has been achieved in recent years, both experimentally and theoretically.

This book is devoted to some of the major aspects of current research in the field of ion-ion and electron-ion collisions. The relevance of the work to astrophysics and thermonuclear fusion is dealt with in the two first lectures. Electron-ion collisions are introduced in two theoretical lectures and developed in three specific lectures on excitation, ionization, and recombination. Ion-ion collisions are treated in three lectures, one dealing with theory and two with experimental aspects. The physics of ion traps and of flowing afterglows are the subjects of the two last lectures.

This book is one of the achievements of the NATO Advanced Study Institute on the Physics of Ion-Ion and Electron-Ion Collisions, held in Baddeck, Nova Scotia, Canada, September 13-26, 1981. Another--less visible but nevertheless considerable--achievement was the intense interaction between lecturers and students during two full weeks of continuous and unrestricted dialogue, in which many durable collaborations were established. We want to thank all the lecturers, not only for the high quality of their lectures but for their total involvement, which was impressive and of immense value to the students.

It wouldn't be fair not to mention J.N. Bardsley, whose manuscript unfortunately could not be included in this book, but who contributed equally with the other lecturers to the success of the Institute.

We are much indebted to Parks Canada, for the kind hospitality received at the Alexander Graham Bell Historic Museum, where the Institute took place. Also, we gratefully acknowledge the help provided by the following:

Canadian Natural Science and Engineering Research (NSERC)
Euratom
The Province of Nova Scotia
Cape Breton Development Corporation (DEVCO)
College of Cape Breton

as well as the support given to us by the Université Catholique de Louvain and the University of Western Ontario.

Finally, we express our gratitude to Marilyn Adams and Nicole Coisman, who took such good care of the administrative work.

The Scientific Affairs Division of NATO in Brussels has earned our appreciation. We would like to express it specifically to its director, Dr. di Lullo, who encouraged us and helped us to organize the Institute.

                                                                F. Brouillard
December 1981   J.W. McGowan

CONTENTS

Electron-Ion and Proton-Ion Collisions in Astrophysics 1
   A. Dalgarno

The Role of Atomic Collisions in Fusion. 37
   D.E. Post

Theory of Electron-Atom Collisions 101
   Y.-K. Kim

Potential Energy Curves for Dissociative Recombination 167
   S.L. Guberman

Electron-Impact Excitation of Ions 201
   D.H. Crandall

Electron-Impact Ionization of Ions 239
   E. Salzborn

Experimental Studies of Electron-Ion Recombination 279
   J.B.A. Mitchell and J.W. McGowan

Theory of Low Energy Ion-Ion Charge Exchange 325
   A. Salin

The Measurement of Inelastic Ion-Ion and Electron-Ion
     Collisions 373
   K.T. Dolder

On the Measurement of Ion(Atom)-Ion(Atom) Charge Exchange. . . 415
   F. Brouillard and W. Claeys

Confinement of Ions for Collision Studies. 461
   J.B. Hasted

Studies of Ion-Ion Recombination Using Flowing Afterglow
     Plasmas 501
   David Smith and Nigel G. Adams

Index 533

ELECTRON-ION AND PROTON-ION COLLISIONS IN ASTROPHYSICS

A. Dalgarno

Harvard-Smithsonian Center for Astrophysics
Cambridge, Massachusetts

INTRODUCTION

Most of the diffuse material in the Universe exists in partly
ionized plasmas. In the interstellar medium, the physical conditions
range from the cold weakly-ionized dense molecular clouds for
which the ionization source is provided by energetic cosmic rays
to the hot highly-ionized dilute galactic corona for which the
ionization source is provided by thermal electrons, heated by
shocks driven by supernova explosions in the disc of the galaxy.
Recombination radiation created following the shock wave produces
X-ray and ultraviolet photons whose absorption above the plane
also contributes to the coronal heating. In planetary atmospheres,
solar radiation is the primary ionization source supplemented at
low altitudes by galactic cosmic rays which become relative more
important for the distant planets.

In the ionized plasmas, the recombination of electrons and
ions determines the ionization structure and the associated
recombination spectrum constitutes a powerful diagnostic probe
of the physical environment in which the emitting species
resides. The recombination processes contribute to the heating
and cooling of the ionized regions and together with subsequent
reactions modify the chemical composition.

1

   In my lectures I will describe some of the diverse variety
of astrophysical phenomena in which collisions of electrons and
protons with ions are significant. I hope to point out areas of
uncertainty where progress depends upon a more detailed understanding of electron-ion processes.

THE EARLY UNIVERSE

   Atomic physics began with the radiative recombination of
electrons and protons

$$e + H^+ \rightarrow H' + h\nu \tag{1}$$

to form neutral hydrogen atoms. In the scenario of the big-bang
cosmology (Weinberg 1972), the universe expanded from an initial
singularity. The earliest moments were controlled by processes
involving the fundamental particles. Nucleosynthesis became
significant once the temperature had fallen below 5×10^9 K and
^{4}He and small amounts of deuterium, ^{3}He and ^{7}Li were produced.
Matter and radiation were maintained in thermal equilibrium by
Thomson scattering of electrons and photons until the temperature
reached values of about 4 000 K. At this point of the evolution,
radiative recombination occured and matter changed rapidly from
a fully-ionized to a mostly neutral state, composed of hydrogen
and helium atoms, some electrons, some protons and many photons.
Table 1 illustrates the sequence of events.

   Following reaction (1), there occurred radiative attachment

$$e + H \rightarrow H^- + h\nu \tag{2}$$

and radiative association

$$H + H^+ \rightarrow H_2^+ + h\nu . \tag{3}$$

The negative ions H^- were destroyed by photodetachment,

$$H^- + h \rightarrow H + e \tag{4}$$

by associate detachment

$$H^- + H \rightarrow H_2 + e \tag{5}$$

and by mutual neutralization

$$H^+ + H^- \rightarrow H + H. \tag{6}$$

Collisions of H^+ and H^- also lead to the formation of H_2^+,

$$H^+ + H^- \rightarrow H_2^+ + e. \text{ *} \tag{6a}$$

---

* This reaction was drawn to my attention by Dr. W. Claeys. The
  cross sections have been measured by Poulaert et al. (1978).

TABLE 1.

------------------------- 10^{12} K

Particle Physics

100 s ------------------------------------ 5×10^9 K

^{4}He, D, ^{3}He, ^{7}Li

Nuclear Physics

1 000 s --------------------------- 5×10^8 K

10^5 years ---------------------- 4×10^3 K

Atomic Physics

Recombination era

$$H^+ + e \quad \rightarrow \quad H' + h\nu$$

H, He, e, H^+, He^+, ν

10^6 years   --------------------

Pre-galactic gas clouds

First generation of stars.

The molecular ions H_2^+ were converted to neutral molecules by reactions with H,

$$H_2^+ + H \quad \rightarrow \quad H_2 + H^+ \tag{7}$$

or were destroyed by dissociative recombination

$$H_2^+ + e \quad \rightarrow \quad H + H \tag{8}$$

and by radiative recombination

$$H_2^+ + e \quad \rightarrow \quad H_2' + h\nu. \tag{9}$$

Progress in determining the variation of the ionization with red shift and the residual electron and molecular abundances is impeded by uncertainties in the rate coefficients of reactions (7) and (8), both of which depend upon the vibrational population of H_2^+.

The rate coefficient of reaction (7) has been measured to be 6×10^{-10} cm^3s^{-1} at a temperature of about 350 K in an experiment in which the vibrational population of the H_2^+ ions is unknown but probably approximates a Franck-Condon distribution (Karpas, Anicich and Huntress 1979). The rate coefficients for specific vibrational levels have not been determined.

The cross sections for reaction (8), dissociative recombination, have been measured over an extensive velocity range by Auerbach et al. (1977) but again for ions in an unknown vibrational distribution. In contrast to reaction (7), theoretical calculations may soon provide reliable estimates of the individual rate coefficients (Giusti, Derkits and Bardsley 1981).

The H_2^+ ions are formed by radiative association, reaction (3). The total rate coefficients have been calculated by Bates (1951) and by Ramaker and Peek (1976) but the values for populating specific vibrational levels were not presented. Presumably the process populates preferentially the high vibrational levels.

The molecular ions are embedded in a blackbody radiation field characterized by a temperature of about 4 000 K. There appears to be no effective sequence of electric dipole transitions between the ground and excited electronic states which could lead to vibrational enhancement. Indeed dipole absorption tends destroy the ions by photodissociation

$$H_2^+ + h\nu \rightarrow H + H^+ \tag{10}$$

in a $1s\sigma_g-2p\sigma_u$ transition. Cross sections for (10) have been calculated by Bates (1952), Buckingham, Reid and Spencer (1952), Dunn (1968) and Argyros (1974). Transitions between rotation-vibration levels of the ground electronic state can occur but because H_2 is homonuclear only by electric quadrupole absorption, an extremely inefficient process. Thus radiative association is the main source of vibrationally excited ions.

The vibrational populations are modified by radiative decay and by vibrational energy transfer in collisions. The lifetimes for radiative decay by electric quadrupole transitions decrease from 1.92×10^6 s for the $v = 1$ level to 4.7×10^5 s for $v = 7$ and 8 and thereafter increase to 1.8×10^8 s for $v = 19$ (Bates and Poots 1953, Peek, Hashemi-Attar and Beckel 1979). Of the collision processes,

$$H + H_2^+(v) \rightarrow H + H_2^+(v') \tag{11}$$

is likely to be the most effective. Its rate coefficient is not known. At the recombination epoch, the particle density was about 10^3 cm^{-3} so that if the rate coefficient is as large as 10^{-10} cm^3 s^{-1}, the long lived high-lying levels will be modified by collisions and will tend to a relative population characterized by the kinetic temperature. If the rate coefficient is less than 10^{-11} cm^3 s^{-1}, the molecules decay radiatively before they undergo any collisions and the ions are mostly in the v = 0 vibrational level.

For H_2^+ (v = 0) ions, reactions (7) and (8) may be slow and the ions may survive until removed by radiative recombination

$$H_2^+ + e \rightarrow H_2' + h\nu. \tag{12}$$

The excited H_2' either predissociates or radiates to a lower discrete or continuum state. It is usually argued that because radiative recombination populates high n Rydberg levels the rate coefficient of (12) is not greatly different from that of reaction (1). However, analogous to dielectronic recombination of complex ions, the rate of (12) may be enhanced by the capture into vibrationally excited Rydberg levels, the same levels which participate in the indirect mechanism for dissociative recombination (cf. Bardsley 1968). Mutual neutralization

$$H_2^+ + H^- \rightarrow H + H + H \tag{13}$$

will also limit the abundance of H_2^+ ions.

This picture of a hydrogen plasma in the recombination era is modified by the presence of helium which is believed to exist in a primeval ratio of helium to hydrogen of about 0.23 by mass. Recombination of helium ions

$$He^+ + e \rightarrow He + h\nu \tag{14}$$

populates the metastable 2^3S state in a fraction 0.75 of the captures. Penning ionization

$$He(2^3S) + H \rightarrow He + H^+ + e \tag{15}$$

and associative ionization

$$He(2^3S) + H \rightarrow HeH^+ + e. \tag{16}$$

occur. The HeH$^+$ molecular ion can also be formed by radiative association

$$He^+ + H \rightarrow HeH^+ + h\nu \tag{17}$$

and by reaction with H_2^+ ions in vibrational levels v $\geq$ 3,

$$H_2^+(v \geq 3) + He \rightarrow HeH^+ + H. \tag{18}$$

Destruction of HeH$^+$ occurs by the reserve reaction

$$H + HeH^+ \rightarrow He + H_2^+ (v \leq 3) \tag{19}$$

but dissociative recombination may be very slow. Dissociative
recombination of H_2^+ is achieved by entry into excited states
of H_2 whose potential curves cross that of H_2^+ at favorable
locations. No such curve exists for HeH and HeH$^+$ ions are
removed instead by reaction (19), by radiative recombination

$$HeH^+ + e \rightarrow He + H + h\nu \tag{20}$$

and by mutual neutralization

$$HeH^+ + H^- \rightarrow He + H + H. \tag{21}$$

These processes illustrated in Fig. 1. set the stage for the
development of condensations in the expanding cooling gas and
together with collision-induced molecular dissociation

$$H + H_2 \rightarrow H + H + H \tag{22}$$

played a major role in bringing about the gravitational collapse
of pre-galactic gas clouds leading to the first generation of
stars in which heavy elements were manufactured. A recent
discussion is given by Yoshii and Sabano (1979).

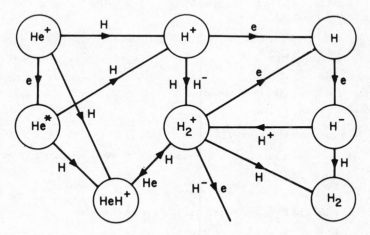

Fig. 1.   The chemistry of the early universe.

These processes also occur in gaseous nebulae and H II
regions with more readily detectable consequences. Recombination
lines in the optical, infrared and radio regions are the
principal features of the emission spectrum.

GASEOUS NEBULAE

A gaseous nebula is a diffuse region of interstellar gas,
ionized by ultraviolet radiation from a central source consisting
of a single star or of several stars. The temperatures vary
between 5 000 K and 15 000 K. Diffuse nebulae have electron
densities varying from 10 cm^{-3} to 10^4 cm^{-3} and are usually strong
emitters of the H I recombination spectrum produced by reaction
(1) as the levels populated by recombination cascade down. The
ionized regions are often several parsecs in diameter and their
masses lie between 10^2 and 10^4 M$_\odot$. A planetary nebula is a
denser form of gaseous nebulae with electron densities between
10^2 and 10^4 cm^{-3} and they are smaller with masses between 0.1
and 1.0 M$_\odot$. The spectrum contains the recombination lines of
H I, He I and He II. In diffuse nebulae, the stars responsible
for the ionization are O or early B-type stars with effective
temperature exceeding 25,000 K. In planetary nebulae, the
central star is highly evolved and is rapidly approaching the
white dwarf stage. Most of the ionized material in a planetary
nebula has been ejected from the star. The effective stellar
temperature often exceeds 10^5 K and singly and doubly ionized
helium are produced by photoionization.

Fig 2 illustrates the infrared spectrum seen towards the
planetary nebula NGC 7027 (Smith, Larson and Fink 1981).
Recombination lines of hydrogen originating in levels with
principal quantum numbers n up to 25 are present. Elaborate
theories have been developed of the radiative recombination
process (cf. Osterbrock 1974) and the theory is precise enough
to be used in conjunction with observation to infer the
reddening due to dust absorption and scattering (see for
example Seaton 1979).

Radio recombination lines produced by emission from high
n levels have been a prolific source of information on
astrophysical plasmas. They have been seen from levels with n
up to 390 in atomic hydrogen and in a wide range of objects.
In addition to diffuse and planetary nebulae, recombination
lines are observed from the diffuse interstellar gas, supernova
remnants, molecular clouds and external galaxies. The emission
line frequencies depend upon the mass of the emitter through the
Rydberg formula.

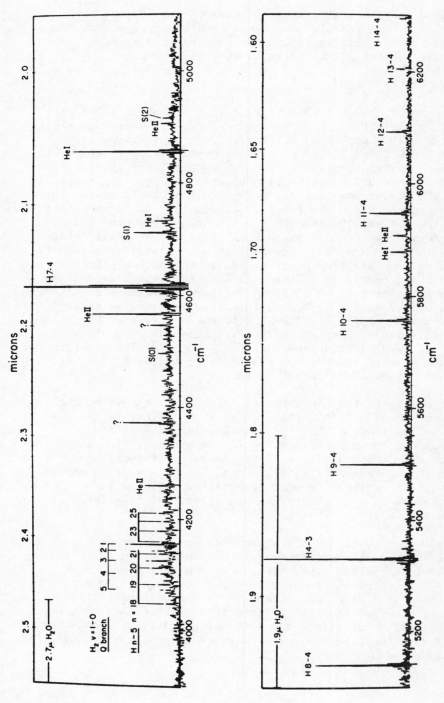

Fig. 2. The infrared spectrum seen towards the planetary nebula NGC 7027 showing recombination lines of hydrogen (from Smith, Larson and Fink 1981).

Thus

$$R_m = \frac{109737.31}{(1+m/M)} \text{ cm}^{-1}$$

for an emitter of mass M, and hydrogen, helium, carbon and a group of heavy elements have been detected. Extensive calculations have been carried out of the level populations in a wide variety of densities and temperatures in plasmas subjected to different external radiation fields. In low density plasmas, the population of the high n levels is affected by absorption of the 2.8° K black body background radiation field. A general computer program for calculating the level populations has been presented by Brocklehurst and Salem (1977). Recent observational studies are contained in the book "Radio Recombination Lines" edited by Shaver (1980).

The spectrum in Fig.2 contains in addition to recombination lines of atomic hydrogen, vibration-rotation lines of molecular hydrogen. The hydrogen molecules can be formed in an ionized gas by the gas phase sequences described earlier and if dust is present by association on the surfaces of the dust particles.

The sequence also produces H_2^+ ions. An additional source arises from photoionization of H_2

$$H_2 + h\nu \rightarrow H_2^+ + e. \tag{23}$$

Because dissociative recombination proceeds rapidly at high electron densities the equilibrium abundance of H_2^+ is small under steady state conditions (Black 1978). However Heap and Stecher (1981) and Feibelman et al (1981) have argued that anomalous extinction towards several planetary nebulae in the spectral region short of 1 500 Å may be attributed to absorption by H_2^+ in the v = 0 state

$$H_2^+(v = 0) + h\nu \rightarrow H + H^+. \tag{24}$$

Heap and Stecher (1981) suggested that the H_2^+ is produced by process (23) occurring as the ionization front advances through the expanding nebula.

The H_2^+ ions created by (23) are vibrationally excited (cf. Ford, Docken and Dalgarno 1975). Radiative decays will occur, terminating in the ground vibrational level. Dissociative recombination may have a similar consequence if it is significantly more rapid for $v \geq 1$ than for v = 0. If the fractional ionization in the shell is small, collisions with neutral hydrogen (11) will tend to produce a vibrational distribution in thermal equilibrium at the gas kinetic temperature.

If the ambient temperature is low, the H_2^+ ions will be driven into the lowest vibrational level, as Heap and Stecher (1981) suggest.

Collisions with helium will occur and a substantial source of HeH^+ may result from reaction (18). A detailed discussion of the formation and destruction of HeH^+ in astrophysical plasmas has been presented by Roberge and Dalgarno (1982) who extended the studies of Black (1978) and Flower and Roueff (1979).

The identification of H_2^+ as the source of the absorption is tentative. It receives some support from a recent detection of CO^+ (Erickson et al. 1981) towards the Orion molecular cloud OMC-1. Because CO^+ reacts rapidly with H_2, as does H_2^+, neither ion can exist in high abundance in cold undisturbed interstellar clouds. A sustained supply of neutral material, continuously exposed to ionizing radiation, appears to be required. The scenario described by Whitworth (1979) may be appropriate.

The spectra of diffuse and planetary nebulae show in addition to the hydrogen and helium recombination spectra emission lines of heavy elements in various stages of ionization. The excitations by electron impact of the metastable states of O III, N II and O II

$$e + X^{m+} \rightarrow e + X^{m+'} \tag{25}$$

are the dominant cooling processes as they transform kinetic energy into radiation energy and are responsible for maintaining the temperature at 5 000 K - 15 000 K. Radiative recombination

$$X^{m+} + e \rightarrow X^{(m-1)+'} + h\nu \tag{26}$$

is the main mode of recombination of the ionized atoms, but charge transfer recombination (cf. Butler, Heil and Dalgarno 1981, Butler and Dalgarno 1981)

$$X^{m+} + H \rightarrow X^{(m-1)+} + H^+ \tag{27}$$

and dielectronic recombination

$$X^{m+} + e \rightarrow (X^{(m-1)+})^* + h\nu \tag{28}$$

(cf. Seaton and Storey 1976) are important supplementary mechanisms. Charge transfer recombination is often the dominant mode for multiply-charged ions in plasmas produced by high frequency sources of ionizing radiation such as quasars and X-ray photo-ionized nebulae (cf. McCray, Wright and Hatchett 1977, Halpern and Grindlay 1980). The possible importance of dielectronic recombination at nebular temperatures was noted by Beigman and

Chichkov (1980) and has been demonstrated by Harrington, Lutz
and Seaton (1981) following a suggestion of Harrington et al.(1980).

Fig.3 reproduces the ultraviolet spectrum of the planetary
nebula NGC 7662, adapted from Harrington et al. (1981). The line
at 2297 Å arises from the $2p^2$ ^1D level of C III in a transition
to the $2s2p$ ^1P level. The $2p^2$ ^1D level is fed by a transition
from the $2p4d$ $^1F^{\circ}$ level which is autoionizing and can be populated
by dielectronic recombination of C IV (Storey 1981).

HIGH TEMPERATURE PLASMAS

Hot plasmas with temperature exceeding 100,000 $\underline{K}$ occur in
solar and stellar coronae, in the galactic coronal gas, in supernova
remnants and stellar wind- induced cavities. Ionization is due to
impact ionization by thermal electrons,

$$e + X^{m+} \quad \rightarrow \quad e + X^{(m + 1)} + e, \tag{29}$$

augmented by photoionization by recombination radiation. At high
temperatures the number of resonance states which are energetically accessible is large and dielectronic recombination controls
the recombination of electrons and ions (Burgess 1964).

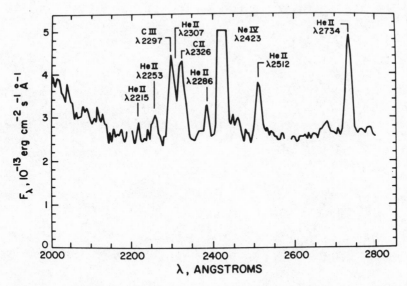

Fig. 3.   The ultraviolet spectrum of the planetary nebula
          NGC 7662 (from Harrington et al., 1981).

The ionization rate coefficient and the recombination rate coefficient are functions of the temperature. At low electron densities, the rates of both are proportional to the electron density and in coronal equilibrium the ionization structure is a function temperature only. At high electron densities exceeding 10^8 cm^{-3}, dielectronic recombination is suppressed by collisional ionization of the excited levels and the ionization structure depends upon the temperature and the density.

Charge transfer reactions involving protons and helium ions must be included for some systems. Baliunas and Butler (1980) have drawn attention to the charge transfer ionization reactions

$$Si^+ + H^+ \rightarrow Si^{2+} + H \tag{30}$$

$$Si^{2+} + He^+ \rightarrow Si^{3+} + He \tag{31}$$

Fig.4, adapted from Baliunas and Butler (1980), compares the

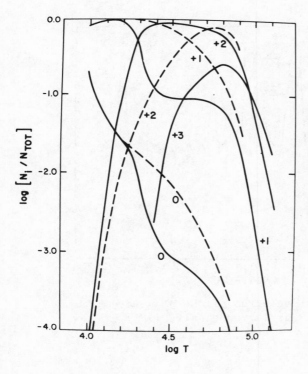

Fig. 4.

The silicon ion distribution in coronal equilibrium calculated with (solid curve) and without (dashed curve) charge transfer ionization and recombination (adapted from Baliunas and Butler 1980).

results of coronal equilibrium calculations with and without
the charge transfer ionization processes and their reserve.

The photons emitted following dielectronic capture into
doubly-excited levels have frequencies close to that of the
resonance level of the ionized system and are called satellite
line. The ratios of the intensities of the satellite lines to
that of the resonance line is a useful measure of the temperature
(Gabriel 1972, Bely-Dubau et al. 1981). Recently Dubau et al.
(1981) have provided data on dielectronic capture of electrons
by the hydrogen-like ion Fe^{25+},

$$Fe^{25+}(1s) + e \quad \rightarrow \quad Fe^{24+}(2pnl)$$

$$\rightarrow \quad Fe^{24+}(1snl) + h\nu \tag{32}$$

The frequency ν is close to the 2p-1s resonance transition of
Fe^{25+}, slightly modified by the spectator electron nl.
Fig.5 (Dubau et al. 1981) is the satellite spectrum at an
electron temperature of 2×10^7 K and a Doppler temperature of
10^7 K. The peak at 1.7917 Å is due to several close-lying n = 2
singlet and triplet satellites.

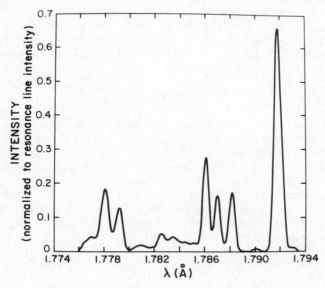

Fig. 5.   The satellite spectrum of iron at an electron
          temperature of 2×10^7 K and a Doppler temper-
          ature of 10^7 K from Dubau et al. 1981.

At high electron densities, the spectrum is modified by collisions of electrons with the doubly-excited states. Collisions with protons are also important since at high temperatures collisions overcome the Coulomb barrier. Protons are particularly effective in changing the angular momentum. The changes modify the radiative lifetime of the doubly-excited level and may reduce the efficiency of dielectronic recombination. The case of C^+ has been studied in detail (Watson, Western and Christensen 1980).

The influence of proton impact on fine-structure excitations in the solar corona was pointed out by Seaton (1963) for the green line of Fe^{13+} arising from the $^2P_{3/2}$ $^2P_{1/2}$ transition. At coronal temperatures, the reaction

$$H^+ + Fe^{13+}(^2P_{1/2}) \rightarrow H^+ + Fe^{13+}(^2P_{3/2}) \tag{33}$$

is more efficient than the electron impact excitation process

$$e + Fe^{13+}(^2P_{1/2}) \rightarrow e + Fe^{13+}(^2P_{3/2}). \tag{34}$$

Considerable attention has been given to proton impact excitation of fine-structure transitions (cf. Kastner and Bhatia 1979, Faucher, Masnou-Seeuws and Prudhomme 1980, Doyle, Kingston and Reid 1980). Proton impacts however are important in general if the energy defect of the process is small compared to the mean thermal energy kT and no change in spin multiplicity is involved. When proton impacts are the dominant source of excitation, the resulting emission intensity is a measure of the proton temperature, which may differ from the electron temperature.

SHOCKED REGIONS OF THE INTERSTELLAR GAS

Shock waves are an ubiquitous phenomenon in interstellar space. A shock propagating faster than the velocity of sound compresses, heats, dissociates and ionizes the gas. The directed energy of the shock is converted in the shock front to thermal energy and for an ionized gas the post-shock temperature T_s is about 14 V_s^2 K where V_s is the shock velocity in km s^{-1}. The recombination of electrons and ions takes place by radiative and dielectronic processes and the recombination radiation may be absorbed by the gas behind and ahead of the shock. The radiation emitted by the shocked gas provides detailed evidence on the characteristics of the shock and its origin (cf. McKee and Hollenbach 1981).

Slow shocks with velocities less than 25 km s^{-1} do not

dissociate molecular species (Hollenbach and McKee 1981) but
by heating the medium they modify the chemistry, endothermic
reactions with H_2 playing a determining role. Elitzur and
Watson (1978, 1980) have suggested that the observed interstellar
CH^+ is produced by the endothermic reaction

$$C^+ + H_2 \quad \rightarrow \quad CH^+ + H \tag{35}$$

in gases heated to several thousand degrees by slow shocks
associated with expanding H II regions. The CH^+ ions are removed
by the reverse reaction, by photodissociation and by dissociative
recombination

$$CH^+ + e \quad \rightarrow \quad C + H. \tag{36}$$

The rate coefficient for (36) has been measured by Mul et al.
(1980), who argued that the CH^+ ions are probably in the lowest
vibrational state. In the shocked gas, although the temperature
is high, the density is low. The molecule is heteronuclear so
that electric-dipole transitions are rapid enough to drive the
ions into the $v = 0$ level and indeed into the lowest rotational
level. Little evidence is available on the rotational dependence
of dissociative recombination.

Other molecular ions which are of interest as measures of
the appropriateness of the shock model are SH^+, HCl^+ and, if the
silicon depletion is not too severe, SiH^+.

Molecular observations of regions where there is clear
dynamical evidence of supersonic motions show an increase by an
order of magnitude in the ratio of HCO^+/CO compared to that in
interstellar clouds (Dickinson et al. 1980, DeNoyer and Frerking
1981). Earlier theoretical calculations by Iglesias and Silk
(1978) of the molecular composition behing a 10 km s^{-1} shock
propagating in a dense molecular cloud had suggested that the
HCO^+/CO ratio would be reduced. The assumed shock environment
may be inappropriate or the chemical model may be inadequate.
Indeed qualitative considerations do suggest an increase in
HCO^+.
The enhanced OH and H_2O resulting from

$$O + H_2 \quad \rightarrow \quad OH + H \tag{37}$$

$$OH + H_2 \quad \rightarrow \quad H_2O + H \tag{38}$$

increases the production of HCO^+ through the reactions

$$C^+ + H_2O \rightarrow HCO^+ + H \tag{39}$$

$$C^+ + OH \rightarrow CO^+ + H \tag{40}$$

$$CO^+ + H_2 \rightarrow HCO^+ + H \tag{41}$$

and the efficiency of removal of HCO^+ by electrons in dissociative recombination

$$HCO^+ + e \quad \rightarrow \quad H + CO \tag{42}$$

is diminished at high temperatures. An additional source of HCO^+ is associative ionization

$$O + CH \quad \rightarrow \quad HCO^+ + e. \tag{43}$$

However HCO^+ is removed by

$$HCO^+ + H_2O \rightarrow \quad H_3O^+ + CO. \tag{43a}$$

INTERSTELLAR CLOUDS

Most of the mass of the galaxy resides in dense molecular clouds. The clouds are opaque to visible radiation and radiation from interstellar molecules in the radio frequency spectrum provides a unique probe of the physical environment in which the initial stages of star formation occur. The clouds are bombarded by energetic cosmic rays which produce ionization. Ion-molecule reactions then modify the chemical composition and dissociative recombination of electrons and molecular ions is a crucial step in the molecular formation and destruction schemes.

The molecules CH and C_2H are widely distributed. A possible scheme for their formation is initiated by the production of CH_2^+ by radiative association

$$C^+ + H_2 \quad \rightarrow \quad CH_2^+ + h\nu \tag{44}$$

(Black and Dalgarno 1973). The CH_2^+ ions undergo dissociative recombination

$$CH_2^+ + e \quad \rightarrow \quad CH + H, \tag{45}$$

forming CH, but more probably chemical reaction,

$$CH_2^+ + H_2 \quad \rightarrow \quad CH_3^+ + H. \tag{46}$$

The abstraction sequence terminates with CH_3^+. The CH_3^+ ions are removed by dissociative recombination

$$CH_3^+ + e \quad \rightarrow \quad CH_2 + H \tag{47}$$

$$CH_3^+ + e \quad \rightarrow \quad CH + H_2. \tag{48}$$

In diffuse clouds ultraviolet photons dissociate CH_2,

$$CH_2 + h\nu \quad \rightarrow \quad CH + H \tag{49}$$

but in dense clouds the branching ratio for reactions (47) and (48) is a critical unknown parameter in the calculation of the abundances of CH.

The CH_3^+ ion may be converted to CH_5^+ by a radiative association mechanism

$$CH_3^+ + H_2 \rightarrow CH_5^+ + h\nu \tag{50}$$

and the unknown products of the dissociative recombination of CH_5^+ constitute a major uncertainty. Presumably methane is one product,

$$CH_5^+ + e \rightarrow CH_4 + H. \tag{51}$$

The methyl radical is another possibility,

$$CH_5^+ + e \rightarrow CH_3 + H + H, \tag{52}$$

which could produce formaldehyde through the reaction

$$CH_3 + O \rightarrow H_2CO + H \tag{53}$$

Alternatively CH_5^+ can react with O according to

$$CH_5^+ + O \rightarrow H_3O^+ + CH_2, \tag{54}$$

and

$$CH_5^+ + O \rightarrow CH_3O^+ + H_2. \tag{55}$$

The latter reaction may lead to formaldehyde by the dissociative recombination of CH_3O^+,

$$CH_3O^+ + e \rightarrow H_2CO + H. \tag{56}$$

Diatomic carbon, C_2, and ethynyl, C_2H, are produced by the sequence

$$C^+ + CH \rightarrow C_2^+ + H \tag{57}$$

$$C_2^+ + H_2 \rightarrow C_2H^+ + H \tag{58}$$

$$C^+ + CH_2 \rightarrow C_2H^+ + H \tag{59}$$

$$C_2H^+ + e \rightarrow C_2 + H \tag{60}$$

$$C_2H^+ + H_2 \rightarrow C_2H_2^+ + H \tag{61}$$

$$C_2H_2^+ + e \rightarrow C_2H + H \tag{62}$$

$$C_2H_2^+ + e \rightarrow C_2 + H + H \tag{63}$$

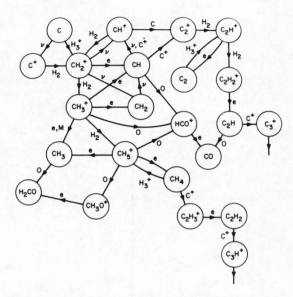

Fig. 6a.   The interstellar ion-molecule carbon chemistry.

but it should be emphasized that there is little evidence that the
end products of dissociative recombination are those postulated
in the chemistry. A lengthy list of reactions has been compiled
by Prasad and Huntress (1980). Figures 6a, 6b and 6c are summary
diagrams of some of the reactions, affecting the distribution of
carbon, oxygen and nitrogen amongst molecules.

The carbon chemistry is the basic component of ion-molecule
schemes for the formation of the complex organic species (cf.
Huntress and Mitchell, 1979, Schiff and Bohme 1979).

Similar uncertainities about dissociative recombination
affect the oxygen chemistry. The final step in the formation
of OH in dense clouds is the reaction

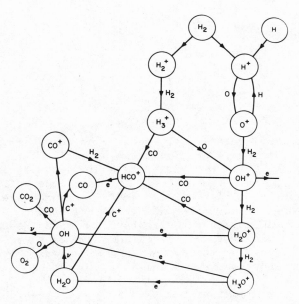

Fig. 6b.   The interstellar ion-molecule oxygen chemistry.

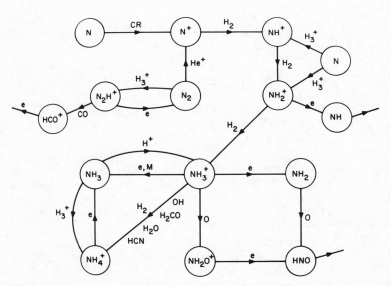

Fig. 6c.   The interstellar ion-molecule nitrogen chemistry.

$$H_3O^+ + e \quad \rightarrow \quad OH + H_2 \tag{64}$$

and of H_2O is the reaction

$$H_3O^+ + e \quad \rightarrow \quad H_2O + H. \tag{65}$$

Several molecular ions have been detected in interstellar clouds. In addition to CH^+, measured by optical absorption in diffuse clouds, CO^+, HCO^+, HCS^+ and N_2H^+ have been identified by emission of radio frequencies. A possible detection of $HOCO^+$ has been reported though HOCN may be a more probable identification (Thaddeus, Guélin and Linke 1981). All are removed by dissociative recombination and by charge transfer to metals. Reactions with H_2 convert CH^+ to CH_2^+ and CO^+ to HCO^+ and neither CH^+ nor CO^+ exists in cold dense clouds. The remaining ions provide significant information about the sources of ionization.

The deuterated ions DCO^+ and N_2D^+ have been observed. Their abundances relative to HCO^+ and N_2H^+ are much larger than the expected D/H ratio and substantial fractionation must have occurred. The major sources of HCO^+ and DCO^+ are the reactions

$$H_3^+ + CO \quad \rightarrow \quad HCO^+ + H_2 \tag{66}$$

$$H_2D^+ + CO \quad \rightarrow \quad DCO^+ + H_2. \tag{67}$$

When in equilibrium, the ion densities are related by the equation

$$\frac{n(DCO^+)}{n(HCO^+)} \simeq \frac{n(H_2D^+)}{3n(H_3^+)} \frac{\alpha_{HCO^+}}{\alpha_{DCO^+}} \tag{68}$$

where α_{HCO} and α_{DCO} are the respective rate coefficients for the dissociative recombination of HCO^+ and DCO^+. An accurate value for the ratio $\alpha_{HCO^+}/\alpha_{DCO^+}$ for ions in their lowest energy level is needed but presumably the ratio is of the order of unity. Expression (68) asserts that the fractionation of DCO^+ is directly related to the fractionation of H_2D^+.

The fractionation of H_2D^+ is achieved by the exchange reaction

$$H_3^+ + HD \quad \overset{\leftarrow}{\underset{\rightarrow}{}} \quad H_2D^+ + H_2 + \Delta E. \tag{69}$$

The backward and forward rates are related at low temperatures by (cf. Adams and Smith 1981)

$$\frac{k_\rightarrow}{k_\leftarrow} = \exp (\Delta E/kT). \tag{70}$$

Since ΔE is about 140 K, considerable fractionation is possible in a cold interstellar cloud. The fractionation is modified by other more rapid modes of destruction of H_2D^+, which may be removed by dissociative recombination at a rate $\alpha n_e s^{-1}$ and by chemical reactions with constituents such as CO and O at a rate $\Sigma k_x n(x) s^{-1}$. Equating the formation and destruction rates of H_2D^+, we obtain

$$k_\rightarrow n(HD) n(H_3^+)$$
$$= n(H_2D^+) (k_\leftarrow n(H_2) + \alpha n_e + \Sigma k_x n(x)) \tag{71}$$

so that

$$\frac{n(H_2D^+)}{n(H_3^+)} = \frac{1}{f} \frac{n(HD)}{n(H_2)} \tag{72}$$

where

$$f = k_\rightarrow \exp(-\Delta E/kt) + \frac{\alpha n_e}{n(H_2)} + \Sigma k_x \frac{n_x}{n_{H_2}} \tag{73}$$

The exponential term is usually negligible – the reverse reaction is very slow in most clouds – and the fractionation is controlled by the fractional ionization and by the fractional content of heavy neutral material, parameters which influence the coupling of a collapsing cloud to its surroundings through the action of the interstellar magnetic field and which may be critical to the dissipation of the rotational angular momentum of the cloud and to star formation. The fraction of material distributed between the gas phase and grains directly affects molecular formation and the cooling of interstellar clouds.

A measure of the (D)/(H) ratio at different locations in the galaxy has cosmological significance. To derive it from molecular observations, the fractional ionization and element content must be known.

If a value for the cosmic abundance ratio (D)/(H) is adopted,
the measured ratios can be used to derive estimates of the
fractional ionization and of the heavy element depletion. In
clouds, ratios exceeding 0.1 have been measured. The actual
value is uncertain partly because of foreground absorption in
cooler regions. If we adopt the (D)/(H) ratio of 2×10^{-5} suggested by measurements in the local neighborhood, we may infer
that the fractional ionization is less than 5×10^{-7} and the
fractional neutral content is less than 5×10^{-5}.

The fractional ionization also affects the abundance ratio
of N_2D^+/N_2H^+ but it derives from the same H_2D^+/H_3^+ ratio,
according to the sequences

$$H_3^+ + N_2 \quad \rightarrow \quad N_2H^+ + H \tag{74}$$

$$H_2D^+ + N_2 \quad \rightarrow \quad N_2D^+ + H \tag{75}$$

and equation (71) applies. Fractionation was first observed for
DCN/HCN. Its chemistry is more involved but consists of sequences
that lead to H_2CN^+ and $HDCN^+$ followed by dissociative recombination and the fractional ionization is again needed if the (D)/(H)
ratio is to be derived.

It may be possible to identify some more complex molecular
system where fractionation occurs through radiative association
of some neutral or ionic species X with H_2 and HD,

$$X + H_2 \quad \rightarrow \quad XH_2 + h\nu \tag{76}$$

$$X + HD \quad \rightarrow \quad XHD + h\nu \tag{77}$$

Presumably (77) is more rapid than (76), enhancement occurs and
the fractionation depends upon the ratio of the rate coefficients.
In any event if radiative association is significant in building
the complex molecules, the fractionation should be considerable
and the deuterated versions readily detectable.

COMETS

The chemistry of cometary comae has some similarities to
the chemistry of interstellar clouds in that ion-molecule
reactions occupy a central position (cf. Oppenheimer 1975). The
major constituent is probably H_2O, replacing hydrogen, but for
some comets CO_2 may be the major parent molecule. As the parent
molecules stream away from the cometary nuclei they are ionized
and dissociated by solar ultraviolet radiation and by the solar
wind interaction and their products undergo chemical reactions.
A review of laboratory data on many of the reactions has been
presented by Huntress et al. (1980).

Fig. 7 is an illustration of the chemistry initiated by the interactions of solar radiation with water vapour. Dissociative recombinations of OH^+, H_2O^+ and H_3O^+ form an integral component of the chemistry.

Dissociative recombination produces energetic atomes and molecules. The kinetic energy is rapidly degraded into thermal energy by elastic collisions and dissociative recombination may be an important heat source for the neutral gas. Many of the species identified in cometary material were detected by observations of resonant or fluorescent scattering of solar photons and the line profiles provide data on the velocity distribution of the atom or molecule. The disposition of the recombination energy between initial energy and kinetic energy of the end products is usually unknown but would be valuable in identifying the excitation sources.

The excited end products may radiate and dissociative recombination may be a large source of emissions. Oppenheimer (1975) suggested that C_2 forms in the coma largely by dissociative recombination of C_2H^+,

$$C_2H^+ + e \rightarrow C_2 + H \tag{78}$$

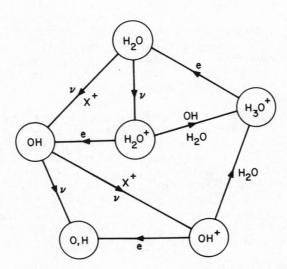

Fig. 7.   The chemistry of a water-dominated cometary coma. The constituent X is any of H, O, CO_2 and N_2.

Recent ab initio calculations suggest that the C_2 is formed preferentially in triplet states Melius, Oppenheimer and Kirby (1981)
and (78) may be responsible for the observed emission in the Swan
bands. The Swan band intensity exceeds the intensity in the
Mulliken system which can be populated by fluorescence originating
in the ground state of the molecule.

Neutron atomic carbon is also seen in comets through lines
at 1561 Å, 1657 Å and 1931 Å. The 1931 Å line originates in the
metastable ^1D state of carbon which may be populated by the
dissociative recombination of CO^+,

$$CO^+ + e \rightarrow C + O. \tag{79}$$

PLANETARY ATMOSPHERES

Chemistries based upon ion-molecule reactions were first
advanced to explain the ionization balance of the terrestrial atmosphere. In the early analyses of the electron content, Bates
and Massey (1946, 1947) argued that radiative recombination

$$O^+ + e \rightarrow O' + h\nu \tag{80}$$

proceeds too slowly to explain the anomalously rapid recombination
inferred from the ionospheric data and they concluded that the
probable sequence was charge transfer of O^+ to form a molecular
ion followed by dissociative recombination.

The mean features of the chemistry of the terrestrial ionosphere are illustrated in Fig. 8. Solar ultraviolet radiation
ionizes the major constituents O, O_2 and N_2. The O^+ ions reacts
chemically with N_2 and O_2 producing NO^+ and O_2^+ which are removed
by dissociative recombination. The N_2^+ ions are removed in part
by dissociative recombination and in part by reaction with oxygen
atoms again producing NO^+.

The actual chemistry is a good deal more complicated.
Reactions with the minor neutral constituents NO and N enter the
scheme and metastable and vibrationally excited systems have large
effects. In addition to the ground ^4S state, O^+ ions are produced
by photoionization of O in the metastable ^2D and ^2P states. The
metastable states can react with N_2 to produce N_2^+ increasing the
source of N_2^+ and decreasing the source of NO^+. The behaviour of
the metastable states of O^+ is still not fully resolved.

Empirical estimates of the recombination coefficient for

$$NO^+ + e \rightarrow N + O \tag{81}$$

may be derived from analysis of ion composition data.

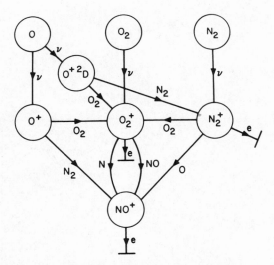

Fig. 8.   The main chemical paths in the terrestrial
          atmosphere.

Fig. 9, reproduced from Oppenheimer et al (1977), is a comparison
of the measured values of the daytime NO^+ density with theoretical
model values calculated using the recombination coefficients
measured in the laboratory by Walls and Dunn (1974), and Huang et
al. (1975). The measured values of Walls and Dunn (1974) which may
not be reliable at low temperatures (cf. Torr and Torr 1979), are
to be preferred, presumably because their experiments involved NO^+
ions in the v = 0 vibrational level. In the atmosphere vibrationally excited NO^+ ions decay in a time of about 10^{-2} s and the recombination time is at least 10 s. Confirmation is provided by the
analyses of noctural data on NO^+ (Torr et al. 1976a) when fewer
mechanisms exist for pupulating vibrationally excited levels. A
dependence of the recombination coefficient on vibrational population can be invoked to explain the differences between the two
sets of laboratory data but appears to be inconsistent with the
measurements of Mul and McGowan (1979) which involve an unknown
vibrational distribution and yield a recombination coefficient in
agreement with that obtained for v=0 ions. Theoretical arguments
have been advanced (Cunningham, O'Malley and Hobson 1981) which
suggest that for the atmospheric ions there should be only a weak
dependence of the recombination coefficient on vibration.

     Similar analyses may be applied to atmospheric measurements
of O_2^+ though loss by reactions with N and with NO

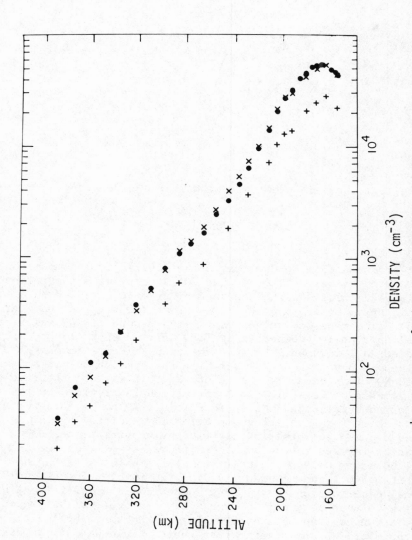

Figure 9. The measured NO$^+$ densities (°) and theoretical densities calculated with the dissociative recombination coefficients of (x) Walls and Dunn (1974) and (+) Huang et al. (1975).

$$O_2^+ + N \quad \rightarrow \quad NO^+ + O \tag{82}$$

$$O_2^+ + NO \quad \rightarrow \quad NO^+ + O_2 \tag{83}$$

and gain by reactions of N^+,

$$N^+ + O_2 \quad \rightarrow \quad N + O_2^+ \tag{84}$$

and of $O^+(^2D)$

$$O^+(^2D) + O_2 \rightarrow O + O_2^+ \tag{85}$$

cause some complications.

Because of atom interchange (Bates 1955)

$$O + O_2^+(v) \quad \rightarrow \quad O_2^+(v'< v) + O \tag{86}$$

the atmospheric ions are probably all in the $v = 0$ level, except at high altitudes. The laboratory ions are not and it appears that the recombination coefficient of O_2^+ is insensitive to the vibrational population.

Recent work by Torr and Torr (1981) appears to modify earlier conclusions about the agreement between laboratory and atmospheric determinations of the total recombination coefficient and they suggest that metastable $O_2^+(a^4\Pi_u)$ ions may be present and behave differently from $O_2^+(X^2\Pi_g)$ ions. The metastable ions are certainly produced in the atmosphere (Dalgarno and McElroy 1963) and can participate in chemical reactions not accessible to the ground state ions. However

$$O + O_2^+(a^4\Pi_u) \quad \rightarrow \quad O_2^+(X^2\Pi_g) + O \tag{87}$$

is probably rapid. Metastable ions may be present in the laboratory experiments and not in the atmosphere.

Considerable uncertainty attends the interpretation of atmospheric measurements on N_2^+, stemming in part from the reaction

$$O^+(^2D) + N_2 \quad \rightarrow \quad O + N_2^+ \tag{88}$$

and the quenching of $O^+(^2D)$ by collisions, and in part from the differences which may exist in the chemistry of vibrationally

excited N_2^+ ions. Initial attempts to invoke a substantial
variation of the rate coefficient of dissociative recombination

$$N_2^+ + e \quad \rightarrow \quad N' + N'' \tag{89}$$

with vibrational level were criticized by Biondi (1978) who
suggested alternatively that vibrationally excited N_2 ions
react more rapidly with O than do ions in the $v = 0$ level.
Subsequently laboratory studies by Zipf (1980) have shown
that the specific dissociative recombination coefficients for
$v = 0$, 1 and 2 are nearly equal. Adopting the laboratory value
(Mehr and Biondi 1969, Mul and McGowan 1979, Zipf 1980), Torr,
Richards and Torr (1982) have concluded from an analysis of
Atmosphere Explorer data that vibrationally excited N_2^+ ions are
removed by charge transfer

$$N_2^+ + O \quad \rightarrow \quad N_2 + O^+. \tag{90}$$

In the atmospheres of Mars and Venus, CO_2 is the major
neutral gas. Its photoionization produces CO_2^+ which is removed
by dissociative recombination

$$CO_2^+ + e \quad \rightarrow \quad CO + O \tag{91}$$

and by reaction with O

$$CO_2^+ + O \quad \rightarrow \quad CO + O_2^+ \tag{92}$$

and

$$CO_2^+ + O \quad \rightarrow \quad CO_2 + O^+ \tag{93}$$

followed by

$$O^+ + CO_2 \quad \rightarrow \quad O_2^+ + CO. \tag{94}$$

The O_2^+ can react with minor species such as NO but mostly it
undergoes dissociative recombination. As a consequence, O_2^+ is
the major ionic component in the ionospheres of Mars and Venus.

On Jupiter and Saturn, H_2 and He are the major neutral
gases. Minor constituents include CH_4, NH_3 and C_2H_6. The H_2^+
ions produced by photoionization and by cosmic rays react
with H_2 to produce H_3^+ ions. The H^+ ions diffuse down and

participate in three-body reactions to again produce H_3^+ ions
or in charge transfer reactions with minor species. If H_2 is
vibrationally excited the reverse of reaction (7) may occur.
The H_3^+ ions can undergo dissociative recombination or form H_5^+
in a further three-body reaction (cf. Atreya, Donahue and Waite
1979, Atreya and Waite 1981).

Reactions of vibrationally excited H_3^+ and H_5^+ with methane
lead to CH_5^+. Ionization fragments of CH_4 also lead to CH_5^+ in two
and three-body reactions with H_2. The CH_5^+ ion recombines to
form CH_4 and CH_3 or it may react with ammonia and ethane to
produce NH_4^+ and $C_3H_7^+$ according to

$$CH_5^+ + NH_3 \quad \rightarrow \quad CH_4 + NH_4^+ \tag{95}$$

$$CH_5^+ + C_2H_6 \quad \rightarrow \quad C_3H_7^+ + CH_4 \tag{96}$$

Capone et al. (1979) list a series of possible reactions leading
to the formation of various hydrocarbons including propane by
dissociative recombination of $C_3H_9^+$,

$$C_3H_9^+ + e \quad \rightarrow \quad C_3H_8 + H \tag{97}$$

Propane has been tentatively identified as a constituent of
Saturn's atmosphere and of Titan's atmosphere (McGuire et al.
1981, Hanel et al. 1981).

The predominant constituent of Titan's atmosphere is
molecular nitrogen (Tyler et al. 1981, Broadfoot et al. 1981) and
methane is a major secondary component. Small amounts of ethane,
acetylene, ethylene and hydrogen cyanide have been detected
(Hanel et al. 1981).

Fig.10 illustrates some aspects of the expected ion chemistry. Complex hydrocarbons can be formed by further reactions with
CH_4 (Capone et al. 1980).

Ion-electron recombination processes may produce measurable
emission in planetary atmospheres. There appears to be but one
example of emission resulting from radiative recombination though
there are several cases of dissociative recombination. The radiative recombination of O^+ with electrons

$$O^+ + e \quad \rightarrow \quad O' + h \tag{98}$$

produces emission lines at 1304 Å and 1356 Å. There lines have
been observed as part of a tropical ultraviolet nightglow.

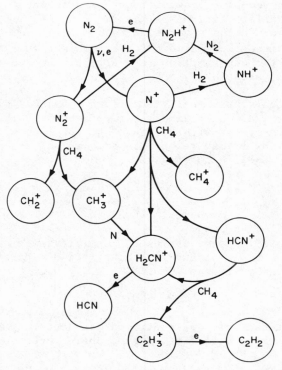

Fig. 10.   Ion-molecule chemistry in the atmosphere of Titan.

Fig.11 reproduces the altitude profile measured with a rocket-borne spectrometer (Brune et al. 1978). Theoretical analysis leads to the conclusion that radiative recombination (98) is the major source of the emission with a possible contribution of about 25% from the ion-ion mutual neutralization (cf. Hanson 1970)

$$O^+ + O^- \quad \rightarrow \quad O' + O \qquad (99)$$

(Brune et al. 1978).

Radiative recombination coefficients for populating the upper levels of the 1356 Å and 1304 Å transitions have been derived from the airglow measurements (Brune et al. 1978). The values of respectively 7.8×10^{-13} cm^3 s^{-1} and 4.9×10^{-13} cm^3 s^{-1} at about 700 K are consistent with theoretical predictions (Julienne et al. 1974).

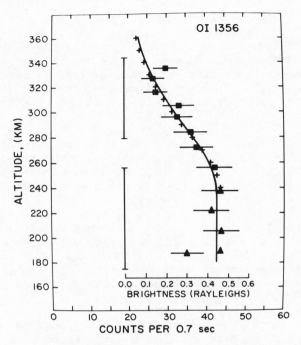

Fig. 11.   The altitude profile of the terrestrial oxygen
           tropical nightglow at 1356 Å (from Brune et al.
           1978).

     In the terrestrial atmosphere,   dissociative recombination
of O_2^+ with electrons has long been recognized as a source of
the oxygen green and red lines at 5577 Å and 6300 Å though
there is continuing discussion about the efficiences with which
the excited ^1S and ^1D levels of oxygen are populated (cf. Torr
and Torr 1982).

     For the red line, a recent analysis by Link, McConnell and
Shepherd (1981) of the nightglow data of Hays et al. (1978) leads
to an effective recombination coefficient of 2.1×10^{-7} cm^3 s^{-1}
which is in close agreement with a measurement by Zipf (1970).
Because the vibrational distributions of the recombining ions
are unlikely to be the same in the atmospheric and the experiment
correlations, the agreement appears to require that the recombination coefficient for $O(^1D)$ production be independent of the
vibrational level.

     For $O(^1S)$ production, Zipf (1979, 1980) has shown that the
recombination coefficient varies with the vibrational level of
O_2^+. Because of deactivation of $O_2^+(v)$ by atom interchange

enhancement of the vibrational population in the atmosphere at the altitudes where dissociative recombination is an important source of $O(^1S)$ atoms is negligible. To resolve the discrepancy between the laboratory data of Zipf (1980) and the airglow data it appears necessary to assume that the fraction of recombinations which terminate in the $O(^1S)$ state increases substantially as the electron temperature increases from 300 K to 1000 K (Bates and Zipf 1980).

Dissociative recombination of O_2^+ is a source of the red and green lines on Mars and Venus and dissociative recombination of CO_2^+ contributes to the CO Cameron band emission (cf. Fox and Dalgarno 1979, 1981). Of greater significance, dissociative recombination is a source of energetic atoms.

Observational evidence for the existence of nonthermal atomic oxygen at high altitudes in the terrestrial atmosphere has been obtained by Yee and Hays (1980) and Yee, Meriwether and Hays (1980). Dissociative recombination of O_2^+ is the major source of the energetic atoms. A similar oxygen corona is predicted for Venus (Nagy et al. 1981).

On Mars dissociative recombination of N_2^+ may be responsible for the loss of nitrogen during the lifetime of the planet (Brinkman 1971, Yung et al. 1977, Fox and Dalgarno 1980). If the N_2^+ is vibrationally excited, the loss rate will be enhanced. The Viking mission established that the fractional abundance of nitrogen is 2.5% and that the $^{15}N/^{14}N$ isotope ratio is a factor of 1.62 ± 0.16 greater than the terrestrial value. These observations are most readily interpreted by postulating the existence of a substantial initial reservoir of nitrogen which has diminished in time as the nitrogen escaped. Dissociative recombination of N_2^+ produces nitrogen atoms which have enough energy to escape the gravitational attraction of the planet. Other processes also provide a supply of energetic atoms. A difference in the escape rates for ^{14}N and ^{15}N occurs because of diffusion separation in the atmosphere so that at the exobase the ratio of $^{14}N^{15}N$ to $^{14}N_2$ relative to that in the bulk atmosphere is 0.82. To reproduce the measured isotope ratio requires a minimum initial reservoir of about 4×10^{22} cm^{-2} (cf. Fox and Dalgarno 1980). Nitrogen must also have escaped from the atmosphere of Titan (Strobel and Shemansky 1981, Smith et al. 1981).

Acknowledgments

I am grateful to Dr. D. Smith and Dr. N.G. Adams for their comments on these notes. The research upon which this study is based was supported in part by the National Science Foundation through its Astronomy and its Atmospheric Science Grant programs.

REFERENCES

Adams, N.G. and Smith, D., 1981, Ap. J. 248, 373.
Argyros, J.D., 1974, J. Phys. B7, 2025.
Atreya, S.K., Donahue, T.M. and Waite, J.H., 1979, Nature 280, 795.
Atreya, S.K. and Waite, J.H., 1981, Nature 292, 682.
Auerbach, D., Cacak, R., Caudano, R., Gaily, T.D., Keyser, C.J.,
    McGowan, J.W., Mitchell, J.B.A. and Wilk, S.F.J., 1977,
    J. Phys. B 10, 3797.
Baliunas, S. and Butler, S.E., 1980, Ap. J. Lett. 235, L45.
Bardsley, N.J., 1968.
Bates, D.R., 1951, MNRAS 111, 303.
Bates, D.R., 1952, MNRAS 112, 40.
Bates, D.R., 1955, Proc. Phys. Soc. 68A, 344.
Bates, D.R. and Massey, H.S.W., 1946, Proc. Roy. Soc. A 187, 261.
Bates, D.R. and Massey, H.S.W., 1947, Proc. Roy. Soc. A 192, 1.
Bates, D.R. and Poots, G., 1953, Proc. Phys. Soc. A66, 784.
Bates, D.R. and Zipf, E.C., 1980, Planet Spa. Sci. 28, 1081.
Beigman, I.L. and Chichkov, B.N., 1980, J. Phys. B13, 565.
Bely-Dubau, F., Dubau, J., Faucher, P. and Gabriel, A.H., 1981,
    MNRAS, in press.
Biondi, M.A., 1978, Geophys. Res. Lett. 5, 661.
Black, J.H., 1978, Ap. J. 222, 125.
Black, J.H. and Dalgarno, A., 1973, Astrophys. Lett. 15, 79.
Broadfoot, A.L., Sandel, B.R., Shemansky, D.E., Holberg, J.B.,
    Smith, G.R., Strobel, D.F., McConnell, J.C., Kumar, S.,
    Hunter, D.M., Atreya, S.K., Donahue, T.M., Moss, H.W.,
    Bertaux, J.L., Blamont, J.E., Pomphrey, R.B. and Linick, S.,
    1981, Science 212, 206.
Brinkman, R.T., 1971, Science 174, 944.
Brocklehurst, M. and Salem, M., 1977, Comp. Phys. Comm. 13, 39.
Brune, W.H., Feldman, P.D., Anderson, R.C., Fastie, W.G. and
    Henry, R.C., 1978, Geophys. Rev. Lett. 5, 383.
Buckingham, R.A., Reid, S. and Spence, R., 1952, MNRAS 112, 382.
Burgess, A., 1964, Ap. J. 139, 776.
Butler, S.E., Heil, T.G. and Dalgarno, A., 1980, Ap. J. 241, 442.
Capone, L.A., Dubach, J., Whitten, R.C., Prasad, S.S. and
    Santhanam, K., 1980, Icarus 44, 72.
Capone, L.A., Dubach, J., Whitten, R.C. and Prasad, S.S., 1979,
    Icarus 39, 443.
Cunningham, A.J., O'Malley, T.F. and Hobson, R.M., 1981, J. Phys.
    B14, 773.
Dalgarno, A. and McElroy, M.B., 1963, Planet. Spa. Sci. 11, 727.
DeNoyer, L.K. and Frerking, M.A., 1981, Ap. J. Lett. 246, L37.
Dickinson, D.F., Kuiper, E.N.R., Dinger, A.S. and Kuiper, T.B.H.,
    1980, Ap. J. Lett. 237, L43.
Doyle, J.E., Kingston, A.E. and Reid, R.G.H., 1980, Astron. Ap.
    90, 97.
*Butler, S.E. and Dalgarno, A., 1980, Ap. J. 241, 838.

Dubau, J., Gabriel, A.H., Loulerque, M., Steeman-Clark, L. and
      Volonte, S., 1981, MNRAS 195, 705.
Dunn, G.H., 1968, Phys. Rev. 172, 1.
Elitzur, M. and Watson, W.D., 1978, Ap. J. Lett. 222, L141.
Elitzur, M. and Watson, W.D., 1980, Ap. J. 236, 172.
Erickson, N.R., Snell, R.L., Loren, R.B., Mundy, L. and Plambeck,
      R.L., 1981, Ap. J. Lett. 245, L83.
Faucher, P., Masnou-Seeuws, F. and Prudhomme, M., 1980, Astron.
      Ap. 81, 137.
Feibelman, W.A., Boggess, A., McCracken, C.W. and Hobbs, R.W.,
      1981, Astron. J. 86, 881.
Flower, D.R. and Roueff, E., 1979, Astron. Ap. 72, 361.
Ford, A.L., Docken, K.K. and Dalgarno, A., 1975, Ap. J. 195, 819.
Fox, J.L. and Dalgarno, A., 1979, J. Geophys. Res. 84, 7315.
Fox, J.L. and Dalgarno, A., 1981, J. Geophys. Res. 86, 629.
Fox, J.L. and Dalgarno, A., 1980, Planet. Spa. Sci. 28, 41.
Gabriel, A.H., 1972, MNRAS 160, 99.
Giusti, A., Derkits, C. and Bardsley, J.W., 1981, XII International
      Conference on Electronic and Atomic Collisions : Abstracts
      (Ed. S. Datz) p. 482.
Hanel, R., Conrath, B., Flasar, F.M., Kunde, V., Maguire, W.,
      Pearl, J., Pirragla, J., Samuelson, R., Herath, L., Allison,
      M., Cruickshank, D., Gautier, D., Gierasch, P., Horn, L.,
      Koppany, R. and Ponnamperuma, C., 1981, Science 212, 192.
Halpern, J.P. and Grindlay, J.E., 1980, Ap. J. 242, 1041.
Hanson, R.B., 1970, J. Geophys. Res. 75, 4343.
Harrington, J.P., Lutz, J.H. and Seaton, M.J., 1981, MNRAS 195,
      21P.
Hays, P.B., Rusch, D.W., Roble, R.G. and Walker, J.C.G., 1978,
      Revs. Geophys. Space Phys. 16, 225.
Harrington, J.P., Lutz, J.H., Seaton, M.J. and Strickland, D.J.,
      1980, MNRAS 191, 13.
Heap, S.R. and Stecher, T.P., 1981, in The Universe at Ultra-violet
      Wavelengths : the First Two Years of IUE, NASA CP-2171.
Hollenbach, D. and McKee, C.F., 1981, Ap. J. Lett. 241, L47.
Huang, C.M., Biondi, M.A. and Johnsen, R., 1975, Phys. Rev. A11,
      901.
Huntress, W.T., McEwan, M.J., Karpas, Z. and Anicich, V.G., 1980,
      Ap. J. Suppl. 44, 481.
Huntress, W.T. and Mitchell, G.F., 1979, Ap. J. 231, 456.
Iglesias, E.R. and Silk, J., 1978, Ap. J. 226, 851.
Julienne, P.S., Davis, J. and Oran, E., 1974, J. Geophys. Res., 79,
      250.
Karpas, Z., Anicich, V. and Huntress, W.T., 1979, J. Chem. Phys.
      70, 2877.
Kastner, S.O. and Bhatia, A.K., 1979, Astron. Ap. 71, 211.
Link, R., McConnell, J.C. and Shepherd, G.G., 1981, Planet. Spa.
      Sci. 29, 589.

Maguire, W.C., Hanel, R.A., Jennings, D.E., Kunde, V.G. and
    Samuelson, R.J., 1981, Nature 292, 683.
McCray, R.A., Wright, C. and Hatchett, S., 1977, Ap. J. Lett. 211,
    L29.
McKee, C.F. and Hollenbach, D., 1981, Ann. Rev. Astron. Ap. 18,
    219.
Mehr, F.J. and Biondi, M.A., 1969, Phys. Rev. 181, 264.
Melius, C.F., Oppenheimer, M. and Kirby, K., 1981, private
    communication.
Mul, P.M., Mitchell, J.B.A., D'Angelo, V.S., Defrance, P., McGowan,
    J.W. and Froelich, H.R., 1981, J. Phys. B14, 1353.
Mul, P.M. and McGowan, J.W., 1979, J. Phys. B12, 1591.
Nagy, A.F., Cravens, T.E., Yee, J-H. and Stewart, A.I.F., 1981,
    Geophys. Res. Lett. 8, 629.
Oppenheimer, M., 1975, Ap. J. 196, 251.
Oppenheimer, M., Dalgarno, A., Trebino, F.P., Brace, L.H.,
    Brinton, H.C. and Hoffman, J.H., 1977, J. Geophys. Res. 82,
    191.
Peek, J.M., Hashemi-Attar, A-R. and Beckel, C.L., 1979, J. Chem.
    Phys. 71, 5382.
Poulaert, G., Brouillard, F., Claeys, W., McGowan, J.W. and Van
    Wassenhove, G., 1978, J. Phys. B 11, L671.
Prasad, S.S. and Huntress, W.T., 1980, Ap. J. Suppl. 43, 1.
Ramaker, D.E. and Peek, J.M., 1976, Phys. Rev. A 13, 58.
Schiff, H.I. and Bohme, D.K., 1979, Ap. J. 232, 740.
Seaton, M.J., 1963, MNRAS 127, 191.
Seaton, M.J., 1979, MNRAS 187, 785.
Seaton, M.J. and Storey, R.J., 1976, in atomic Processes and
    Applications (eds. P.G. Burke and B.L. Moiseiwitsch),
    North Holland : Amsterdam.
Shaver, P.A., 1980, Radio Recombination Lines (Reidel : Holland).
Smith, G.R., Strobel, D.F., Broadfoot, A.L., Sandel, B.R.,
    Shemansky, D.E. and Holberg, J.B., 1981, J. Geophys. Res., in
    press.
Smith, H.A., Larson, H.P. and Fink, U., 1981, Ap. J. 244, 835.
Storey, P.J., 1981, MNRAS 195, 27P.
Summers, H.P., 1977, MNRAS 178, 101.
Thaddeus, P., Guelin, M. and Linke, R.A., 1981, Ap. J. Lett.,
    246, 141.
Torr, D.G. and Torr, M.R., 1979a, J. Atmos. Terr. Phys. 41, 787.
Torr, D.G., Torr, M.R., Walker, J.C.G., Brace, L.H., Brinton, H.C.,
    Hanson, W.B., Hoffman, J.H., Nier, A.O. and Oppenheimer, M.,
    1976a, Geophys. Res. Lett. 3, 209.
Torr, D.B., Torr, M.R., Walker, J.C.G., Nier, A.O., Brace, L.H.
    and Brinton, H.C., 1976b, J. Geophys. Res. 81, 5578.
Torr, D.G., Richards, P.G. and Torr, M.R., 1982, J. Geophys., Res.
    in press.
Torr, M.R. and Torr, D.G., 1981, Planet Spa. Sci., in press.
Torr, M.R. and Torr, D.G., 1982, Revs. Geophys. in press.

Tyler, G.L., Eshleman, V.R., Anderson, J.D., Levy, G.S., Lindal,
     G.F., Wood, G.E. and Croft, T.A., 1981, Science 212, 201.
Walls, F. and Dunn, G.H., 1974, J. Geophys. Res. 79, 1911.
Watson, W.D., Western, C.R. and Christensen, R.B., 1980, Ap. J.
     240, 956.
Weinberg, S., 1972, Gravitation and Cosmology (Wiley : New York).
Whitworth, A., 1979, MNRAS 186, 59
Yee, J.H. and Hays, P.B., 1980, J. Geophys. Res. 85, 1795.
Yee, J.H., Meriwether, J.W. and Hays, P.B., 1980, J. Geophys.
     Res. 85, 3396.
Yoshii, Y. and Sabano, Y., 1979, Publ. Astron. Soc. Japan 31,
     505.
Yung, V.L., Strobel, D.F., Kong, T.Y. and McElroy, M.C., 1977,
     Icarus 30, 26.
Zipf, E.C., 1970, Bull. Am. Phys. Soc. 15, 418.
Zipf, E.C., 1979, Geophys. Res. Lett. 6, 881.
Zipf, E.C., 1980, Geophys. Res. Lett. 7, 645.
Zipf, E.C., 1980, J. Geophys. Res. 85, 4232.

THE ROLE OF ATOMIC COLLISIONS IN FUSION

D. E. Post

Princeton University
Plasma Physics Laboratory
Princeton, New Jersey 08544

ABSTRACT

Atomic physics issues have played a large role in controlled fusion research. A general discussion of the present role of atomic processes in both magnetic and inertial controlled fusion work is presented.

INTRODUCTION

Atomic collision processes play a large role in both magnetic and inertial confinement fusion research. There are four principal areas of fusion research where atomic processes are important: (1) the hot central plasma where the fusion reactions are expected to occur, (2) the plasma edge where the plasma interacts with the external environment, (3) plasma heating methods used to produce the hot plasma, and (4) diagnostic techniques used to measure the physical properties of the plasma. In magnetic fusion research, atomic processes such as line radiation and charge exchange have contributed significantly to the energy and particle balance in the hot center of the plasma. Atomic processes involving low Z ions, hydrogen and impurity molecules, and ion-solid collisions are important at the plasma edge. Neutral beam heating has been crucial to the success of the tokamak and mirror programs. Many of the techniques used to diagnose magnetic fusion experiments rely on electron and ion collision processes.

Atomic physics is even more important in inertial confinement research. A knowledge of equations of state and opacities is crucial to designing fusion target pellets and to understanding their behavior. The absorption and stopping of laser light and energetic ions in the outer layers of an imploding pellet, the design and construction of high energy lasers and ion beams, and the use of spectroscopy and laser scattering to diagnose pellet implosions all rely on atomic collisions of one type or another.

These lectures will introduce all of these issues. I will first cover the general issues in fusion such as fusion reactions, confinement requirements, and general reactor schemes. In the second section, I will begin to discuss magnetic confinement and introduce the experimental and theoretical plasma physics issues in magnetic confinement. The third section will cover the role of impurities in the energy balance in magnetic fusion experiments. The fourth section will concentrate on the atomic collision issues important at the plasma edge. The fifth section will discuss plasma heating, and the sixth section will cover plasma diagnostics. The seventh section will cover inertial confinement. The eighth section will discuss the role of plasma physics in atomic physics, and the ninth will list useful review articles.

It is useful to first describe the conditions (electron and ion temperature, density, etc.) that characterize inertially and magnetically confined plasmas. These conditions can be derived from a few simple assumptions.

Most currently envisioned fusion schemes rely on the nuclear reaction $D + T \rightarrow n + He + 17.5$ MeV. The neutron has 14.1 MeV of kinetic energy. The mean free path for neutrons in a plasma is much larger than the dimensions of commonly discussed fusion plasmas, and thus the neutron energy is absorbed in a "blanket" structure surrounding the fusion chamber. The charged alpha particle is confined and used to heat the plasma directly, that is to resupply the heat losses of the fusing plasma. The reaction rates of deuterium-tritium (DT) fusion (Greene, 1967) are large only for a temperature $T \gtrsim 10$ keV (Fig. 1). This sets a lower limit on the temperature required at the center of a fusion plasma. A hot plasma will try to cool in a characteristic loss time τ_E due to various energy loss processes including conduction, convection, expansion, radiation, etc. A second condition for a fusion plasma can be introduced from the requirement that the alpha particles produced from the fusion reaction deposit their heat in the plasma to balance the energy losses. A simple energy balance equation for a DT plasma in which the alpha heating exceeds the energy loss rate is

$$\frac{3}{2} k \frac{\partial(\sum_i n_i T_i)}{\partial t} \approx \frac{3}{2} k \frac{(n_D T_D + n_T T_T + n_e T_e)}{\tau_E} \lesssim n_D n_T \langle \sigma v \rangle_{DT} E_\alpha \quad , \quad (1)$$

where the sum is over the different ion species and the electrons, τ_E is the plasma energy confinement time, n_i and T_i are the density and temperature of plasma species i, $\langle \sigma v \rangle_{DT}$ is the fusion reaction rate of deuterium-tritium (DT) (Fig. 1), and E_α is the kinetic energy of the alpha particle (3.5 MeV). Assuming that $n_D = n_T = (n/2)$ and that $T_D = T_T = T_e$ and denoting n_e by n, one obtains the condition $n\tau_E \gtrsim (12kT/\langle \sigma v \rangle E_\alpha)$ for ignition. $n\tau_E$ for DT is a strong function of temperature (Fig. 2), and has a minimum at $T \sim 25$ keV, with $n\tau_E \gtrsim 1.5 \times 10^{14}$ sec/cm^3. Thus one goal of controlled fusion research is the attainment of $T \gtrsim 10$ keV and $n\tau_E \gtrsim 10^{14}$ sec/cm^3.

We can generalize these considerations to include radiation losses explicitly. Taking bremsstrahlung as an example, we can equate bremsstrahlung and other losses to the alpha heating

$$\frac{3nT}{\tau_E} + bn^2\sqrt{T} = \frac{1}{4}n^2\langle \sigma v \rangle E_\alpha \, ,$$

or

$$n\tau_E = \frac{3T}{(1/4)\langle \sigma v \rangle E_\alpha - bT^{1/2}} \, , \qquad (2)$$

for ignition $\left(b = 3.336 \times 10^{-19} \text{ eV/(sec-cm}^3\text{(keV)}^{1/2}) \right)$, (Book, D. L, 1980). Ignition can only occur when the denominator of Eq. (2) is

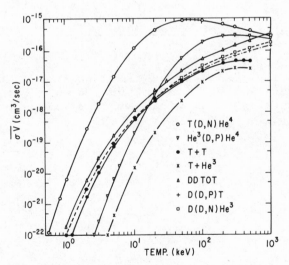

Fig. 1.   Reaction rates for a variety of fusion reactors averaged over a Maxwellian ion temperature distribution (Greene 1967). (762346).

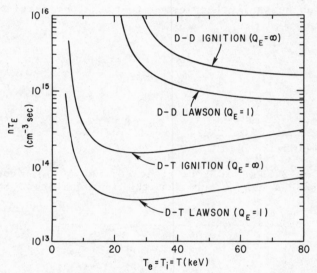

Fig. 2. $n\tau_E$, the product of the plasma density and energy confinement time for ignition and energy breakeven (Lawson criteria) for DT and DD plasmas (Jassby and Towner). (773873)

greater than zero. The temperature at which this occurs is called the ignition temperature.

Ignition is a stringent criterion requiring very good confinement. We can relax the confinement requirements by recovering some of the lost energy and using it to heat the plasma and derive what is known as the "Lawson criterion" (Lawson, 1957). If η is the efficiency with which we can recover the lost energy and convert it to plasma heating, we can write

$$\frac{3nT}{\tau_E} + bn^2 T^{1/2} = \frac{1}{4} n^2 <\sigma v> E_\alpha$$

$$+ \eta \left(\frac{3nT}{\tau_E} + bn^2 T^{1/2} + \frac{1}{4} n^2 <\sigma v> E_n \right) , \qquad (3)$$

where E_α is the energy of the alpha particles (3.5 MeV), and E_n is the energy of the neutrons (14.1 MeV). This yields a "Lawson" condition for $n\tau_E$ of the form

$$n\tau_E = \frac{3T \ (1 - \eta)}{(1/4) \ <\sigma v> \ (E_\alpha + \eta E_n) - (1 - \eta) \ bT^{1/2}} . \qquad (4)$$

Another way of characterizing the economic potential of a fusion system is to define an energy multiplication factor Q_E = Energy extracted/Energy input. Then

$$Q_E = \eta \frac{P_I + P_F}{P_I} , \qquad (5)$$

where P_F is the fusion power plus the power generated by neutron reactions in the material surrounding the reactor, and P_I is the power one has to inject into the plasma to balance the energy losses. $Q_E = 1$ corresponds to the Lawson condition in which the reactor is able to run itself, and $Q_E = \infty$ corresponds to ignition. Another commonly defined Q is $Q_P = P_F/P_I$, the power multiplication factor.

The DT reaction requires a source of tritium which is usually assumed to be produced by n + Li reactions (breeding) in a blanket surrounding the burning DT plasma. This complicates reactor designs. There is therefore an advantage to use only deuterium. The reaction chain would be (Rose and Clark, 1961)

$$D + D \rightarrow n(2.45 \ MeV) + He \ (0.82 \ MeV)$$

$$D + D \rightarrow p(3.0 \ MeV) \ + T(1. \ MeV)$$

$$_3\text{He} + D \rightarrow p(14.7 \text{ MeV}) + {}_4\text{He}(3.6 \text{ MeV})$$

$$T + D \rightarrow n(14.1 \text{ MeV}) + {}_4\text{He}(3.5 \text{ MeV}) \ .$$

The major disadvantage of this scheme is that the reaction rates are a factor of 10 to 100 lower than DT (Fig. 1) except at very high temperatures. The lower reaction rates for DD require better confinement (higher $n\tau_E$) (Fig. 2) and higher operating temperatures than DT. Fusion with DD requires temperatures of about 40 keV and $n\tau_E \sim 2 \times 10^{15}$ sec/cm^3 and is thus much harder than fusion with DT. Consequently all designs for fusion experiments in this century are based on DT.

The word "scientific breakeven" is often used with regard to the next generation of fusion experiments. It usually refers to $Q_p \approx 1$; i.e., the fusion power and the heating power are roughly equal. While $Q_p \approx 1$ represents an important milestone, it still falls short of the Q_p of ten or more needed for an economic pure fusion reactor.

At the moment there are two approaches to producing the temperatures and confinement needed for fusion. One approach began in about 1952 and uses the pressure of magnetic fields to confine the plasma. The plasma is heated by energetic neutral atom beams, ion beams, internal currents, or radio frequency waves. The other approach, at least for peaceful uses, began sometime thereafter (mid 1960's), and is basically an attempt to scale down a hydrogen bomb. The basic idea is to use a laser beam, ion beam, or electron beam to heat the surface of a pellet (Nuckolls et al., 1972). The heated surface of the pellet "boils off," accelerating the outer layers of the pellet inward. The spherical convergence of the pellet compression would produce a high density and high temperature plasma in the center of the pellet for a long enough time to ignite deuterium and tritium. The central pellet plasma would be confined by its own inertia thus leading to the term "inertial confinement."

MAGNETIC CONFINEMENT, THEORY AND EXPERIMENTS

We can use very general considerations to derive the physical conditions that characterize the atomic processes we have to consider for magnetic fusion. Since a temperature of 10 keV is far hotter than can be tolerated by any material, the plasma must be kept out of contact with any containment vessel. The idea of magnetic confinement is to use the pressure of high strength magnetic fields (1-10 Tesla) to confine the plasma. The plasma pressure is limited by the requirement that it cannot be much greater than the magnetic field pressure. Since most engineering structures can support, at best, pressures of about 1000 atmospheres, it is reasonable to expect the plasma pressure to be on the order of 1000 atmospheres or smaller. Since the pressure is the product of the density and temperature ($p = nT$), a temperature of 10 keV and a pressure of 1–1000 atmospheres implies densities ranging from $6 \times 10^{13} \text{cm}^{-3}$ up to $6 \times 10^{16} \text{cm}^{-3}$. Realistic systems would be expected to operate with densities of 10^{13} to 10^{15}cm^{-3}. For the confinement condition $n\tau_E \sim 10^{14} \text{sec/cm}^3$, these densities imply $\tau_E \sim 0.1$ to 1 sec.

A plasma can be confined because it consists of electrons (charge $q_e = -e$) and ions (charge $q_z = Ze$) which are affected by magnetic and electric fields. The charged particles in a magnetic field B will execute gyro-orbits of frequency $\omega = (qB/mc)$ and radius $\rho_L = v_\perp/\omega$, where $v_\perp$ is velocity perpendicular to B (see Chen, 1974). The charged particles will move freely along the field. The magnetic field will do no work on the charged particles since the work done by the field on a particle with charge q is $W = \vec{F} \cdot \vec{v} = q(\vec{E} + (\vec{v}/c) \times \vec{B}) \cdot \vec{v} = e\vec{E} \cdot \vec{v}$. Thus a magnetic field provides confinement in two directions (perpendicular to the field). The problem is then how to provide confinement in the third direction along the field lines.

There are, in general, two approaches to magnetic confinement (Fig. 3). The first approach is to provide confinement by bending the field lines around into a torus so that the lines are essentially endless. The second is try to stop the end losses by reflecting the particles from the ends of the field lines. The first approach is termed toroidal or closed confinement, and the second is called linear or open confinement. Examples of toroidal confinement systems are tokamaks, stellarators, and toroidal pinches (see Teller, 1981). Mirror machines and linear pinches are examples of open confinement systems (Teller, 1981). Confinement along the field lines is essential, since the ion velocity $(v \sim (E/m)^{1/2})$ is about 2×10^8 cm/sec. Without confinement along the field lines a confinement time of one second would require a 2000 km long reactor, much beyond what is reasonable.

There are three requirements that any magnetic fusion experiment must meet. The particles must be on confined orbits. The magnetic field created by the external and internal currents must be in a quasi-stable equilibrium (forces must balance in a stable way). The loss rate of particles and energy due to collisions and instabilities must be small.

The net motion of a particle in a magnetic field is given by the motion of the guiding center (the motion of the center of the gyro-orbit). In an inhomogeneous and curved magnetic field, the motion of a particle in a gyration is non-uniform with the result that there is a small net motion of the guiding center of the

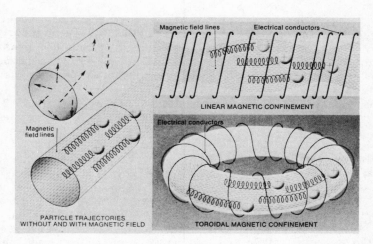

Fig. 3.   Schematic drawing of charged particle confinement by a
          magnetic field.   (786452)

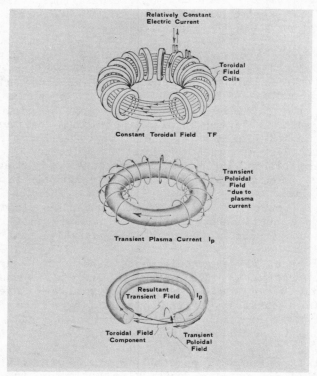

Fig. 4.   Schematic illustration of a tokamak showing the toroidal
          and poloidal magnetic fields. (754023)

particle after one gyration. These small net motions add and result in a "drift" of the particle guiding center. A second feature of guiding centers is that there are constants of the motion of the guiding centers (adiabatic invariants) which are useful in describing the particle orbits. One constant is the magnetic moment μ

$$\mu = \frac{W_\perp}{B} \, ,$$

where $W_\perp$ is the kinetic energy perpendicular to the magnetic field B. For a fixed total kinetic energy $E = W_\perp + W_\parallel$, where $W_\parallel$ is the energy parallel to the magnetic field, conservation of μ implies that there is a maximum magnetic field $B = (E/\mu)$ in which the particle orbit can exist. Thus particles can be confined along a field line by making the fields small in the middle and large at both ends (Fig. 3). The particles bounce away from the high field region, hence the term mirror. Of course only part of the particles are confined, those where $(W/W_\perp) < B_{end}/B_{center}$ ($= R_n$, the mirror ratio) with W and $W_\perp$ being evaluated at the middle of the field line. Particles with $(W/W_\perp) > R_n$ escape and are lost. Thus scattering events which change the angle of the particle velocity with respect to the field line can change $W/W_\perp$ and cause previously confined particles to be lost. In a conventional mirror this turns out to be the dominant loss mechanism.

A new type of mirror machine is being investigated, the tandem mirror (Fowler and Logan, 1977). It is basically a long solenoid with a conventional mirror at each end. This design overcomes some of the problems with the high loss rate of a standard mirror.

An open confinement scheme has a confinement time of about an ion-ion scattering time, which is fairly short. A closed confinement scheme, such as a tokamak or stellarator, bends the field lines around in a torus so that they remain in the plasma forever. The particle orbits, in both cases, are described by drifts and the conservation of μ. Orbits for mirrors and toruses are reviewed in Siambus, et al. (1965) and Furth and Rosenbluth (1969).

In a simple toroidal field, the particles would drift vertically due to the curvature and gradients of the toroidal field. Since ions and electrons would drift in opposite directions, a charge separation would develop, creating an electric field in the vertical direction. Another drift, the $\vec{E} \times \vec{B}$ drift, would cause the plasma to expand outward, and there would be no equilibrium. Equilibrium can be obtained if the magnetic field

lines are made helical (Fig. 4) so that the particle drifts average
to zero. The current in a tokamak is usually induced by making the
plasma the secondary winding of a transformer. In a tokamak, the
helical pitch of the field is then provided by an internal
current. In a stellarator, it is provided by external helical
windings.

The study of how the forces produced by the plasma pressure
and the forces produced by the magnetic and electric fields balance
in a plasma is called magnetohydrodynamics (MHD). The magnetic and
electric fields are produced by internal and external currents and
fields. An important quantity in MHD is the ratio of the plasma
pressure to the magnetic field pressure, $\beta = 2nT/(B^2/8\pi)$. As β
increases, the plasma typically becomes more unstable. Theoretical
estimates of the upper limit on β in tokamaks are four to seven
percent (some estimates go as high as twenty percent). Experiments
have reached two and one-half percent. Mirror experiments have
reached betas of fifty percent and even one-hundred percent. The
pressure, nT, is fixed by the fusion power density needed. Raising
the magnetic field strength raises the cost of the magnets and thus
the cost of a reactor. There is thus a great incentive to use as
high a β as possible, which is a clear advantage mirror machines
may have over tokamaks. However, both types of machines have a
large volume which contains magnetic field but no plasma, so a
detailed comparison requires detailed designs. A good review of
MHD equilibrium and stability issues for mirrors is Taylor (1963)
and for toruses in Solovev and Shafranov (1967).

The biggest advantage of toroidal systems is that their
particle and energy losses should be diffusive. A collision which
changes the particle velocity should only cause the guiding center
to move from the original field line to an adjacent one. Thus
energy would be lost through conduction and convection
perpendicular to the magnetic field.

In a tokamak, the helical field lines map out closed surfaces
called flux surfaces. The plasma temperature and density are
reasonably constant along the field lines, so the energy is lost by
diffusion perpendicular to the flux surfaces from the center to the
plasma edge. A naive estimate of the thermal diffusivity across
the field lines is given by $D \sim (r_L^2/\tau_i)$ where r_L is the gyroradius
and τ_i is the ion-ion scattering time (Spitzer, 1963). We can
write a simple energy balance equation for a cylinder

$$\frac{\partial E}{\partial t} = \frac{1}{r} \frac{\partial}{\partial r} \left(rD \frac{\partial E}{\partial r} \right),$$

where r is the distance from the center of the plasma, and E is the
plasma energy density. This can be approximated as

$$\frac{E}{\tau_c} \approx D \frac{E}{(a/2)^2} \, ,$$

or

$$\tau_c \approx \frac{a^2}{4D} = \tau_i \frac{a^2}{4r_L^2} \, ,$$

where a is the minor radius of the torus and τ_c is the confinement time. Thus the confinement time has been increased by the factor $a^2/4r_L^2$ compared to a mirror ($\tau_c \sim \tau_i$ for a mirror). However, things are, as always, a little more complicated. Many of the particles are "mirror trapped" in the toroidal field since it increases toward the center of the torus, and the correct gyroradius to use is not the one for the toroidal field, but the one for the poloidal field (the field the short way around the torus) set up by the plasma current. Since $B_p \sim 0.1 \times B_T$, the factor $a^2/4r_L^2$ is not as large as one might at first expect. However, it is still $\gtrsim 50$ for most tokamak and stellarator experiments.

The measured loss rates are, unfortunately, greater than the rates calculated using more exact versions of the classical theory above. These "anomalous" losses are probably due to small scale turbulence which either increases the effective step size across the field lines between collisions, or decreases the collision time. The physics of this loss process is the focus of much of the experimental and theoretical work in the fusion community. Baldwin (1977) presents a review of classical and anomalous losses in mirrors. The theory of classical losses for toruses is reviewed in Hinton and Hazeltine (1976) and for anomalous losses in Tang (1978).

Table I lists the major tokamak experiments in the United States. I will concentrate on atomic physics issues in tokamaks, since they are the best developed magnetic confinement experiments. It should be noted that mirror machines offer the promise of superior MHD properties, and have some reactor design advantages as well. If they succeed in achieving the good confinement of tokamaks, they should have similar atomic physics problems. A recent review of mirror research is given in Cohen (1980). Two recent reviews of tokamaks are given in Rawls (1979) and Furth (1975).

To date tokamaks have achieved central temperatures of about 7 keV, densities of $10^{14} - 10^{15}$ cm^{-3}, and $n\tau_E \sim 3 \times 10^{13}$ sec/cm^3. Mirror experiments (TMX) have achieved temperatures of about 200 eV, densities of about 5×10^{13} and $n\tau \sim 7 \times 10^{10}$ sec/cm^3 (Coensgen

Table 1 (Rawls, 1979). Design Parameters and "Typical" Plasma
Properties for Major U.S. Tokamaks

Tokamak	a (cm)	R_0 (cm)	B_T^a (kG)	I_p^a (kA)	$T_e(0)^b$ (eV)	$T_i(0)^b$ (eV)	n_e^b (cm^{-3})	τ_E^b (ms)
ST	13	109	50(40)	130(70)	2500	600	4×10^{13}	10
ATC[c]	17	90	20(15)	100(60)	1100	250	1.5×10^{13}	5
ORMAK	23	80	25	340(170)	1500[d]	1800[d]	3×10^{13}	10
ALCATOR A	10	54	120(80)	400(200)	1200	800	6×10^{14}	20
DOUBLET II-A	14×40	66	10(8)	320(150)	400	200	3×10^{13}	4
PLT	45	130	45(32)	1400(550)	4000[d]	7000[d]	8×10^{13}	100
ISX	25	90	18(12)	160(110)	1300[d]	1800[d]	6×10^{13}	20
DOUBLET III	45×150	143	40(20)	5000(2000)	1000	400	1×10^{14}	60
ALCATOR C	17	64	140(70)	1000(400)	900	700	5×10^{14}	30
PDX	45	145	24(24)	500(500)	3000[d]	6000[d]	1×10^{14}	80

[a] Design values (normal operating values in parentheses).

[b] Maximum obtained - not necessarily in same discharge.

[c] Before compression.

[d] With neutral beam heating.

Table 2. Examples of the United States Tokamak System

	PLT	→ TFTR	→ FED(estimated)
R (major radius)	140 cm	250 cm	520 cm
a (minor radius)	40 cm	85 cm	130 cm
Date	1978	1984	1991
Cost (millions of 1981 dollars)	30	350	1500

et al., 1980). The growth in the United States tokamak fusion program is illustrated in Table II. Similar sized tokamaks to TFTR are being constructed throughout the world (JT-60, Japan; JET, Europe; and T-15, USSR). The mirror program is building a large experiment at Livermore, California, costing about $200 million, to be ready in about 1984. The United States magnetic fusion budget is about $450,000,000 a year, and the worldwide budget is about 10^9/year.

IMPURITIES

Ideally, the plasma in a fusion experiment consists entirely of hydrogen ions and electrons. Unfortunately, this is often not the case. Ions of elements used in the construction of the vacuum vessel, walls and limiters, and lighter elements usually associated with high vacuum systems are often present in significant quantities. These "impurity" ions degrade the fusion potential of the plasma. They contribute to the plasma pressure without adding significantly to the fusion yield. Since the MHD stability limits are usually set by a maximum $\beta = P/(B^2/8\pi)$ where $P = \sum n_i T_i$, summed over the plasma constituents, impurities "dilute" the reacting plasma reducing the fusion yield for a given β. Fully stripped impurities also increase the bremsstrahlung losses per ion since these losses are proportional to Z^2.

Dilution and enhanced bremsstrahlung are usually the major concern for fully stripped ions. However, partially stripped ions can lose appreciable energy through line radiation. At densities of $10^{14} cm^{-3}$, the plasma is optically thin, and the radiation goes directly from the center of the plasma to the walls thus short circuiting the plasma confinement (Fig. 5).

Energy losses from the plasma edge, such as radiation from partially stripped low Z ions like carbon or oxygen and losses due to the charge exchange of hot ions and cold atoms, do not degrade the confinement since the plasma energy has already reached the edge. In fact, low Z impurity radiation may help reduce the high Z impurity level by reducing the power that strikes the wall and limiters as charged particles. Energetic particles can produce high Z impurities by sputtering wall or limiter materials.

The major energy losses come from electron impact excitation of bound electrons which decay radiatively (Jensen, et al., 1977). From the requirement that $n \sim 10^{14}/cm^3$ and $T \sim 10$ keV, we can specify the conditions appropriate for the atomic processes in the central plasma. The radiative lifetime for most excited states (Einstein $A \sim 10^{10}$ sec^{-1} => $\tau_{rad} \sim 10^{-10}$ sec) is short compared to the electron excitation collision time ($\upsilon \sim 10^5$ sec^{-1} => $\tau_{ee} \sim 10^{-5}$ sec). This condition in which the ions are in the ground state before excitation is usually referred to as coronal. One can then

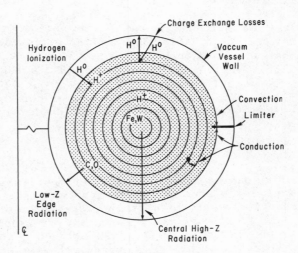

Fig. 5.   Schematic diagram of energy flows in a tokamak.   The
          center of the torus is on the left.   (782058)

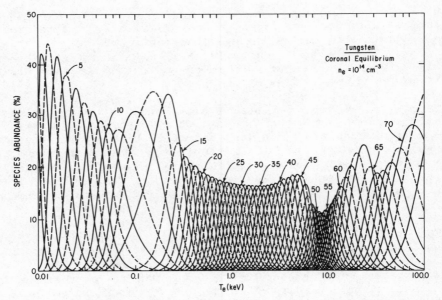

Fig. 6.   Charge state distribution as a function of electron
          temperature for tungsten in coronal equilibrium (Post,
          et al., 1979).   (782223)

write an ionization balance equation of the form

$$\frac{\partial n^i}{\partial t} = R^{i+1 \to i} \, n^{i+1} + I^{i-1 \to i} \, n^{i-1}$$

$$- R^{i \to i-1} \, n^i - I^{i \to i+1} \, n^i \,, \tag{6}$$

where n^i is the density of charge state i, $R^{i+1 \to i}$ is the frequency for recombination from charge state $i + 1$ to i, and $I^{i \to i+1}$ is the frequency for ionization from charge state i to $i + 1$. It is useful to study the "coronal equilibrium" case with $\partial n^i / \partial t = 0$ for all i from 0 to Z. Then $R^{i+1 \to i} \, n^{i+1} = I^{i \to i+1} \, n^i$. If we know the recombination and ionization rates, we can solve for the charge state distribution.

The ionization rate can be written as $n_e \langle \sigma v \rangle$ where n_e is the electron density, and $\langle \sigma v \rangle$ is the reaction rate for ionization averaged over a Maxwell-Boltzmann electron distribution for a given electron temperature. The rates are usually summed over all the bound electrons. $\langle \sigma v \rangle$ for a given charge state is then only a function of the electron temperature.

Ionization cross sections are usually evaluated theoretically and benchmarked by a few measurements. Commonly used rates were recently reviewed by De Michelis and Mattioli (1981). As an example of the rates that are typically used by plasma modelers, I will describe the rates used in our code at Princeton (Post et al., 1977). In our calculations, we use an amalgam of cross sections from Jordan (1970); Jacobs (1972); Sampson and Golden (1971); Lotz (1967), and supplemented by those of Younger (1981).

The dominant recombination processes are radiative and dielectronic. Again we use general prescriptions. In general, dielectronic recombination is dominant when the ion has three or more bound electrons. Radiative recombination is important when dielectronic recombination is small (0, 1, or 2 bound electrons). Three body recombination is extremely small except at very low temperatures and high densities (1 eV and $10^{14} cm^{-3}$). All of the rates (except three body) are roughly proportional to the electron density. Dielectronic recombination becomes small at high densities but at $10^{14} cm^{-3}$ it is roughly proportional to the electron density. Thus we can write $R^i, I^i = n_e \langle \sigma v \rangle^i$ with the $\langle \sigma v \rangle^i$ being dependent only on the electron temperature.

Using the above rates we can compute the equilibrium fraction of each charge state for any given element as a function of electron temperature for any element. The general nature of our formulae allows us to tackle a high Z element such as tungsten (Post et al., 1979) (Fig. 6) as well as the lower Z elements such as C, O, and Fe.

The next step after determining the ionization balance is to determine how much energy is lost through the excitation of bound electrons through electron impact collisions and their subsequent radiative decays. We use primarily the prescriptions of Seaton (1972) with suitable modifications. For systems with many bound electrons (> 15), $\Delta n = 0$ transitions are extremely important. Using the calculated charge state distribution one can then compute the radiative power loss density as

$$P = n_z n_e \sum_{i=0}^{z} f_i \sum_{\substack{all \\ excitations}} \Delta E^i \langle \sigma v \rangle^i , \tag{7}$$

where n_z is the impurity density, n_e is the electron density, f_i is the fractional abundance of charge state i, and the sum is over all excitations of ions with charge i. ΔE^i is the excitation energy and $\langle \sigma v \rangle^i$ is the excitation rate. Polynomial fits of these cooling rates have been tabulated (Post et al., 1977).

The line radiation is, of course, largest when there are many bound electrons (three or more). In general, an ion will strip out until the ionization potential is of the order of the electron temperature (Fig. 7). Thus lower Z impurities will have little line radiation at high temperatures. Higher Z impurities will have larger power loss rates, not only due to having higher ionization potentials and higher excitation energies, but also because they have many more bound electrons with many possible excitations. Thus the cooling rate for tungsten is about sixty times that for iron and about five hundred times that for oxygen (Fig. 8) (Jensen, 1977).

Although dielectronic recombination had long been recognized as an important recombination mechanism for astrophysical plasmas (Burgess, 1965; Cox, 1969), it was only applied to magnetic fusion work in 1975-1976 (Merts et al., 1976; Jensen et al., 1977). Until then, tungsten limiters and metal rails, inserted to keep the plasma from hitting the walls, were widely used in tokamaks because of their excellent refractory properties. Estimates of the tungsten density in tokamaks were about 0.1-0.3%. If one neglects dielectronic recombination, tungsten strips down to an average charge of + 40 at a temperature of 1 keV, and the radiation is not extremely large. Including dielectronic recombination reduces the average charge to about twenty-five at an electronic temperature of 1 keV, with an increase in the radiation factor of about ten (Jensen et al., 1977). At this level tungsten could radiate and indeed was radiating most of the energy contained in tokamaks such

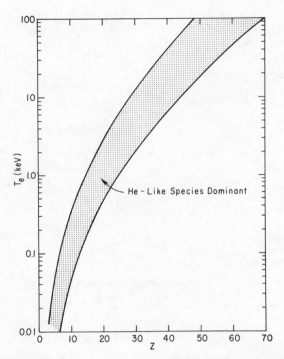

Fig. 7. The temperature range for which helium-like ions of
atomic number Z are the dominate charge state. For a
given Z, hydrogen-like and fully stripped ions pre-
dominate at temperatures above the band. Ions with
three or more bound electrons are dominate below the
band. Only ions below the band can radiate strongly.
(792394)

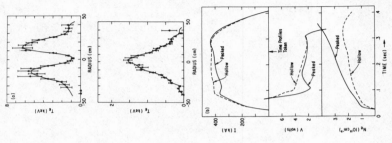

Fig. 9. Hollow electron temperature profiles in the PLT tokamak caused
by large amounts of radiation
from tungsten ions (Hsuan, et al.,
1978). (786412)

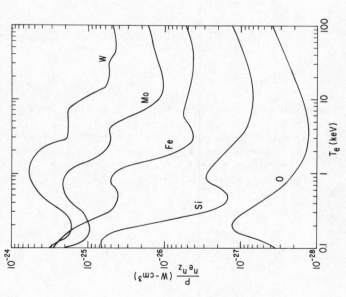

Fig. 8. Emissivities of various elements in
coronal equilibrium (Jensen, et al.,
1977). (772182)

as ORMAK (Isler et al., 1977) and PLT (Hinnov and Mattioli, 1978). PLT even had hollow electron temperature profiles due to the large radiation losses (Hsuan et al., 1978) (Fig. 9). Attempts to heat PLT with the injection of high energy neutral beams only brought more tungsten into the discharge, resulting in the heating power being radiated away with no temperature increase. Calculations were done of the expected tungsten lines (Isler et al., 1977) (~ 50 Å), and these lines were measured. The suspicions about tungsten which had already been reinforced by the power loss calculations were finally confirmed. The tungsten limiters were replaced with carbon limiters, and PLT was subsequently heated to about 7 keV (Eubank et al., 1979). Very high Z materials are no longer used in tokamaks.

Many tokamaks are heated by the injection of beams of high energy neutral atoms (20-40 keV). The presence of these neutral atoms can lead to an additional recombination mechanism through the reaction

$$H^o + A^{+q} \rightarrow H^+ + A^{q-1} \; .$$

This process is particularly important for systems with just a few bound electrons since radiative recombination is inherently small and dielectronic recombination is very small in these cases due to a small number of allowed excitations. The ionization balance equilibrium equation has to be modified (Krupin et al., 1979; Hulse et al., 1980; Puiatti et al., 1981)

$$\frac{n_{i+1}}{n_i} = \frac{I^{i \rightarrow i+1}}{R^{i+1 \rightarrow i} + (n_n/n_e) \, v\sigma^{i+1 \rightarrow i}} \; , \tag{8}$$

where I and R are the usual ionization and recombination rates (dielectronic and radiative), n_n is the neutral density, v is the neutral velocity, and $\sigma^{i+1 \rightarrow i}$ is the charge exchange cross section. The main effect of this process is to postpone or eliminate the burnout of an ion to non-radiative states as higher temperatures are approached. Modest values of n_n/n_e can significantly affect the ionic balance (Fig. 10) (Hulse et al., 1981). The enhancement of the lithium-like and more highly recombined states can lead to an enhancement in the total impurity radiation (Hulse et al., 1981) (Fig. 11). The effect has only been a major term in the power balance in beam heated tokamaks where the plasma has a fairly high impurity content already, and high power, low energy beams (leading to a high neutral density, n_o) are used.

Clearly, the presence of impurities will increase the plasma energy losses, and thus make fusion more difficult. We can rewrite

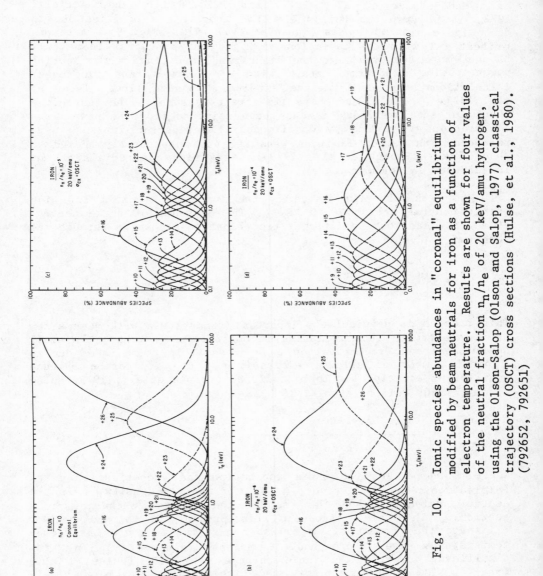

Fig. 10. Ionic species abundances in "coronal" equilibrium modified by beam neutrals for iron as a function of electron temperature. Results are shown for four values of the neutral fraction n_n/n_e of 20 keV/amu hydrogen, using the Olson–Salop (Olson and Salop, 1977) classical trajectory (OSCT) cross sections (Hulse, et al., 1980). (792652, 792651)

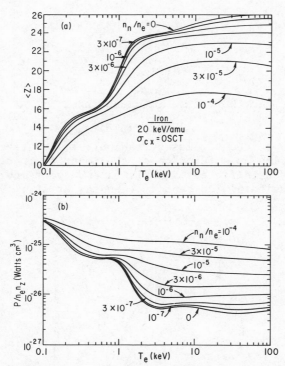

Fig.   11.   Iron neutral-beam modified "coronal" equilibrium.
             Hydrogen neutrals at 20 keV/amu are considered, using
             the Olson-Salop (Olson and Salop, 1977) classical
             trajectory cross sections.   The curves are parameterized
             by the neutral fraction n_n/n_e. Shown are (a) the
             average charge state $<Z>$, and (b) the overall radiation
             rate coefficient $P/n_e n_Z$, both as functions of the
             electron temperature (Hulse, et al., 1980).   (792653)

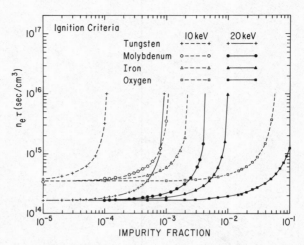

Fig. 12.   The $n_e\tau_E$ due to plasma transport losses required for
            ignition as a function of the impurity fraction n_Z/n_i
            for four different impurities and two temperatures. n_i
            is the total ion density (Jensen, et al., 1977).
            (772101)

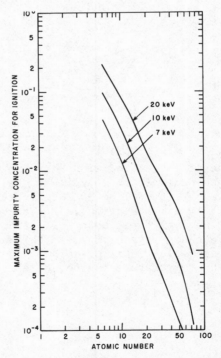

Fig. 13.   Maximum allowed impurity concentration versus Z for
            ignition at various plasma temperatures $T_e = T_i$,
            assuming zero non-radiative losses (Jensen, et al.,
            1978). (773294)

the energy balance Eq. (2) to include line radiation and beam heating in the form (Jensen et al., 1977; Jensen, et al., 1978)

$$\frac{(3/2)\ (n_e T_e + \underset{ions}{\Sigma}\ n_i T_i)}{\tau_E} + R = P_b + P_\alpha\ , \tag{9}$$

where R is the radiation loss, P_b is the beam heating, P_α is the alpha heating, T_e is the electron temperature, and T_i is the ion temperature. For a DT plasma, $P_\alpha = 0.2\ QP_b$ where $Q = P_f/P_b$. P_f is the fusion power and Q is the fusion power multiplication factor. Equation (9) can be written as

$$n_e \tau_E = \frac{(3/2)\ [(1+f\bar{Z})T_e + (1+f_z)\ T_i]}{H - f_z L_z - L_H}\ , \tag{10}$$

where $H(Q) = P_\alpha/n^2$, f_z is the fraction of the ions that have charge Z, and $\bar{Z}$ is the average charge state. L_z is the radiation rate (in units such as watts cm^3) per free electron per ion, and L_H is the hydrogen bremsstrahlung rate (in units such as watts cm^3). As one can see from Eq. (10) (analgous to Eq. 2) as f_z is increased, the $n_e \tau_E$ required for a given Q (or H) increases. At a large enough f_z, the denominator of Eq. (9) becomes zero and even perfect confinement is not good enough.

At such a point the radiation losses equal the fusion and beam heating. Since the radiation rates are largest for high Z and decrease (in general) as T_e is increased, higher Z raises the $n_e \tau_E$ required and higher T_e lowers it (Fig. 12). We can define a fatal fraction as the f_z at which the denominator of Eq. (10) becomes zero. This then sets an upper limit on the impurity concentration that can be tolerated. If we take as our criteria that the plasma must be ignited, we find that the fatal fraction scales as $\sim Z^{-2.5}$ (Jensen et al., 1978). Less than 10^{-4} tungsten can be tolerated as compared to 10^{-1} carbon (Fig. 13).

Although impurities have played and continue to play an important role in the central plasma, it is always a bad influence by spoiling confinement. Eliminating high Z impurities is viewed as one of the key problems in fusion work.

However, the key questions of the economic and physics viability of fusion reactors still hinge on questions of MHD stability (maximizing β) and plasma transport. Impurities are viewed as a nuisance which hopefully will go away. Atomic processes do not contribute usefully to better confinement or better stability, and most plasma physicists do not particularly care about improving our knowledge of electron impact collision physics in hot plasmas. For them it is enough to know that tungsten is very bad and carbon is not "too" bad. However better knowledge of atomic collision physics in hot plasmas is still crucially important since we do not know how well we will be able to control impurities. In the other three areas of fusion research, the plasma edge, heating methods, and diagnostics, atomic processes play a beneficial role, and most of the qualitatively new applications of atomic processes will probably be in these areas.

PLASMA EDGE

Atomic processes play an important role in the behavior of the tokamak edge plasma. The particle balance (Fig. 14) in a tokamak consists of neutral hydrogen atoms and molecules going into the plasma and undergoing ionization and charge exchange. Some of the hydrogen atoms gain enough energy by charge exchanging with hot plasma ions to increase their mean free path for ionization [$\lambda = (v_0/n_e \langle \sigma v \rangle)$] to penetrate to the center before they are ionized (Hughes et al., 1978). The plasma ions then diffuse across the flux surfaces until they reach the edge where they recombine at the wall or at a limiter. The ions and neutral atoms strike the wall and limiter and can sputter off material so that some of it enters the plasma and becomes an impurity. Light impurities such as oxygen are also desorbed from the wall due to electron and ion impacts with the wall.

The central portion of a fusion plasma is quite hot (1-5 keV in current tokamaks and greater than 10 keV in a future reactor experiment). Low Z impurities are completely ionized there and do not radiate strongly. The edge plasma near the vacuum vessel walls is generally much cooler with temperatures typically in the 10-50 eV range and sometimes as low as 1-10 eV (McCracken and Stott, 1979). At such low temperatures, atoms of low Z elements such as carbon and oxygen are usually only partially ionized, and thus can radiate strongly. In addition, the plasma edge is usually quite turbulent, and the ions there are poorly confined. Thus ions can recombine by contact with solid structures such as the wall or limiter, and be reintroduced into the edge plasma as neutrals. This "recycling" of impurity ions leads to the production of low ionization states of the impurity ions born from the ionization of the "recycling" neutral impurity atoms. If these ions can diffuse in time scales short compared to ionization times, particularly the time required to ionize an atom from a neutral atom to a helium-like ion, ions with excitation energies much less than the electron temperature can exist. This lowering of average charge state below "coronal equilibrium" can lead to a substantial enhancement of the radiation rate compared to "coronal equilibrium." Enhancement of the recombination due to charge-exchange recombination with the recycling neutral hydrogen can augment this radiation enhancement as well (Hogan, 1981). For example, if a neutral carbon atom is instantaneously deposited into a $10^{13} cm^{-3}$, 50 eV plasma, it will

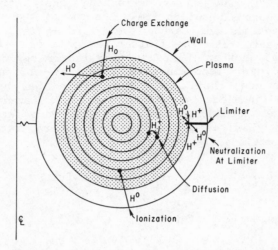

Fig. 14.   Schematic diagram of the particle flows in a tokamak. .
             The center of the torus is on the left.   (782057)

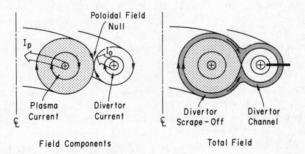

Fig. 15.   The magnetic field geometry for a poloidal divertor. A
             null in the poloidal field is created by a coil carrying
             a current parallel to the plasma current. Field lines
             outside this null point encircle both the main plasma
             and the divertor coil. The plasma edge is thus "di-
             verted" to a region away from the main plasma. In
             addition to the fields shown here, there is also a
             strong toroidal magnetic field.   (803288)

radiate about 10 keV before reaching the helium-like state appropriate to coronal equilibrium. The sum of the single electron ionization potentials for carbon from 0 to +4 is 150 eV. If the carbon +4 ion then hits the limiter, recombines and reenters the plasma, it loses another 10 keV in radiated energy before reaching q = +4. Ordinary recombination times are often larger than transport times near the edge. This problem has been studied by solving time dependent equations for the transport of the ionization stages for carbon and oxygen including recombination and ionization (Hawryluk et al., 1979) as a generalization of Eq. 6. The relevant equations are

$$\frac{\partial n^i}{\partial t} = \frac{1}{r}\frac{\partial}{\partial r}\left(r\Gamma^i\right) + R^{i+1\to i}n^{i+1} + I^{i-1\to i}n^{i-1}$$

$$- n^i\left(R^{i\to i-1} + I^{i\to i+1}\right),\qquad\qquad (11)$$

where Γ^i is the impurity flux $\left(\Gamma^i = D\left(\partial n^i/\partial r\right)\right)$, for example.

As we have said before, impurity radiation from the edge may be beneficial. If we examine the energy balance in a tokamak (Fig. 5), we see that while radiation from the center is detrimental to energy confinement, radiation from the edge is not, since the energy would, in any event, be quickly lost from the edge by convection and conduction. In fact losing energy from the edge by having high energy particles strike the wall or limiters can lead to the sputtering of metallic impurities which then get to the plasma center and radiate from there. Losing the energy from the edge by photons does not lead to much sputtering, and is a very attractive way of removing the thermal plasma energy losses. Low Z elements such as oxygen and carbon often work well in current tokamaks since they are fully stripped at the plasma center and can radiate strongly at the edge.

The interaction of the edge plasma with the vacuum vessel and limiters is important because it can lead to the introduction of unacceptable levels of moderate Z and high Z impurities into the plasma. The most commonly studied mechanism for impurity production is chemical and physical sputtering of metal surfaces by ions and neutrals. Extensive data in the form of the yield per atom as a function of the projectile energy has been obtained and catalogued (McCracken and Stott, 1979; Roth et al., 1979). While it is expected that sputtering will be the dominant impurity production mechanism on future large experiments, it has not been fully established that it is the dominant mechanism on current experiments.

In addition to the impurity control problem, another problem with fusion reactors is exhausting the helium ash which results from the DT fusion reaction. On an INTOR-sized experiment (INTOR, 1980), a reactor feasibility experiment, the helium is produced at the rate of one percent (of total particles)/sec. Thus helium must be exhausted and replaced with D and T if the fusion reactor is to burn for longer than about twenty seconds.

One approach to solving the impurity control and helium pumping problems for INTOR has been to use a poloidal divertor (Fig. 15), a device in which the edge plasma is swept away from the main plasma to a chamber where it can be controlled. A large number of complicated atomic processes take place when a plasma strikes a metal surface. These effects have been modeled (Heifetz et al., 1981; Petravic et al., 1981) including a large number of reactions (Table III). The plasma flows into the divertor (Fig. 16), strikes the neutralizer plate, and recombines. The resulting neutrals move around the divertor, undergoing charge exchange until they are either ionized or go down the pump or return to the main plasma. The reflection properties of the 1-500 eV neutrals and ions are a crucial ingredient in the model. What is used now is a collection of theoretical and experimental results (Eckstein and Verbeek, 1979). There is almost no data on low energy collisions (1-30 eV), and the theoretical models are very poor in that energy range.

A key feature of the above mentioned models is the number of molecules formed by low energy hydrogen atoms striking a surface. The assumption is made that in steady state the atoms that are not all reflected as energetic atoms, but some of them penetrate significantly into the crystal lattice and are trapped, eventually diffuse to the surface, and are desorbed as molecules. The molecules undergo reactions 3, 4, and 5 in Table III. If they are ionized (H_2^+), they further undergo reactions 5a and 5b. For moderate electron temperatures the molecule is first ionized, then dissociated into H^+ and H^O (Fig. 17). This leads to a very localized ionization source near the neutralizer plate, which in turn leads to lowering the plasma temperature and raising the plasma density in the divertor (Petravic et al., 1981). This mode of divertor operation has the potential of reducing the sputtering by reducing the energy of the ions to below the sputtering threshold. It has also been observed experimentally on D-III (Mahdavi et al., 1981; Nagami et al., 1980).

An additional function of a divertor for a reactor experiment would be to provide helium pumping. Early estimates of the pumping speeds needed for reactor experiments were large (500,000 ℓ/sec) and caused a serious design problem (INTOR, 1980). It was proposed that a poloidal divertor could enhance the pumping of helium compared to hydrogen (Shimomura et al., 1979). The argument was

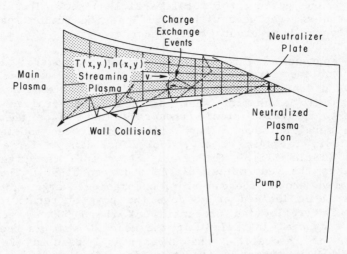

Fig. 16.   Schematic representation of the operation of a poloidal
            divertor. In addition to the fields shown here, there
            is also a strong toroidal magnetic field.   (81P0177)

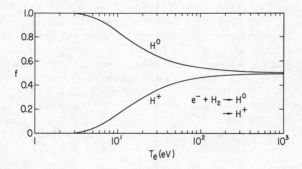

Fig. 17.   Equilibrium fraction of H^0 and H^+ produced from $H_2 + e^-$
            $\rightarrow H^0$, H^+ following molecular ionization and dissociation
            channels (Heifetz, et al., 1981).   (802112)

Table 3. Reactions Used in Divertor Model (Heifetz, 1981)

(1)	$H^o + H^+ \rightarrow H^+ + H^o$
(2)	$e + H^o \rightarrow H^+ + 2e$
(3)	$e + H_2^o \rightarrow 2H^o + e$
(4)	$e + H_2^o \rightarrow H^o + H^+ + 2e$
(5)	$e + H_2^o \rightarrow H_2^+ + 2e$
(5a)	$e + H_2^+ \rightarrow 2H^o$
(5b)	$e + H_2^+ \rightarrow H^o + H^+ + e$
(6)	$He^o + He^+ \rightarrow He^+ + He^o$
(7)	$He^o + He^{++} \rightarrow He^{++} + He^o$
(8)	$He^o + H^+ \rightarrow He^+ + H^o$
(9)	$e + He^o \rightarrow He^+ + 2e$

Table 4.   Neutral Beam Systems

Energy	Ion	Power	Experiment	Date
40 keV	D^o, H^o	3 MW	PLT	(1978)
50 keV	D^o, H^o	8 MW	PDX	(1981)
80 keV	D^o, H^o	7 MW	D-III	(1983)
120 keV	D^o	20 MW	TFTR	(1984)
150 keV	D^o	60 MW	FED	(1990)

that helium would slow down due to collisions with the wall, and be quickly ionized, whereas hydrogen could recover the energy lost by wall collisions by charge exchange with the hot plasma ions, and thus keep a long ionization mean free path. Detailed calculations (Callen et al., 1980; Seki et al., 1980; Heifetz et al., 1981) showed that this was not the case. Charge exchange of hydrogenic ions and neutrals can reduce the directional transport of hydrogen relative to helium because it "isotropized" the hydrogen neutrals. The mean free path for charge exchange for hydrogen is about one-half the mean free path for ionization.

However, "helium enrichment" is not necessary since, if a divertor can be operated in a high density regime (Petravic et al., 1981; Mahdavi et al., 1981; Nagami et al., 1980), the neutral pressure at the pumping duct can be as high as 0.1 torr, and large gas throughputs can be obtained with modest pumping speed systems.

Another area that is beginning to receive attention is the transport of molecular impurities such as CH_4 in the plasma edge (Langer, 1981). The production of methane by the chemical sputtering of carbon limiters and steel surfaces by hydrogen is thought to be a significant source of carbon impurities in current experiments. The chain of events is

$$e^- + CH_4 \rightarrow CH_3^+ + H + e^-$$

$$e^- + CH_3^+ \rightarrow CH_2^+ + H + e^-$$

$$e^- + CH_2^+ \rightarrow CH^+ + H + e^-$$

$$e^- + CH^+ \rightarrow C + H \quad .$$

The original CH_4 is very slow, and ionizes at the very edge of the plasma. It continues to be dissociated by repeated electron collisions from the 5-20 eV edge plasma. However, CH_x^+ is an ion and can begin to thermalize with the relatively hot (5-20 eV) ions in the edge plasma. Thus a 0.05 to 0.1 eV CH_4 can possibly produce a significant fraction of energetic C^o (about 10-20 eV) which can then penetrate more deeply into the plasma than a 0.05-0.1 eV C^o. Calculations for a variety of edge profiles show that the effect is significant and produces an enhancement in the neutral penetration for C^o of 2-3 cm on the average with a significant tail penetrating up to 10 cm further than C^o which was launched from the wall with an energy characteristic of the wall temperature. Detailed calculations of this type will become more common as more data

becomes available. A good understanding of the plasma edge will
grow in importance as the fusion community begins to design
realistic reactor level experiments which require the handling of
large heat and particle fluxes. Essential to that understanding
will be detailed knowledge of the hydrogen and impurity collision
processes in low temperature (1-200 eV) edge plasmas with densities
in the $10^{10} - 10^{14} cm^{-3}$ range.

PLASMA HEATING

The third area of fusion research involving atomic processes
is plasma heating. Since a DT plasma requires a temperature of
about 10 keV for heating by alpha particles to be effective, the
plasma must first be heated to about 10 keV. Most fusion
experiments have large internal currents, which supply power to the
electrons in the plasma by drawing the current against the plasma
resistance. However, since the plasma resistivity is proportional
to $T_e^{-3/2}$ (Spitzer, 1963), this method is most effective at low
temperatures. Other heating methods include compressing the
plasma, and launching radio waves into the plasma. However, the
most successful heating method used for tokamaks and mirrors so far
has been the injection of high energy (20-60 keV/amu), high current
(20-100 amps) neutral hydrogen beams into the plasma. The neutral
atoms cross the confining magnetic field and are "ionized" by
collisions with the plasma (Fig. 18). The energetic ions are
confined by the magnetic field and heat the background plasma
electrons and ions by elastic coulomb collisions (Jassby, 1977).

The beam atoms are "ionized" by electron and ion impact
ionization and by charge exchange (Fig. 19) (Freeman and Jones,
1974). Electron ionization is a relatively small fraction of the
stopping cross section. Below 40 keV/amu, charge exchange is
dominant, and above 40 keV/amu proton impact ionization is
dominant. Above about 80 keV/amu the stopping cross section is
approximately inversely proportional to the beam energy.

Some early work (Moriette, 1974) suggested that the ion impact
ionization cross section, σ_q, for $H^0 + A^{+q} \rightarrow H^+ + A^{+q} + e^-$ should
scale as $\sigma_q = q^2 \sigma_H$. Given the impurity levels of the then current
experiments, it was feared that this effect would raise the
stopping cross section so that neutral beams would be ionized at
the plasma edge and not heat the center (Hogan and Howe, 1976;
Duchs et al., 1977). Furthermore, it was feared that the intense
heating of the edge would lead to a large impurity influx, thus
stopping the beam even more effectively. For this reason among
others, there was great interest in the actual values of the
electron loss cross sections for hydrogen and multiply charged
ions. As a result of intense theoretical (Olson and Salop, 1976;
Olson and Salop, 1977) and experimental work (Phaneuf et al., 1978;
Berkner et al., 1978; Bayfield et al., 1976), the dominant

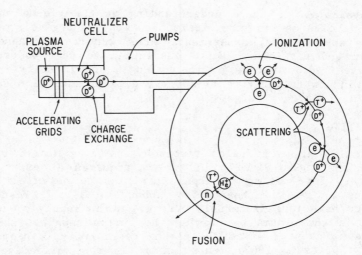

Fig. 18.   Neutral injection by the acceleration of H$^+$ and
              neutralization by charge exchange, and the subsequent
              stopping of the beam in the plasma by coulomb scattering
              and nuclear reactions (Jassby, 1977).   (753441)

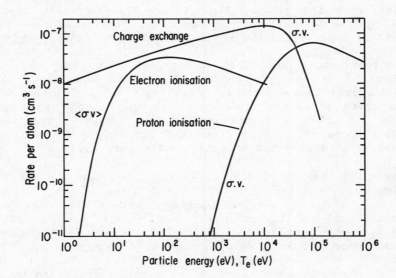

Fig. 19.   Reaction rates for atomic hydrogen for electron impact
              ionization, proton impact ionization, and charge
              exchange with protons as a function of lab hydrogen
              energy.   The energy scale for the electron ionization
              rate is the electron temperature (Freeman and Jones,
              1974).   (762150)

collision process was discovered to be charge exchange instead of ionization. The scaling was $\sim Z^{1.2}$, instead of Z^2. The results have been put in a general formula (Olson et al., 1978) for the combined processes of charge transfer and ionization for $1 \leq q \leq 50$ and 50 keV/amu $< E <$ 5,000 keV/amu:

$$\sigma_{loss} = 4.6 \; q \; \frac{(1 - e^{-y})}{y} \times 10^{-16} cm^2 \; ,$$

where

$$y = \frac{E}{32q}$$

and E is in keV/amu. With this scaling, impurities have very little effect on neutral beam penetration (Fig. 20) for beams of 100 keV/amu or less. This is a consequence of the Z scaling of σ_q for a neutral beam atom. The mean free path would be

$$\frac{1}{\lambda} = n_e \frac{\langle \sigma v \rangle}{v_o} + n_H \sigma_H + n_q \sigma_q$$

$$\approx n_e \frac{\langle \sigma v \rangle}{v_o} + n_H \sigma_H + n_q q \sigma_H$$

$$\approx n_e \left(\frac{\langle \sigma v \rangle}{v_o} + \sigma_H \right) \; ,$$

where $\langle \sigma v \rangle$ is the electron impact ionization rate, v_o is the neutral velocity, n_H is the hydrogen ion density, and n_q is the density of ions of charge q. This is the same answer obtained for the case with no impurities. However, if $\sigma_q \approx Z^2 \sigma_H$, then

$$\frac{1}{\lambda} \approx n_e \left(\frac{\langle \sigma v \rangle}{v_o} \right) + Z \; \sigma_H \right) \approx \frac{Z}{\lambda_H} \; ,$$

for typical cases, and since Z can be as large as three or four, impurities could lead to very poor penetration of a neutral beam into a tokamak plasma.

The charge exchange of hydrogen ions with partially ionized impurity ions, $H^+ + A^{+q} \rightarrow H^o + A^{+q+1}$, could serve as an important loss mechanism for fast beam injected ions if the cross sections were greater than $10^{-19} cm^2$ and the density of partially ionized impurity ions were large ($\sim$ 0.1 - 1% of n_e). However, estimates of these cross sections seem to be lower (in the $10^{-21} cm^2$ range).

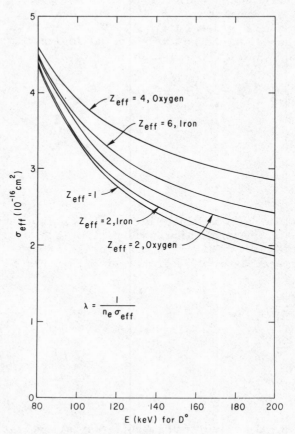

Fig. 20. Effective trapping cross section versus neutral hydrogen
beam energy, defined as

$$\sigma_{eff} = \frac{Z-Z_{eff}}{Z-1} \left(\sigma_{cx}^{H} + \sigma_{p-ion}^{H} \right) + \frac{<\sigma V>_{e-ion}}{Z(Z-1)} \left(\sigma_{cx}^{imp} + \sigma_{ion}^{imp} \right)$$

for a deuterium neutral beam traversing a deuterium
plasma with an impurity of charge Z. σ_{cx}^{H} is the hydrogen charge exchange cross section, σ_{P-ion}^{H} is
the proton impact ionization cross section, σ_{cx}^{imp}
is the impurity-hydrogen charge exchange cross section,
and σ_{ion}^{imp} is the ion impact ionization cross section.
The cross sections are taken from (Freeman and Jones,
1974) and Olson et al., 1978).(792225)

Understanding atomic processes is also crucial to understanding neutral beam systems. Neutral hydrogen beams are made by forming a low temperature plasma (1-10 eV), extracting the protons and accelerating them to $\sim$ 20-60 keV/amu while keeping the beam space charge neutralized, and then neutralizing the protons by charge exchange in a gas cell of H_2. The unneutralized ions are then bent out of the beam by magnets and the neutral beam proceeds on to the plasma (Fig. 18).

The atomic physics of molecular hydrogen plays a key role in the ion source. H^+, H_2^+, and H_3^+ are all present (Raimbrult and Girard, 1981). The H_2^+ and H_3^+ are undesirable, and considerable effort has been expended to reduce their concentration. They are formed from $H_2 + e^-$ and $H_2^+ + H_2$ reactions, and the successful approaches for reducing H_2^+ and H_3^+ concentrate on keeping the H^+ in the ion sources away from the walls to reduce the H_2 density and avoiding the production of H_2^+.

The reason H_2^+ and H_3^+ are undesirable is that they are extracted and accelerated and neutralized along with H^+. Thus the beam can have an appreciable fraction of neutrals with one-half and one-third of the extraction and acceleration voltage. These lower energy neutral atoms do not penetrate to the center of the plasma as well as the full energy atoms, and thus are not as effective in heating the plasma.

The fraction of an H^+ (or D^+ or T^+) beam that can be neutralized by charge exchange in a gas cell declines rapidly above 50 keV/amu (Fig. 21), chiefly due to the decline in the charge exchange cross section H^+ (fast) + H^O → H^O (fast) + H^+ compared to the atom impact ionization cross section H^O (fast) + H^O → H^+ (fast) + ?. This renders H^+ based neutral beam systems unfeasible above about 100 keV/amu due the low conversion efficiency for turning H^+ beams into H^O beams. It may be possible to recover the power in the unneutralized component of the H^+ beam, or inject it through a second neutralizer cell. Such schemes, however, lead to significant complications in any practical beam system, and have not yet even been demonstrated to work on any large beam system. A fairly large column density of neutral hydrogen (on the order of $10^{16}/cm^2$) has to be maintained in the neutralizer gas cell. Keeping this neutral gas out of the main vacuum system for the plasma requires extensive high speed cryogenic hydrogen vacuum pumping systems. These difficulties increase as the beam energy is raised and in practice a large neutral beam at energies of 80-100 keV/amu is a very complicated and inefficient system, involving separators, energy recovery, and huge pumping speeds (20 megaliters of liquid He cryopumping for INTOR (INTOR, 1980)) (Fig. 22) (Jassby, 1977). However, neutral beams do have the advantage that they have had the most success in heating tokamaks (Eubank et al., 1979) and mirrors (Coensgen et al., 1980). A summary of the

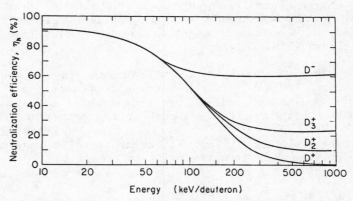

Fig. 21. Equilibrium neutralization efficiency for D^-, D^+, D_2^+, D_3^+ in a gas cell of H_2 as a function of the energy per deuteron (Jassby, 1977). (763705)

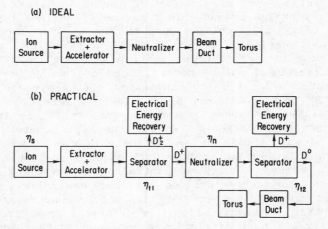

Fig. 22. Components of (a) ideal and (b) practical neutral-beam injector systems for a tokamak reactor (Jassby, 1977). (763692)

parameters of typical tokamak neutral beam systems is given in Table IV.

However, the complicated operation and low efficiency of H^+ based neutral beams, especially at the energies above the 100 keV desirable for large experiments to achieve good penetration and heating, has led to interest in beam systems based on negative ions. H^- beams can be neutralized with about sixty percent efficiency at high energies in a H_2 gas cell (Fig. 21). The chief difficulty has been the construction of high current H^- sources (10 amps or greater). A wide variety of approaches are under investigation (Sluyters, 1980). The most successful have been the use of double charge exchange of H^+ with alkalai and alkaline earth vapors, $H^+ + Na \rightarrow H^- + ...$, and the conversion, on an alkalai coated surface, of H^+ to H^-. Another promising approach is the volume production of H^- in an energetic molecular hydrogen gas (Bacal, 1981). The exact details are not clear but it is thought that the volume production process involves the electron impact dissociation of vibrationally excited H_2 into H^o and H^-. High energy beams would be useful for mirror machines as well as tokamaks.

Marginal gains in the neutralization efficiency of H^+ beams might be obtained, allowing one to push the maximum feasible energy of such beams up to ~ 120 keV/amu, by the use of He or Ne as a neutralizer gas instead of H_2 (Grisham and Post, 1981a). One gains efficiency at the cost of increasing the pumping difficulty with the use of helium.

The neutralization of H^- by photodetachment has received some study (Sluyters, 1980) but looks marginally acceptable due to high laser powers required, and the relative inefficiency of these lasers.

The search for better beams has led to a new idea, the injection of high energy neutral or singly charged atoms of Li, B, C, O, Na, Si, etc. (Grisham et al., 1981b; Dawson and MacKenzie, 1980). In the neutral beam approach (Fig. 23), negative ions of Li, C, O, or Si would be accelerated to 1-2 MeV/amu in an RF accelerator, then neutralized by either gas collisions or photo-detachment, and finally injected into the tokamak or mirror experiment (Post et al., 1981a) where they ionize in steps until they are fully stripped. The ions then heat the plasma by coulomb collisions. It appears that it may be feasible to make 500 mA to 2 amp O^- sources, for example, and the necessary RF accelerator technology is being developed for other purposes. The key issues are the formation of the negative ions Li^-, C^-, O^-, and Si^-, the one electron loss cross sections of the negative ions for gas collisions and photodetachment, and the electron loss cross sections for all charge states of the atoms at relative velocities

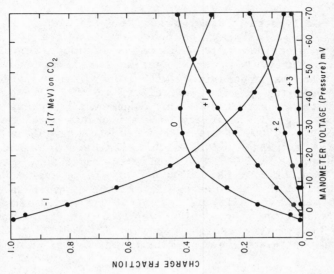

Fig. 24.   The charge state fraction of a
7 MeV Li⁻ beam after passing
through a CO_2 gas cell as a function of pressure in the gas cell
for an experiment at the Brookhaven National Laboratory (Grisham, et al., 1981c). (81P0121)

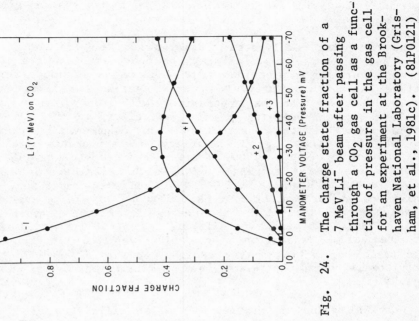

Fig. 23.   Outline of a high energy neutral
injection system based on negative
ions (Grisham, et al., 1981b).
(802298)

of ~ 10^9 cm/sec.   Recent experiments (Grisham et al., 1981c;
McCullough et al., 1981) have shown that about fifty percent of a
10^9 cm/sec Li^- beam can be neutralized in a gas cell (Fig. 24) but
the maximum neutral fractions for C^-, O^-, and Si^- are thirty,
twenty-five, and twenty percent.   The Brookhaven (Grisham et al.,
1981c) results give a hint that the cross sections for the loss of
two or more electrons in a single collision may be large.   The
injection of positive ions depends on their being injected on
unconfined orbits and being ionized in the plasma to a higher
charge state, thereby shrinking the gyroradius, and changing the
unconfined orbit to a confined one.   This effect is also important
for the neutral light atom beam.   The calculations (Grisham, et
al., 1981b) (Fig. 25) of this process involve a succession of one
electron loss collisions.   Significant multi-electron loss cross
sections would change the penetration and heating calculations.
Also uncertain is the role of two step processes in which an atom
or ion is first excited by a collision then ionized by a second
collision resulting in a possible enhancement in the ionization
rate.

The chief advantage of the light atom neutral beam approach
lies in the very large beam energy that could be used (1-2
MeV/amu).   The upper limit on the beam energy is set by the
requirement that the neutral atoms be ionized in the plasma, and
the ions thus formed be confined while they slow down and heat the
plasma.   The high energy would reduce the current required, thereby
making possible a relatively simple DC beam system with small
pumping requirements.   One oxygen beam could deliver 1 amp at 30
MeV for a total power of 30 MW.   With such low currents, the
impurity contamination would be smaller (0.1% per sec) than the
helium ash buildup.   Since the beams are low Z, the central
radiation would be low.   Normal H^+ beams typically have a 0.5-1%
contamination level of oxygen and other impurities.   Due to the
high energies of the low Z neutral beams compared to 50-100 keV/amu
for conventional H^+ based beams, the injected neutral oxygen
current per megawatt of beam power is lower for a 1 MeV/amu neutral
oxygen beam than for a 75 keV/amu H^+ based H^o neutral beam.

There are problems of impurity production and hydrogen ion
density increases in tokamaks during RF heating.   These are
probably caused by the RF degrading the confinement of ions at the
plasma edge.   If loss of confinement is the problem, the poorly
confined ions strike the limiter and walls, knocking impurities off
and desorbing hydrogen from the walls.   While this may be a serious
problem, it is not intrinsically different from the general plasma
wall interaction problem.

DIAGNOSTICS

The field of plasma diagnostics has made as much or more

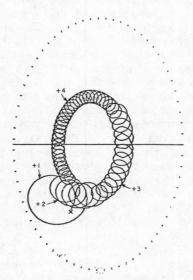

Fig. 25.   Gyro-orbits for a 32 MeV neutral oxygen atom injected
             into a large plasma. The view is a cross section of a
             tokamak. The neutral oxygen atom is ionized at x, and
             successively ionized to +8 (we only show +1, +2, +3, and
             +4) while it drifts toward the plasma center (Grisham,
             et al., 1981b). (81X0685)

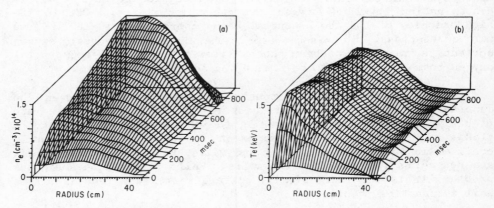

Fig. 26.   Time dependent electron temperature and density profiles
             in PLT measured by Thomson scattering (Bretz, et al.,
             1978).

progress in the last ten to fifteen years as other areas of fusion
research. Now, much more than in the past, basic plasma parameters
(n_e, T_e, T_i) are measured by a variety of methods. Many of the
techniques now can provide spatially resolved and time dependent
data during a single discharge (Fig. 26). The plasma parameters
quoted for fusion experiments can be relied upon, a marked change
from fifteen to twenty years ago. Each important parameter is
usually measured by several techniques (Rawls, 1979). A review of
diagnostic techniques has been done by the TFR group (DeMichelis,
1978), and I will just summarize the general field and discuss
several atomic physics issues. Table V lists many of the
quantities measured, the techniques commonly employed, and a few
techniques being developed for use in the next generation of
tokamaks. I will discuss laser techniques, then neutral particle
diagnostics, particle beam techniques, and UV, visible, and x-ray
spectroscopy.

The most fundamental parameters in a plasma are T_e, n_e, and
T_i, the electron temperature, electron density, and the ion
temperature respectively. The basic diagnostic for n_e and T_e uses
Thomson scattering. Light from a Q-switched ruby laser is
collected after being scattered by 90° by the plasma electrons
(Bretz, 1978). The doppler broadening of the scattered light gives
the electron temperature and the scattered intensity gives the
electron density. The phase shift of a microwave beam, due to the
index of refraction for microwaves caused by free electrons,
provides another measurement of the electron density. Other
interferometry techniques using HCN (Vernon et al., 1977), alcohol
(Wolfe, 1976), and CO_2 (Baker and Lee, 1978) lasers have recently
been developed primarily for use with high electron density
experiments.

Just as Thomson scattering and microwave interferometry are
the fundamental diagnostics for T_e and n_e, neutral particle
analyzers are the primary tool for measuring the ion temperature.
The technique is to use charge exchange of the confined hot ions
with hydrogen neutral atoms (either background or introduced by a
beam) to neutralize the hot ions and allow them to escape the
confining magnetic field. Once outside the plasma, the neutrals
can be ionized and analyzed. The ion temperature can then be taken
from the slope of the energy spectrum of the hot neutrals.

A second use of neutral particle diagnostics is the
measurement of the fast ion velocity distribution produced either
from RF heating of the plasma or from the slowing down of the
injected fast ions which are introduced during neutral beam
heating. These measurements are important since they provide
information about the details of how the plasma heating technique
is working.

Table 5. Plasma Diagnostics

Plasma	Quantity	Present Techniques
Electron density profile	$n_e(r)$	Thomson scattering, microwave interferometry, laser interferometry.
Electron temperature profile	$T_e(r)$	Thomson scattering, cyclotron radiation, x-ray spectra.
Ion temperature profile	$T_i(r)$	Charge-exchange spectra, doppler broadening of impurity lines including forbidden times, neutron emission.
Ion density profile (hydrogen and impurities)	$n^q(r)$	Line spectra (visible, UV, x-ray), charge-exchange recombination.
Neutral beam ion distribution	$f_{beam}(v)$	Charge-exchange spectra.
Electron velocity distribution	$f_e(v)$	x-ray
Plasma current profile	$j(r)$	Orbit shift of injected beam, Zeeman splitting, Faraday rotation.
Magnetic instabilities	MHD	Loops, x-ray
Plasma turbulence	Fluctuations	Probes, CO_2 laser scattering, microwave scattering
Radiation losses	$P_{rad}(r)$	Bolometers, spectroscopy
Edge plasma, plasma-wall interaction		Electric probes, ion doppler broadening hydrogen light, surface analysis.
Electrostatic potential	$\phi(r)$	Ionization of heavy probe beams

Sample future work would be better versions of above plus:

Ion temperature profiles	$T_i(r)$	Forbidden lines of injected impurities for $T_e > 3$ keV.
Impurity distribution	$n^{+q}(r)$	Multichannel spectrometers
Fast alpha distribution	$f_\alpha(v)$	Double charge-exchange with neutral Li beam

Extensions of this idea to the use of double charge exchange to measure energetic He^{++} distributions have been proposed. The first technique of this type would involve the use of neutral helium beams of 20-100 keV to neutralize the He^{++} ions used in ICRF $_3He^{++}$ minority heating (Post et al., 1981b). In the second technique of this type, the distribution of fast alpha particles would be measured by double charge exchange between the fast alpha particles and a high energy neutral lithium beam ($\sim$ 6 MeV) (Post et al., 1981c). The double charge exchange cross section of Li and He^{++} has recently been measured by McCullough et al. (1981) (Fig. 27). Three remaining issues relevant to this diagnostic technique involve the atomic physics of Li and He. A substantial and known portion of the neutral lithium beam must penetrate into the center of the plasma. The importance of two-step processes (successive collisions between the 6 MeV Li^{0} beam and the plasma) leading first to an excited state of Li^{0} and then to the ionization of the loosely bound excited state, must be evaluated. The double charge exchange cross section for excited Li^{0} and He^{++} has to be evaluated. The excited state distribution of He^{0} produced by the double charge exchange must be known, since only the ground state and low lying excited states of H_e^{0} have a good chance of escaping the plasma before being ionized again.

High energy heavy ion beams have also been used to measure the electrostatic potential in the plasma (Jobes and Hosea, 1973). In the ST tokamak, a 100 keV Th^{+} beam was injected into the plasma, where some of it was ionized to Th^{++}. By detecting the Th^{++} beam current as a function of position and energy, the electrostatic potential at the point of the second ionization was determined. Similar techniques are used on mirror experiments, where the electrostatic potential is a crucial parameter.

A neutral lithium beam has been used on Pulsator (McCormick et al., 1977) to measure the poloidal field by the Zeeman splitting of the collisionally excited resonance line (2s - 2p, 6708 Å). The polarization of the unshifted component of the Zeeman triplet line is used to measure the angle between the toroidal field and the poloidal field, from which the current can be inferred.

Spectroscopy (visible, UV, and x-ray) is one of the major diagnostic tools for fusion plasma research. It is used both to measure the plasma conditions (n_e, T_e, T_i, etc.) and to measure the plasma constituents. Its main use in measuring plasma parameters is to measure the plasma ion temperature by measuring the line broadening due to doppler shifts from the thermal motion of the radiating ions (DeMichelis, 1978). Two new uses and extensions of this technique have recently been made.

The first has been to use the doppler shift to measure the rotation velocity of a tokamak plasma. The PLT plasma has been

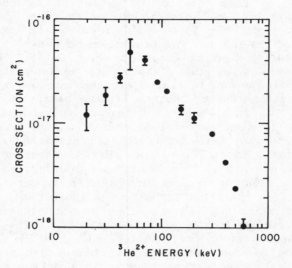

Fig. 27.   Cross sections for charge exchange of He^{++} and Li^{0} as a
            function of energy (McCullough, et al., 1981).

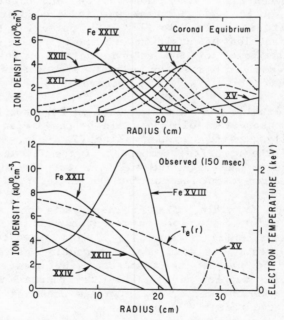

Fig. 28.   Observed iron charge state distribution in PLT compared
            with coronal equilibrium distribution (Hinnov) and
            (Stodiek, et al, 1980).   (803942)

observed to rotate at speeds of about 10^7 cm/sec (Suckewer, 1981) during high power neutral beam heating when the beams are injected in only one direction. This measurement was done using forbidden lines of FeXX. The second application of spectroscopy has been to use forbidden lines of highly ionized ions to measure the ion temperature in the hot central plasma. It is usually difficult to obtain sufficient resolution for the short wavelength resonance lines (1-100Å) of the highly stripped ions present in the center of a hot discharge to measure doppler shifts and broadening. However, the use of forbidden lines with wavelengths in the near UV (eg., 2665 Å for FeXX where optical techniques are useful) has made it easier to use doppler broadening measurements in very hot plasmas. A good summary is given by Suckewer (1981). The technique is being extended to higher electron temperatures (3-5 keV) by the use of Kr and Mo injected into the plasma in trace amounts (Cohen, 1981).

The slope of the energy spectrum of the x-ray continuum radiation is often used to measure the electron temperature. The continuum radiation (usually at energies of 2-5 x T_e) is due to radiative recombination and bremsstrahlung, and the spectrum has a temperature dependence of (DeMichelis and Mattioli, 1981)

$$\frac{1}{T_e^{1/2}} \exp\left[-\left(h\nu/T_e\right)\right] .$$

For $h\nu \gg T_e$, the x-ray intensity spectrum gives T_e.

The presence of high energy electrons (runaways) accelerated by the toroidal electric field can be measured by the enhancement of the high energy x-ray spectrum above the recombination and bremsstrahlung spectrum.

Another technique for measuring T_e is to use the intensity ratio of a dielectronic satellite line to the K_α-line of helium-like FeXXV using a high resolution crystal spectrometer at 1-2 Å (Bitter et al., 1979). This ratio is proportional to $\exp(E_i/T_e)/T_e$ where E_i is the ionization energy of lithium-like iron. The electron temperature thus deduced is accurate enough to confirm other measurements of the electron temperature.

The major use of spectroscopy, however, has been to measure the constituents of the plasma. The line emission of different impurity charge states allows the measurement of both the relative position and the absolute density of the impurity ions as a function of time. By comparing the observed positions of each impurity charge state with calculated positions assuming various transport models, the plasma transport can be studied (Fig. 28). Deducing the plasma transport rates requires highly accurate ionization, recombination, and excitation rates. It also requires including the effects of charge exchange recombination, since charge exchange recombination can shift the ionization equilibrium.

The measurement of the plasma transport and impurity transport in general is crucial since impurity radiation losses pose a severe threat to fusion reactor experiments. Much of the transport analysis has been complicated by the lack of knowledge of the behavior of the fully stripped impurity. Techniques to measure the fully ionized impurity density are under development. They involve the measurement of the distribution of the excited state radiation of hydrogen-like ions produced by charge exchange of the fully stripped ions and hydrogen neutrals (Isler, 1977; Afrosimov et al., 1978; Isler et al., 1981). The development of this technique requires a good understanding of the excited state distribution of the ion produced by the charge exchange. With a more complete picture of all of the charge states, progress in understanding impurity transport will hopefully follow.

Finally, spectroscopic measurement of a few of the dominant resonance lines of an impurity is used in conjunction with models for all the line transitions to determine the total impurity line radiation energy losses.

Another technique used to measure the energy losses due to radiation and neutral particles is bolometry, which measures the total energy falling on a detector. It provides a total loss rate, which is checked by the spectroscopic measurements.

The ion temperature is also measured by the neutron emission, since the fusion reaction rate is a strong function of T_i ($\propto T_i^{3-4}$ at 1-3 keV).

INERTIAL CONFINEMENT

The invention of lasers and high energy particle beams in the 1960's opened the possibility of compressing small pellets of DT and other fusionable materials to high enough densities and temperatures so that the fusion alphas would be stopped in the pellet and thus ignite the pellet in a very much scaled down version of a hydrogen bomb. The basic idea is that energy (laser, beams) is focused on the pellet surface (Fig. 29). The surface is rapidly heated and ablated. The continued ablation drives the pellet material inward with a spherically convergent shock wave. The central plasma is compressed, hopefully adiabatically. The central pressure (PV^γ = const) increases until the pressure of the compressed plasma equals the pressure of the driving ablator. Then the pellet center expands and the pressure drops. The ignition conditions are that T exceed a few keV, and ρR be greater than 1 gm/cm^2. ρR is the product of the the compressed fuel density and

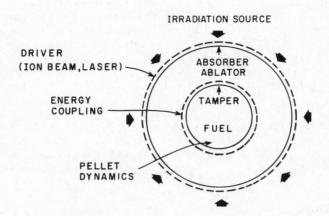

Fig. 29.   Schematic illustration of an inertial confinement
             pellet. (81P0191)

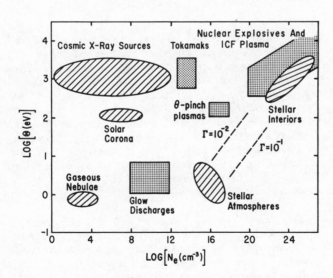

Fig. 30.   Density and temperature regimes of some natural and man-
             made plasmas. The two dashed lines show (N_e, Θ-values
             corresponding to different constant values of the
             Coulomb parameter Γ for a hydrogen plasma (taken from
             Weisheit, 1982). (802419)

pellet radius required for alpha containment. Typical pellet dimensions (consistent with yields less than one ton of TNT) are in the range of 0.01 - 1 cm for the uncompressed pellet and 10^{-1} to 10^{-3} cm respectively for the compressed pellet. Since the temperature must be in the 1-10 keV range for fusion to occur, the thermal velocity of the particles is $\sim 10^{8}$ cm/sec. The pellet will thus fly apart in 10^{-9} - 10^{-10} seconds. This time can roughly be interpreted as a confinement time, τ_c, so $n_e \tau_c \sim 10^{14}$ sec/cm^3 implies that $n_e \sim 10^{23}$ - 10^{25}/cm^3. Thus the conditions for inertial confinement fusion are $T >$ a few keV, $\tau_c \sim 10^{-9}$ sec, and $n_e \sim 10^{24}$ cm^{-3}.

Atomic physics is somewhat more intimately involved in the physics of inertial confinement than in magnetic confinement, but the issues can still be roughly divided into those affecting the central plasma, the edge plasma, heating and diagnostics (Table VI).

Conditions in the central plasma depend crucially on the equation of the state of the plasma in the pellet and the opacity of the central pellet plasma. The equation of state is given by

$$P = \frac{\rho}{A} \left(z^* + 1 \right) T,$$

where P is the pressure, ρ is the mass density, A is the atomic weight of the plasma ion, z^* is the number of free electrons per ion, and T is the temperature. The key quantity to calculate is z^*, which is also the effective charge of the ion. The heat capacity also requires z^*. The other item of interest in calculating the hydrodynamics is the opacity. The transport of thermal x-rays is an important factor in the behavior of the imploding and exploding pellet. Particularly important is line radiation from moderate Z and high Z materials used in multi-layered pellet designs. These data are included as essential ingredients in large hydrodynamic calculations such as LASNEX (Zimmerman and Kruer, 1975).

The high densities ($\sim 10^{24}$/cm^3) and short times ($\sim 10^{-9}$ sec) encountered in inertial confinement open up a class of atomic phenomena that are different than conventional astrophysical and laboratory plasmas. The interatomic spacing, defined as the ion sphere radius, R_i, (Weisheit, 1982), where

$$\frac{4 \pi R_i^3}{3} = \frac{<z_i>}{N_e} \quad ; \quad R_i = \left(\frac{1.61 \times 10^{24} <z_i> \text{cm}^{-3}}{n_e} \right)^{1/3} a_o ,$$

Table 6.   Inertial Confinement Atomic Physics Issues
(adapted from (Hauer, 1981)

ICF Related Process	Atomic Processes Involved	Theory and Modeling	Diagnostic Techniques
Absorption of laser light	Ionization and recombination dynamics in presence of large density gradients and electric fields	Rate equation models coupled to hydro codes or other plasma modeling ionic species and energy spectrum	(1) Spatially resolved spectral line observation (2) Charged particle
Transport and deposition of fast ions and electrons	(1) Collisional ionization and excitation (2) bremsstrahlung (3) electron scattering	Stopping power formulations needed for input to comprehensive hydro modeling	Inner shell and bremsstrahlung radiation used as indicator of particle deposition
Thermal, particle, and radiation transport	(1) Bound contributions to absorption (2) thermal and electrical conduction in partially ionized plasma	(1) Opacity calculations needed for radiation treatment in comprehensive models (2) Calculations of transport coefficients	(1) Line radiation from layered targets as indicator of thermal transport (2) Absorption spectra
Production of high density imploded plasma	(1) Quantum effects in the equation of state (EOS) of dense plasmas (2) line radiation from higher Z plasmas	(1) Thomas Fermi Dirac EOS as starting point (2) spectral modeling with rate equations coupled to hydro codes	(1) Observation in dense plasma of (a) line profiles (b) line shifts (c) continuum edge (2) high intensity shock studies

Table 7.   Large United States Inertial Confinement Experiments

Name	Type	Wavelength	Site	Date	Power	Energy
Helios	CO_2	10μ	LASL	1978	30 TW	6 kJ
Shiva	Ni glass	1μ	LLL	1978	30 TW	6 kJ
Antares	CO_2	10μ	LASL	1982	50 TW	40 kJ
Nova	Ni glass	1μ	LLL	1985	60 TW	30 kJ

where $a_o = 0.529 \times 10^{-8}$ cm (the Bohr radius), and $\langle Z_i \rangle$ is the average number of free electrons per ion, is the order of the size of the orbits for bound electrons. This implies that the pressure of nearby ions will affect the energy levels, collision rates, etc., of an ion. Binary collision models which treat an ion as an isolated system undergoing an impact collision will have to be modified. A measure of the importance of this effect is the ion coupling parameter given by the ratio of the coulomb potential at the ion sphere radius divided by the temperature (Dewitt et al., 1973)

$$\Gamma = \frac{\langle Z_i e \rangle^2}{R_i T} = \frac{\langle Z_i \rangle^{5/3}}{T(eV)} \left(\frac{n_e}{8 \times 10^{19} \text{cm}^{-3}} \right)^{1/3} \ .$$

When $\Gamma \gg 1$, coulomb interactions control the particle motions, but when $\Gamma \ll 1$ thermal motions are dominate. ICF (inertial confinement fusion) plasmas lie in the $10^{-2} \lesssim \Gamma \lesssim 1$ range (Fig. 30).

The first step in calculating collision rates, line positions, etc., is to calculate the potential a bound electron sees on a partially stripped ion. The contribution due to adjacent ions is usually done with a modified Debye-Huckel model (Weisheit, 1982). The contribution of the bound electrons is typically computed using a Thomas-Fermi type model.

There is appreciable Stark broadening of emission lines due to the transient electric fields produced by the thermal motion of the ions and electrons (Hooper, 1968).

The first major effect of nearby ions is that the ionization continuum is lowered. Electrons of the nth quantum shell (n = principal quantum number) have an average orbit radius

$$r_n = a_o \frac{n^2}{Z^*} \ .$$

The density effects are significant (More, 1982) when $r_n \gtrsim R_i$ or $n_i \gtrsim 1.6 \times 10^{24}$ cm^{-3} (Z^*/n^2). Electron wave functions of adjacent ions overlap, leading to a broadening of atomic states into energy bands and "pressure ionization." The energy levels are also displaced due to the screening by "free" electrons of the core potential seen by a bound electron.

Screening affects not only the energy levels but also the eigenfunctions of bound electrons. Thus screening changes the strengths of radiative transitions from the values for isolated ion (Fig. 31). The strength of the screening can be characterized by

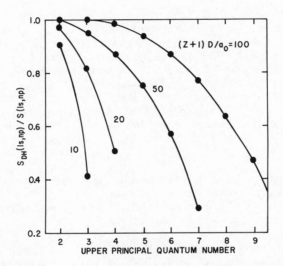

Fig. 31.   The ratio S_{DH}/S of Debye-Huckel-screened to unscreened
transition strengths for the principal series in a
hydrogenic ion of <u>nuclear</u> charge $(Z + 1)e$ for various
normalized screening lengths $(Z + 1) D/a_0$. As an aid
to the eye, curves have been drawn to connect values
for a given normalized screening length (taken from
Weisheit, 1982).   (802418)

the Debye length D,

$$D = \left[4\pi e^2 \sum_i z_i^2 n_i / T_i\right]^{1/2}$$

$$= \left\{\sum_i z_i^2 n_i / \left(1.97 \times 10^{22} cm^{-3} T_i (eV)\right)\right\}^{-1/2} a_o \ .$$

In general, the transition strengths tend to decrease rapidly as the shielding length becomes short (Weisheit, 1982).

Screening can significantly reduce the cross sections for electron impact excitation near threshold. For comparison Weisheit (1982) shows the $^4S \rightarrow 2_p$ excitation cross section of the isoelectronic configurations (Fig. 32) O^+ and N. The energy is scaled by the threshold energy of each transition. The finite range of the N^o coulomb potential, illustrating the effects of screening, causes the nitrogen cross section to vanish near the threshold whereas the O^+ potential is long range and the cross section rises steeply near the threshold. The same effect occurs for the 1s $\rightarrow$ 2s transition in one electron atoms as the screening is increased.

Dielectronic recombination rates are also reduced due to the removal of high n orbitals as illustrated by Fig. 33 which uses unscreened radiative and collisional rate coefficients (Weisheit, 1975).

The edge plasma is extremely important to inertial confinement since the energy needed to compress the pellet is deposited there. Typical densities run from $10^{14} - 10^{22} cm^{-3}$ and typical temperatures run from 10 eV to 1 keV. The key issues are the absorption and stopping of laser light and the absorption and stopping of fast ions and electrons. Magnetic fields in the megagauss range are generated and can affect the collision rates. As in the magnetic fusion case, the interaction of the pellet debris with the reactor vessel is important. The emphasis is on the survival of wall to pulsed heat loads and particle loads (about 1,000 joules/cm^2 in 10^{-9} seconds).

The exact mechanisms for the absorption of laser light are not understood. The light propagates through the low density plasma at the pellet edge until it reaches the "critical" density where the light frequency equals the plasma frequency $\nu_p = 8.97 \times 10^3$ $n_e^{1/2}$/sec (Spitzer, 1962) (about $10^{19} cm^{-3}$ for CO_2 lasers, about $10^{21} cm^{-3}$ for Nd glass).

Ion beams are stopped by collisions with the partially ionized ions and electrons in the pellet. Most experimental data has been

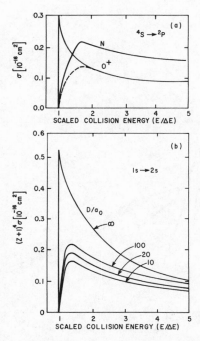

Fig. 32.   (a) The computed electron impact excitation cross
section [108] for the transition $(2p^3)^4S \rightarrow (2p^3)^2P$
in N and O$^+$. When the nitrogen cross section is scaled
to agree at high energies with the oxygen cross section,
the dashed curve shows its low energy behavior. For
nitrogen, the threshold energy is $\Delta E = 3.58$ eV; for
oxygen, it is $\Delta E = 5.02$ eV. (b) The computed electron
impact excitation cross section for the transition 1s
$\rightarrow$ 2s in a one-electron ion of nuclear charge $(Z + 1)e$.
The Debye-screened ion results are the Born calculations of Hatton et al., (1981), and the unscreened ion
result $(D \rightarrow \infty)$ is the Coulomb-Born calculation of Oh,
et al. (1978). The threshold energy is $\Delta E = 10.2$
$(Z + 1)^2$ eV (taken from Weissheit, 1982). (802423)

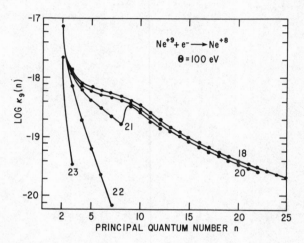

Fig. 33.   Rate coefficients $\{\kappa_9(n) \quad [cm^3 sec^{-1}]\}$ for dielectric
recombination to excited states (1snℓ) of Ne^{+8} in a
100 eV plasma. The logarithm of the plasma electron
density $N_e(cm^{-3})$ is shown by each curve) at densities
$N_e \geq 10^{20} cm^{-3}$, continuum lowering removes all but the
few states indicated (taken from Weisheit, 1982).
(81T0028)

gathered only for cold materials. For cold materials the stopping is due primarily to ionization and excitation of bound electrons. Stopping power calculations involve adding the contributions of the bound electrons of the partially ionized atoms and the stopping due to the free electrons (Hauer, 1981). A portion of the absorbed laser energy gives rise to a high energy electron population. As with heavy ions, fast electrons are stopped by impact ionization and excitation collisions with bound electrons, and scattering from free electrons.

The main approaches to inertial confinement are defined by the type of beam used to compress the pellets (lasers, light ions, heavy ions, and electrons). The original approach was to use high energy lasers (Nuckolls et al., 1972). The primary difficulties with lasers are their low electrical efficiency and poor coupling of the laser power to the pellet. A facet of the coupling problem is the generation of high energy electrons which penetrate to the pellet core and heat it before the final compression stage, thus increasing the power required to reach the required high density. Current experiments indicate that the laser-plasma coupling is improved, and the hot electron preheat is reduced as the laser wavelength is reduced. Large CO_2 lasers (10μ) are being used at the Los Alamos National Laboratory. Neodymium glass lasers (1μ) are the most common, with programs at Lawrence Livermore National Laboratory, KMS, Inc., and the University of Rochester, in the United States, at the Lebdeev Institute in USSR, at Osaka University in Japan, at LeMeil in France, and at the Max Planck Institute in Garching, West Germany. There is intense interest in going to shorter wavelengths using exotic lasers such as KrF lasers or by frequency doubling. Livermore is constructing a frequency doubled laser ($\sim 0.5\mu$) beginning with a Neodymium glass laser ("Novette").

Given the apparent simplicity of the stopping of high energy ions in a plasma, the short range of the ions, and the possibility of high electrical efficiency for ion beams, there is an active program to use both light and heavy ion beams for compressing pellets. The light ion beam research is being done in the United States at Sandia Laboratories in Albuquerque, New Mexico, and at the Kurchatov Institute in Moscow. The heavy ion beam fusion work is being done at the Argonne National Laboratory, and the Lawrence Berkeley Laboratory in the United States, and the Rutherford Laboratory in the United Kingdom.

One candidate beam is ~ 1 GeV Cs^+. Recent work on the reaction $Cs^+ + Cs^+ \rightarrow Cs^0 + Cs^{+2}$, $Cs^+ + Cs^{++}$, etc., has had a bearing on the feasibility of such beams, since these reaction could destroy the beam.

Since lasers involve electronic transitions in molecules and atoms, atomic processes will continue to play a large role in the search for high power, efficient lasers. The thrust toward shorter wavelengths (< 1μ) will be an important research area. Significant improvements in both these areas will have to be made for lasers to be feasible as drivers for a inertial confinement fusion reactor. At this moment ion beams seem more promising for reactor applications than lasers. Thus there will be an increasing emphasis on the production and acceleration of efficient, high energy, and high power ion beams.

Diagnostic techniques for inertial confinement experiments rely heavily on atomic processes. Spectroscopy in the UV, x-ray, and visible is used (Yaakobi et al., 1977). One common technique is to seed the DT gas with neon or argon and look at the emission lines for line profile changes, line shifts, and other high density effects (Bristow et al., 1979). Many pellets have layers of high Z materials, and time and space resolved line radiation, and x-ray continua from these high Z layers allows one to determine the time history of the temperature and density of the layers in the compressing and exploding pellet. Spectral information coupled with rate equation modeling of ionization and recombination in the strong temperature and density gradients at the ablation layer where the laser is deposited, provides a measure of the laser plasma interaction. Optical laser probes are used with interferometry, faraday rotation, and polarization to measure the density profiles in the ablating surface of the pellet.

One example of this is the use of K_α radiation produced by the impact excitation of fast electrons as a measure of the power absorbed which goes into the production of fast electrons. The simple ratio of the total stopping power to the stopping power due to K_α excitation is given by (Hauer, 1981).

$$\frac{dE/dx|_{tol}}{dE/dx|_{K_\alpha}} = C \frac{\ln(4E/I)}{\ln(E/E_{K_\alpha})} \quad ,$$

where E is initial electron energy, I is the average bound excitation energy, and E_K is the K_α excitation energy.

This simple analysis is complicated by the heating and stripping of the ions being excited and the subsequent shifts of the emission lines (for A^O, A^{+1}, A^{+2}, A^{+3}, etc.). Resonance line absorption into K shell excitations, the contribution of the free electrons to the stopping power, and the variation of the ratio of K_α line emission to Auger emission with charge state all further complicate the analysis.

Bremsstrahlung radiation is also used as a tool to measure the hot electron temperature and transport using detailed bremsstrahlung cross sections (Pratt et al., 1977). Extensive modeling of these two emissions and their interaction with the plasma is required to recover quantitative information about the behavior of the fast electrons.

The best conditions achieved so far by the inertial confinement program are n ~ 3 × 10^{24} cm^{-3}, T ~ 1 keV, and τ_c ~ 1.5 nsec (Hauer, 1981). Pellets of initial diameters of two-hundred to four-hundred microns have been compressed to ten to twenty microns. The major funding commitment is for large laser systems at Livermore and Los Alamos in the United States (Table VII). The United States budget is $215 million/year.

THE ROLE OF PLASMA PHYSICS IN ATOMIC PHYSICS

The programmatic needs of the fusion community have provided motivation and resources for many of the current research topics in atomic physics. The interest in charge exchange of multi-charged ions with hydrogen was spurred on by the concerns of the fusion community that neutral beams would not penetrate plasmas with even relatively small impurity levels. In addition, fusion research has, as a by-product, produced new experimental and theoretical tools to explore atomic physics in new regimes. Plasmas with temperatures as high as 7 keV and densities up to 3 × 10^{24} cm^{-3} are produced in laboratories. These have led to the identification of the energy levels of highly stripped ions (such as W^{+50}) (Reader and Luther, 1980). Large spectroscopy groups have been set up at the major fusion laboratories with the attendant improvements in spectrometers and detector technology. These groups have observed effects before seen only in astrophysical plasmas before, such as dielectronic satellite lines and recombination edges. The work on excitation-autoionization is of great interest to the fusion community. The work on the atomic physics of dense matter is a qualitatively new area that will almost certainly grow in interest and importance.

A lot of extremely useful work is being done by the various data center groups throughout the world to assemble the data relevant to fusion and put it in a digestible form so that plasma physicists can easily use the data. In the United States there are data centers at the Oak Ridge National Laboratory in Oak Ridge, Tennessee; at the Joint Insitute Laboratory for Astrophysics in Boulder, Colorado, and at the National Bureau of Standards in Washington, D. C. In Japan there are data centers at the Plasma Physics Institute at Nagoya University in Nagoya, and at the Japanese Atomic Energy Research Institute at Tokai. In Europe there are groups at the International Atomic Energy Agency in Vienna, Austria; at Queen's University, Belfast, Northern Ireland,

Table 8. Textbooks and Review Articles

Topic	Reference
Plasma Physics	Chen, (1974); Spitzer, (1962)
Tokamaks	Rawls, (1979); Furth, (1975)
Mirrors	Cohen, (1980)
Impurity Atomic Physics	DeMichelis et al. (1981); Suckewer, (1981)
Plasma Wall Interactions	McCracken, (1979)
Neutral Beam Heating	Jassby, (1977); Raimbrult, (1981)
Tokamak Diagnostics	DeMichelis, (1978); Rawls, (1979)
Dense Matter Atomic Physics	Weisheit, (1982); More, (1982)
Inertial Confinement	Hauer, (1981); Miley, (1979)
Atomic Data Needs of Fusion	Physica Scripta (1981)

and at the University of Paris in France.   In the USSR there is a group at the Kurchatov Institute in Moscow.

REVIEWS

Good textbooks and review articles are essential for people outside a field to efficiently learn what they need to know about the field.   I have provided a short key to textbooks and reviews in Table VIII.

ACKNOWLEDGMENTS

The author is grateful for discussions and contributions to Drs. H. Furth, A. Hauer, E. Hinnov, R. Hulse, D. Meade, and J. Weisheit.   The author is also grateful to Dr. Clifford Singer for discussions and a careful proofreading of the manuscript.

The work done at Princeton is supported by the U. S. DoE Contract No. DE-AC02-76-CHO-3073.

REFERENCES

Afrosimov, V. V., Gordeev, Y., Zimoviev, A., and Korotkov, A., 1978, JETP. Lett. 28, 500.
Bacal, M., 1981, Physica Scripta 23, 122.
Baker, D. R., and Lee, S-T., 1978, Rev. Sci. Instrum. 49, 919.
Baldwin, D. E., 1977, Rev. Mod. Phys. 49, 317.
Bayfield, et al., 1976, Abstracts Fifth International Conference on Atomic Physics, Berkeley, Ca., p. 126.
Berkner, R. N., et al., 1978, J. Phys. B. 11, 5.
Bitter, M., et al., 1979, Phys. Rev. Lett. 43, 129.
Book, D. L., 1980, NRL Plasma Formulary, Office of Naval Research.
Bretz, N., et al., 1978, Appl. Opt. 17, 192.
Bristow, T., et al., 1979, IAEA-CN-37/B-4, Plasma Physics and Controlled Nuclear Fusion, IAEA, Vienna, Austria.
Burgess, A., 1965, Astrophysics J. 141, 1588.
Callen, J. D., et al., 1980, IAEA-CN-38/Y-3, Plasma Physics and Controlled Nuclear Fusion, IAEA, Vienna, Austria.
Chen, F., 1974, "Introduction to Plasma Physics," Plenum Press.
Coensgen, F., et al., 1980, Phys. Rev. Lett. 44, 1132.
Cohen, B. I., 1980, UCRL-80, Lawrence Livermore Laboratory Report, Livermore, CA.
Cohen, S., private communication (1981).
Cox, D., and Tucker, W., 1969, Astrophys. J. 157, 1157.
Dawson, J, and MacKenzie, K., 1980, PPG-470, Center for Plasma Physics and Fusion Engineering, University of California, Los Angeles, California
DeMichelis, C., and Mattioli, M., 1981, Nuclear Fusion 21, 677.
DeMichelis, C., 1978, Nucl. Fusion 18, 5.
Dewitt, H. E., Grasboske, H. C., and Cooper, M. S., 1973,

Astrophys. J. 181, 439.

Duchs, D., Post, D., and Rutherford, P., 1977, Nuclear Fusion 17, 565.

Eckstein, W., and Verbeek, H., 1979, Data on Light Ion Reflection, IPP9/32, Max-Planck-Institut Fur Plasmaphysik, Garching.

Eubank, H., et al., 1979, Phys. Rev. Lett. 43, 270.

Fowler, T. K., and Logan, B. G., 1977, Comm. Pl. Ph. 2, 167.

Freeman, R., and Jones, E., 1974, CLMR-137, Culham Laboratory, Abingdon, England.

Furth, H. P., and Rosenbluth, M. N., 1969, Plasma Physics and Controlled Nuclear Fusion Research, IAEA, Vienna, Austria, I, 821.

Furth, H. P., 1975, Nuclear Fusion 15 (1975), 487.

Greene, S., 1967, UCRL-702522, Lawrence Livermore Laboratory, Livermore, CA.

Grisham, L., and Post, D. E., 1981a, PPPL-1803, Princeton University, (to appear in Nuclear Technology/Fusion).

Grisham, L., Post, D., Mikkelsen, D., and Eubank, H., 1981b, "Plasma Heating with Multi-MeV Neutral Impurity Beams," PPPL-1759 (1981), Princeton University, (to appear in Nuclear Technology/Fusion).

Grisham, L. R., Post, D. E., Johnson, B. M., Jones, K. W., and Barette, J., et al., 1981c, PPPL-1857, to appear in Rev. Sci. Instrum.

Hatton, G. J., Lane, N. F., and Weisheit, J. C., 1981, Phys. Rev. A (in press).

Hauer, A., 1981, "Survey of Atomic Physics Issues in Experimental Inertial Confinement Fusion Research," XII ICPEAC, Gatlinburg, Tennessee.

Hawryluk, R., Suckewer, S., and Hirshman, S., 1979, Nuclear Fusion 19, 607.

Heifetz, D., et al., 1981, "A Monte Carlo Model of Neutral Particle Transport in Divertor Devices," Princeton Plasma Physics Laboratory, submitted to J. Comp. Phys. 178, 218.

Hinnov, E., and Mattioli, M., 1978, Phys. Lett. A 66, 106.

Hinnov, E., private communication.

Hinton, F. L., and Hazeltine, R. D., 1976, Rev. Mod. Phys. 48, 239.

Hogan, J. T., and Howe, H. C., 1976, Journal of Nuclear Materials 63, 151.

Hogan, J. T., 1981, Twelfth ICPEAC, Gatlinburg, Tennessee.

Hooper, C. F., 1968, Phys. Rev. 165, 215.

Hsuan, H., et al., 1978, Proc. Joint Varenna and Grenoble Int. Symp. on Heating in Toroidal Plasmas, Grenoble.

Hughes, M. H., et al., 1978, J. Comp. Phys. 28, 43.

Hulse, R. A., Post, D. E., and Mikkelsen, D. R., 1980, J. Phys. B 13, 895.

INTOR, 1980, International Tokamak Reactor, Zero Phase, International Atomic Energy Agency, Vienna, Austria.

Isler, R. C., Neidigh, R. V., and Cowan, R. D., 1977, Physics Letters 63A, 295.

Isler, R. C., 1977, Phys. Rev. Lett. 38, 1359.
Isler, R. C., Murray, L., and Kasai, S., 1981, XII ICPEAC, Gatlinburg, Tn., p.663.
Jacobs, A., 1972, J. Quant. Spectrosc. Radiat. Transfer 12, 243.
Jassby, D., and Towner, H., private communication.
Jassby, D. L., 1977, Nuclear Fusion 17, 309.
Jensen, R. V., et al., 1977, Nuclear Fusion 17, 1187.
Jensen, R. V., et al., 1978, Nucl. Sci. and Eng. 65, 282.
Jobes, F. C., and Hosea, J. C., 1973, Proceedings of the Sixth European Conference on Controlled Fusion and Plasma Physics, Moscow, p. 199.
Johnson, B., private communication.
Jordan, C., 1970, Mon. Not. R. Astron. Soc. 148, 17.
Krupin, V. A., Marchenko, V. S, and Kakovlenko, S., 1979, JETP Lett. 29, 3895.
Langer, W., 1981, "Transport of Molecular Impurities at the Edge of a Plasma," Princeton Plasma Physics Laboratory (submitted to Nuclear Fusion).
Lawson, J. D., 1957, Proceedings of the Physical Society, London 70B, 6.
Lotz, W., 1967, Astrophys. J. Suppl. 14, 207.
Mahdavi, M., et al., 1981, "Particle Exhaust from Discharges with an Expanded Boundary Divertor," GA-A16334, General Atomic.
McCormick, K., Kick, M., and Olivan, J., 1977, Proceedings Eighth European Conference on Controlled Fusion and Plasma Physics, Prague, 1, 140.
McCracken, G., and Stott, P., 1979, Nuclear Fusion 19, 889.
McCullough, R., et al., 1981, "Electron Loss by Fast Li$^-$ Ions," XII International Conference on the Physics of Electronic and Atomic Collisions, Gatlinburg, Tennessee.
McCullough, R., Goffe, T., Lennon, M., Shah, M., and Gilbody, H., 1981, "Electron Capture by He$^+$ and He^{+2} Ions in Li$^-$ Vapour," XII International Conference on the Physics of Electronic and Atomic Collisions, Gatlinburg, Tennessee, p. 661.
Merts, A. L., Cowan, R. D., and Magee, N. H., Jr., 1976, "The Calculated Power Output from a Thin Iron-Seeded Plasma," LASL Report LA-6220-MS, Los Alamos, N.M.
Miley, G., Ed., "Proceedings of the IEEE Minicourse on Inertial Confinement Fusion," 1979, Montreal, Canada.
More, R. M., 1982, "Atomic Physics in Inertial Confinement Fusion," Applied Atomic Collisions, Volume II, Bederson, B., Barnett, C., and Harrison, M., Academic Press.
Moriette, P., 1974, Proc. Seventh Yogoslav Symposium on the Physics of Ionized Gases, p. 43.
Nagami, M., et al, 1980, IAEA-CN-38/0-2, Plasma Physics and Controlled Nuclear Fusion, IAEA, Vienna, Austria.
Nuckolls, J., et al., 1972, Nature 239, 139.
Oh, S. D., Macek, J., and Kelsey, E., 1978, Phys. Rev. A 17, 873.
Olson, R. E., and Salop, A., 1976, Phys. Rev. A 14, 579.

Olson, R. E., and Salop, A., 1977, Phys. Rev. A 16, 531.

Olson, R. E., Berkner, K. H., Graham, W. G., Pyle, R. V., Schlacter, A. S., and Stearns, J. W., 1978, Phys. Rev. Letters 41, 163.

Petravic, M., et al., 1981, "A Cool, High Density, Regime for Poloidal Divertors," PPPL-1824, Princeton Plasma Physics Laboratory (submitted to Phys. Rev. Letters).

Phaneuf, R. A., et al., 1978, Phys. Rev. A 17, 534.

Physica Scripta 23 (1981).

Post, D. E., et al., 1977, At. Data and Nuc. Data Tables 20, 397.

Post, D. E., et al., 1979,"Plasma Physics and Controlled Nuclear Fusion Research 1978" IAEA-CN-37/F-3, IAEA, Vienna, Austria.

Post, D., Grisham, L., Santarius, J., and Emmert, G., 1981a, "Heavy Atom Neutral Beams for Tandem Mirror End Plugs," PPPL-1786, Princeton University (to appear in Nuclear Fusion).

Post, D., Grisham, L., and Medley, S., 1981b, PPPL-1864 (January 1981), submitted to Nuclear Fusion.

Post, D. E., et al., 1981c, Journal of Fusion Energy 1, 129.

Pratt, R. H., et al., 1977, Atomic Data and Nuclear Data Tables 20, 175.

Puiatti, M. E., Breton, C., DeMichelis, C., and Mattioli, M., 1981, "Impurity Charge Exchange Processes in Tokamak Processes,"Euratom-CEA Association, Fontenay-aux-Roses, Report EUR-CEA-FC 1085.

Raimbrult, P., and Girard, J. P., 1981, Physica Scripta 23, 107.

Rawls, J., 1979, Status of Tokamak Research, DoE/ER-0034 United States Department of Energy, Washington, D.C. (1979).

Reader, J., and Luther, G., 1980, Phys. Rev. Lett. 95, 609.

Rose, D. J., and Clark, M., 1961, Plasmas and Controlled Fusion, MIT Press, and J. Wiley, Inc., Cambridge, MA., and New York, p. 18.

Roth, J., Bohdansky, J., and Ottenberger, W., 1979, Data on Low Energy Light Ion Sputtering, Max-Planck-Institut Fur Plasmaphysik, IPP9/26, Garching.

Sampson D., and Golden, L., 1971, Astrophys. L. 170, 169.

Seaton, M., 1962, "The Theory of Excitation and Ionization by Electron Impact," in Atomic and Molecular Processes, Ed., D. R. Bates, (Academic Press, Inc., New York).

Seki, et al., 1980, Nuclear Fusion 20, 1213.

Shimomura, Y., et al., 1979, JAERI-M-8294, Japan Atomic Energy Research Institute, Tokai, Japan.

Siambus, J. G., and Trivelpiece, A., 1965, Phys. Fluids 8, 2047.

Sluyters, Th., Ed., 1980, Proceedings of the Second International Symposium on the Production and Neutralization of Negative Hydrogen Ions and Beams, BNL-51304, Brookhaven National Laboratory.

Solovev, L. S., and Shafranov, V. D., 1967, Culham Laboratory Report CLM-Trans p.12.

Spitzer, L., 1963, Physics of Fully Ionized Gases, Interscience

(John Wiley and Sons), New York

Spitzer, L., 1963, Physics of Ionized Gases, Interscience (John Wiley and Sons, New York.

Stodiek, W., et al., 1980, "Transport Studies in the Princeton Large Torus," IAEA-CN-38/A-1, Plasma Physics and Controlled Nuclear Fusion Research, IAEA, Vienna.

Suckewer, S., 1981, Physica Scripta 23, 72.

Tang, W. M., 1978, Nuclear Fusion 18, 1089.

Taylor, J. B., 1963, Phys. Fluids 6, 1529.

Teller, E., 1981, Fusion, Academic Press, New York.

Veron, D., Certain, J., and Crenn, J. P., 1977, J. Opt. Soc. Am. 67, 964.

Weisheit, J., 1982, "Atomic Phenomena in Dense Plasmas," Applied Atomic Collisions, Volume II, Bederson, B., Barnett, C., and Harrison, M., Academic Press, New York.

Weisheit, J., 1975, J. Phys. B. 8, 2556.

Wolfe, S. M., 1976, Appl. Opt. 15, 2645.

Yaakobi, B., et al., 1977, Phys. Rev. Lett. 39, 1526.

Younger, S., Phys. Rev. A 23, 1138 (1981).

Zimmerman, G., and Kruer, W., 1975, Comm. Plasma Physics 2, 85.

THEORY OF ELECTRON-ATOM COLLISIONS*

Yong-Ki Kim

Argonne National Laboratory, Argonne, Illinois 60439†
and
Joint Institute for Laboratory Astrophysics
University of Colorado and National Bureau of Standards
Boulder, Colorado 80309

I. INTRODUCTION

The collision of electrons with atoms (including ions as special cases) is not only of interest to atomic physics itself but also has important applications in other fields such as astrophysics, plasma physics, and radiation physics.

Although the main emphasis of this conference is on the inter-action with ions, this lecture will cover interactions with both neutral atoms and ions because:

(a) Electron-atom collisions are better understood, and
(b) More is learned by comparing the similarities and differences between electron-atom and electron-ion collisions.

The systematics peculiar to electron-ion collisions are primarily due to the presence of an attractive potential and infinitely many

---

*Work performed in part under the auspices of the U.S. Department of Energy.
†Permanent Address.

bound states between the colliding particles. We shall use the word "atoms" to include "ions" hereafter unless specified otherwise.

Collisions of electrons with atoms can be classified into two broad categories, slow and fast collisions, depending on the speed of the incident electron compared to the orbital speed of the bound electron participating in the collision.

In a slow collision, the incident electron and the target atom form a compound state, and strong coupling exists among alternative compound states as well as among the various channels of separation after the collision. As a result, both the incident and target particles must be described on equal footing, i.e., the interaction between them cannot be treated as a perturbation compared to the interactions among the bound electrons. The R-matrix and close-coupling methods are examples of theories describing such systems. Slow collisions still pose severe difficulties (and challenges) to theorists.

Fast collisions, on the other hand, are easier to formulate and better understood than slow collisions. The incident electron behaves differently from atomic electrons, and hence it can be described separately from the bound electrons. The collision occurs mostly at large impact parameters, and the interaction is weak. The interaction between the incident electron and the target is represented as an impulse and is treated in the framework of a perturbation theory. For the perturbation treatment to succeed, the collision should be dominated by small impulse—or momentum transfer—compared to the average momenta of the bound electrons. Dipole-allowed transitions clearly belong to this class, and this is the main reason that the Born approximation works better for dipole-allowed transitions than dipole-forbidden ones, as we shall see later.

Also, there is a close relationship between collisions of photons and fast electrons with atoms. The cross sections for various bare incident particles (e.g., e^-, e^+, p, alpha, etc.) can be scaled from each other at high incident energies. These simplifications for fast projectiles are essential features of the first Born approximations.

A major departure of electron-ion collision process from electron-atom collisions occurs in the threshold behavior. For the excitation of a neutral atom by an electron, the cross section vanishes at the threshold, whereas for the excitation of an ion, it starts with a finite value. The excitation cross section for an ion usually has its maximum at the threshold.

In the next few lectures, we shall discuss:

(c) Theory of fast collisions, with emphasis on the systematics
    and relationships to other types of collisions;
(d) Electron-impact ionization including the systematics of second-
    ary electrons;
(e) Relativistic and correlation effects in collision   cross
    sections;
(f) Theory for slow collisions; and
(g) Conclusion.

II.   THEORY OF FAST COLLISIONS: THE BORN APPROXIMATION

A.   Derivation of the Basic Formulas of the FBA

As was mentioned briefly in the Introduction, the first order
perturbation theory is adequate to describe collisions of fast electrons—and other fast charged particles—with atoms, as long as the
collision is dominated by weak interactions.   In this case, the
Coulomb interaction between the projectile and the target H_{int}, can
be treated as a perturbation compared to the sum of the total energy
of the target H_a and the kinetic energy of the incident electron H_e.
In this chapter, we shall derive the basic formulas of the first Born
approximation (FBA) and discuss their qualitative features and rela-
tionships to photon-atom interaction.

1. Preliminaries

We first consider the excitation of an atom from its ground state
0 to a discrete excited state n.   This will allow us to understand
the basic structure of the FBA without being encumbered by mathematical details.

There are three basic forms of the FBA; the plane-wave Born approximation, The Coulomb Born approximation, and the distorted-wave
Born approximation.   They are different in what part of H_{int} is included in the unperturbed Hamiltonian H_0.

For an atom of nuclear charge Z with N electrons, H_a is given by

$$H_a = \sum_{j=1}^{N} \left[\frac{P_j^2}{2m} - \frac{Ze^2}{r_j} + \sum_{k>j} \frac{e^2}{r_{jk}} \right] \quad , \tag{1}$$

where $\vec{p}_j$ and $\vec{r}_j$ are the momentum and position vectors of the jth electron, respectively, $r_{jk} = |\vec{r}_j - \vec{r}_k|$, and e and m are the charge and the rest mass of the electron, respectively. For the projectile electron,

$$H_e = p_e^2/2m \quad , \tag{2}$$

where $\vec{p}_e$ is the momentum of the incident electron. For an inelastic collision, of course, $\vec{p}_e$ will be different before and after the collision. We shall use subscripts i and f for the initial and the final states of the projectile, i.e.,

$$H_{ei} = p_i^2/2m \quad , \tag{3a}$$

and

$$H_{ef} = p_f^2/2m = H_{ei} - E_n \quad , \tag{3b}$$

where E_n is the excitation energy gained by the target atom. The interaction Hamiltonian is given by

$$H_{int} = \sum_{j=1}^{N} \frac{e^2}{r_{ej}} - \frac{Ze^2}{r_e} \tag{4}$$

where r_e is the distance between the projectile and the target nucleus and $r_{ej} = |\vec{r}_e - \vec{r}_j|$. Here we have assumed that the target nucleus is so heavy that the center of mass during the collision coincides with the nucleus. For a heavy projectile, this assumption may be invalid. Then, theory must be derived in the center of mass frame and cross sections must be transformed to the laboratory frame for comparison with experiment.

2. The Plane-Wave Born Approximation (PWBA)

In the simplest—and the most popular—form of the FBA, the unperturbed Hamiltonian of the colliding particles, H_0, is the sum of H_a and H_e. The perturbed Hamiltonian, H', therefore, includes both electron-electron and electron-nucleus interactions in H_{int}, Eq. (4):

$$H_0 = H_a + H_e \quad , \tag{5}$$

$$H' = H_{int} \quad . \tag{6}$$

Since H_0 does not include any interaction between the projectile and the target, the appropriate wave function ψ_0 for H_0 is a product of the atomic wave function ψ_a and that for a free electron ϕ_e, i.e., a plane wave:

$$\Psi_0 = \psi_a \phi_e \quad .\tag{7}$$

The interaction matrix element for the excitation to the state n is

$$M_n = <\psi_n \phi_f |H'|\psi_0 \phi_i>$$

$$= <\psi_n \phi_f |\Sigma_j \; e^2/r_{ej}|\psi_0 \phi_i> - <\psi_n|\psi_0><\phi_f|Ze^2/r_e|\phi_i> \quad .\tag{8}$$

Aside from a normalization factor, we denote the plane wave by

$$\phi_e = \exp(i\vec{P}_e \cdot \vec{r}_e/\hbar) = \exp(i\vec{k}_e \cdot \vec{r}_e) \quad .\tag{9}$$

Then, the projectile-dependent terms in Eq. (8) can be integrated out using the relation[1]

$$\int \frac{e^{i\vec{k}\cdot\vec{r}}}{|\vec{r}-\vec{r}_j|} \; d\vec{r} = \frac{4\pi}{K^2} e^{i\vec{k}\cdot\vec{r}_j} \quad .\tag{10}$$

With Eq. (10) we get

$$M_n = (4\pi^2/K^2)[<\psi_n|\Sigma_j e^{i\vec{K}\cdot\vec{r}_j}|\psi_0> - Z\delta_{n0}] \quad ,\tag{11}$$

where $\vec{K}$ is defined in terms of the momentum transfer,

$$\vec{K}\hbar = \vec{P}_i - \vec{P}_f \quad .\tag{12}$$

The Kronecker delta in Eq. (11) results from the orthonormality of the atomic wave functions, and it contributes only to elastic scattering. Now we can write the cross section in the PWBA using Fermi's golden rule:[2]

$$\frac{d\sigma_n}{d\Omega} = \frac{4m^2 e^4}{(K\hbar)^4} \frac{v_f}{v_i} |Z\delta_{n0} - <\psi_n|\Sigma_j e^{i\vec{K}\cdot\vec{r}_j}|\phi_0>|^2$$

$$= \frac{k_f}{k_i} \frac{4a_0^2}{(Ka_0)^4} |Z\delta_{n0} - <n|\Sigma_j e^{i\vec{K}\cdot\vec{r}_j}|0>|^2 \quad ,\tag{13}$$

where $\vec{v}_i$ and $\vec{v}_f$ are the initial and final velocities of the projectile, and $a_0 = \hbar^2/me^2 = 0.529$ Å is the Bohr radius. Note that Ka_0 is
dimensionless, and hence all terms in the last line of Eq. (13) are
explicitly dimensionless except for the term $4a_0^2$, which carries the
correct dimension of a cross section.

Equation (13) is the cross section for the angular distribution
of the projectile. The relationship between $d\Omega$ and $\vec{K}$ is nontrivial;
it involves both the scattering angle Θ and the energy loss E_n. From
Fig. 1, we have

$$K^2 = k_i^2 + k_f^2 - 2k_i k_f \cos\Theta \ , \tag{14a}$$

and from Eqs. (3a) and (3b),

$$k_f^2 = k_i^2 - 2mE_n/\hbar^2 \ . \tag{14b}$$

Substitution of Eq. (14b) into (14a) yields

$$(Ka_0)^2 = 2(T/R)[1 - E_n/2T - (1 - E_n/T)^{1/2} \cos\Theta] \ , \tag{15}$$

where

$$T = mv_i^2/2 \ , \tag{16}$$

and $R = me^4/\hbar^2 = 13.6$ eV is the Rydberg energy. Note again that
Eq. (15) is presented in a dimensionless form, and T defined by
Eq. (16) reduces to the nonrelativistic kinetic energy of the incident electron. Also, Eq. (13) can be expressed in terms of T and K,
using Eq. (14a):

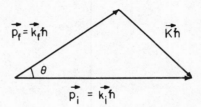

Fig. 1. Definition of the momentum transfer vector.

$$\frac{d\sigma_n}{d(Ka_0)^2} = \frac{4\pi a_0^2}{T/R} \frac{|Z\delta_{n0} - <n|\Sigma_j e^{i\vec{K}\cdot\vec{r}_j}|0>|^2}{(Ka_0)^4} . \tag{17}$$

Equations (14) and (17) are the basic formulas for the PWBA we shall work with.

3. The Coulomb Born Approximation (CBA)

   When the target is an ion, particularly with high ionicity,

$$\zeta = Z - N , \tag{18}$$

the nuclear interaction term, $-Ze^2/r_e$, in H_{int} [Eq. (4)] is more important than the electronic interaction term $\Sigma_j e^2/r_{ej}$. In this case, it may be better to include the nuclear term into the unperturbed Hamiltonian, H_0. Then,

$$H_0 = H_a + H_e - Ze^2/r_e , \tag{19a}$$

$$H' = \Sigma_j e^2/r_{ej} . \tag{20a}$$

Equation (19a), however, still does not contain any interaction term between bound and incident electrons. Thus, the appropriate wave function Ψ_0 for H_0 is still a product of ψ_a and $\phi_e(Z)$; $\phi_e(Z)$ now must satisfy

$$(p_e^2/2m - Ze^2/r_e)\phi_e(Z) = \varepsilon_e \phi_e(Z) . \tag{21a}$$

The solution of Eq. (21a) is the hydrogenic Coulomb function with continuum energy ε_e. The interaction matrix element [Eq. (8)] now contains only the interelectronic term:

$$M_n = e^2 < \psi_n \phi_f|\Sigma_j r_{ej}^{-1}|\psi_0 \phi_i(Z) > . \tag{22a}$$

The CBA uses Eq. (22a) as the interaction matrix element in deriving the cross section formula from Fermi's golden rule again:

$$\frac{d\sigma_n}{d\Omega} = \frac{m^2}{4\pi^2 \hbar^4} \frac{v_f}{v_i} |M_n|^2$$

$$= \frac{k_f}{k_i} \frac{1}{(2\pi a_0)^2} |< \psi_n \phi_f(Z)|\Sigma_j r_{ej}^{-1}|\psi_0 \phi_i(Z)>|^2 . \tag{23}$$

Dimension of the matrix element in Eq. (23) is $(length)^2$ and the correct dimension for the cross section is preserved.

Of course, one can use a screened nuclear charge, e.g., the ionicity ζ instead of the nuclear charge Z in Eq. (21a), in which case
Eqs. (19a)-(22a) should be replaced by

$$H_0 = H_a + H_e - \zeta e^2/r_e \quad , \tag{19b}$$

$$H' = \Sigma_j e^2/r_{ej} - (Z - \zeta)e^2/r_e = \Sigma_j e^2/r_{ej} - Ne^2/r_e \quad , \tag{20b}$$

$$(p_e^2/2m - \zeta e^2/r_e)\phi_e(\zeta) = \varepsilon_e \phi_e(\zeta) \quad , \tag{21b}$$

and

$$M_n = e^2 <\psi_n \phi_f(\zeta)|\Sigma_j r_{ej}^{-1} - N/r_e|\psi_0 \phi_i(\zeta) >$$

$$= e^2 <\psi_n \phi_f(\zeta)|\Sigma_j r_{ej}^{-1}|\psi_0 \phi_i(\zeta)> - Ne^2 \delta_{n0} < \phi_f(\zeta)|r_e^{-1}|\phi_i(\zeta) >. \tag{22b}$$

The last term on the right-hand side (RHS) of Eq. (22b) contributes onl
for elastic scattering. For the CBA with a screened charge ζ, Eq.
(22b) should be used in Eq. (23).

It is not possible to obtain a simple reduction of the electronic
interaction term in Eqs. (22a) and (22b) as was possible for the PWBA
[Eqs. (10) and (11)]. The usual procedure is to expand r_{ej}^{-1} into
products of spherical harmonics, $Y_{\lambda\mu}$:

$$r_{ej}^{-1} = |\vec{r}_e - \vec{r}_j|^{-1}$$

$$= \sum_{\lambda=0}^{\infty} \frac{r_<^\lambda}{r_>^{\lambda+1}} \frac{4\pi}{2\lambda+1} \sum_{\mu=-\lambda}^{\lambda} Y_{\lambda\mu}^*(\theta_e,\phi_e)Y_{\lambda\mu}(\theta_j,\phi_j) \quad , \tag{24}$$

where $r_>$ is the larger of (r_e, r_j) and $r_<$ is the smaller of (r_e, r_j).
For the projectile, we use an expansion into partial waves:

$$\phi_\varepsilon = \sum_{\ell=0}^{\infty} \sum_{m=\ell}^{\ell} R_{\varepsilon\ell}(r_e) Y_{\ell m}(\theta_e,\phi_e) \quad , \tag{25}$$

where $R_{\varepsilon\ell}$ is the continuum radial function with the projectile energy ε.

For discrete excitations, the integral of $Y_{\lambda\mu}(\theta_j,\phi_j)$ in Eq. (24) between the bound-state wave functions,

$$< \psi_n | Y_{\lambda\mu} | \psi_0 > \quad ,$$

will contribute only for a few values of λ owing to the selection rules for angular momentum eigenstates. This in turn will reduce the number of nonvanishing integrals involving the projectile states,

$$< \phi_n | Y_{\lambda\mu}^* | \phi_i > \quad , \tag{26}$$

to a single series sum though both ϕ_i and ϕ_f are represented by infinite series [Eq. (25)].

For ionizing collisions, however, these simplifications do not occur, and the interaction matrix element requires double series summation. In practice, of course, these series are truncated to save computational burden.

As we can see from the above discussion, a different choice of ζ will result in a different cross section. Limited experience with existing theoretical and experimental data suggests that net ionicity defined by Eq. (18) is preferable to the bare nuclear charge Z.

4. The Distorted-Wave Born Approximation (DWBA)

A further improvement on the CBA can be made by replacing a constant ionicity (or screened charge) ζ by a screening function. In practice, we solve a Schrödinger equation for the projectile in the field of bound electrons and the nucleus of the target atom. The screened charge distribution in the target atom can be represented by a local or nonlocal potential —e.g., Hartree-Fock potentials. For brevity, we choose a local potential, $V(r_e)$, to represent the screening of the target nuclear charge by bound electrons. Then, instead of Eqs. (19)-(22), we have

$$H_0 = H_a + H_e - e^2 [Z/r_e - V(r_e)] \quad . \tag{19c}$$

$$H' = \Sigma_j e^2/r_{ej} - e^2 V(r_e) \quad , \tag{20c}$$

$$\{p_e^2/2m - e^2 [Z/r_e - V]\}\phi_\varepsilon(V) = \varepsilon_e \phi(V) \quad , \tag{21c}$$

and

$$M_n = e^2 < \psi_n \phi_f(V) | \Sigma_j r_{ej}^{-1} - V | \psi_0 \phi_i(V) >$$

$$= e^2 < \psi_n \phi_f(V) | \Sigma_j r_{ej}^{-1} | \psi_0 \phi_i(V) > - \delta_{n0} < \phi_f(V) | V(r_e) | \phi_i(V) >.$$

$$(22c)$$

Again, the choice of the screening function explicitly enters in M_n only for elastic scattering. For inelastic scattering, M_n will depend on V indirectly through $\phi_\varepsilon(V)$ in the first matrix element of Eq. (22c). The basic numerical technique to evaluate M_n and hence the DWBA cross section, which is the same form as Eq. (23), is to expand the continuum functions into partial waves and r_{ej}^{-1} into products of spherical harmonics as was done for the CBA.

One can develop different versions of the DWBA by choosing different $V(r_e)$ as well as constructing total wave functions ψ_0 out of ψ_a and ϕ_ε—e.g., with or without antisymmetrization.

5. Generalized Oscillator Strength—The Photon Connection

For an inelastic scattering, Eq. (17) can be written as

$$\frac{d\sigma_n}{d(Ka_0)^2} = \frac{4\pi a_0^2}{T/R} \frac{R}{E_n} \frac{f_n(K)}{(Ka_0)^2} , \qquad (27)$$

where the generalized oscillator strength (GOS), $f_n(K)$, is defined as[3]

$$f_n(K) = \frac{E_n}{R} \frac{|< \psi_n | \Sigma_j e^{i\vec{K} \cdot \vec{r}_j} | \psi_0 >|^2}{(Ka_0)^2} . \qquad (28)$$

Some simple and useful properties of the GOS are listed below.

(a) The GOS is an even function of K—this is obvious from the definition, Eq. (28).

(b) The GOS reduces to the well-known dipole oscillator strength, f_n, at K = 0—we can prove this by expanding $e^{i\vec{K} \cdot \vec{r}}$ in Eq. (28) into power series and by taking the limit $K \rightarrow 0$:

$$e^{i\vec{K}\cdot\vec{r}} = 1 + i\vec{K}\cdot\vec{r} - (\vec{K}\cdot\vec{r})^2/2 + \ldots$$

and

$$\lim_{K\rightarrow 0} | <n|e^{i\vec{K}\cdot\vec{r}}|0> |^2/(Ka_0)^2 = \lim_{K\rightarrow 0}| <n|\vec{K}\cdot\vec{r}|0> |^2/(Ka_0)^2$$

$$= |<n|x/a_0|0>|^2 .$$

In the last step, we took $\vec{K}$ to be parallel to the x axis. The GOS now reduces to

$$\lim_{K\rightarrow 0} f_n(K) = \frac{E_n}{R} | <n|x/a_0|0> |^2 \equiv f_n . \tag{29}$$

(c) Similar to the Thomas-Kuhn-Reiche sum rule for f_n, we have the Bethe sum rule

$$\sum_n f_n(K) = N \quad \text{for all K} , \tag{30}$$

where $\sum_n$ denotes summatiom over all excitations (including ionization and inner-shell excitations) and N is the total number of electrons in the atoms. The proof of the Bethe sum rule is given in the classic article by Bethe[3] and other standard textbooks.[4,5] In this context, the Thomas-Kuhn-Reiche sum rule is a special case (K =0) of the Bethe sum rule. The Bethe sum rule is valid for ions, as long as the number of bound electrons is used for N, not the nuclear charge.

The fact that the GOS reduces to f_n at K=0 provides a very important link between electron-impact and photon-impact cross sections. This connection exists only through PWBA, and hence it can be exploited only when the PWBA is valid—i.e., when the incident electron speed far exceeds those of the bound electrons, particularly the orbital speed of the electron being excited or ionized.

Note that $f_n(K)$ as defined by Eq. (28) is a function of K only and does not depend on the incident energy. Hence, the GOS needs to be calculated only once for a given transition as a function of K. Then, one can calculate the PWBA cross section, $d\sigma_n/d\Omega$ for all incident energies. Again, this simplification occurs only for the PWBA.

In comparing experiment with theory, one can derive an "experimental" GOS by using the measured angular distribution on the left-hand side (LHS) of Eq. (27) and solving it for $f_n(K)$. If "experimental" GOS for different incident energies form a universal curve, it is a good indication that the PWBA is valid for the experiment. As is evident from Eq. (28), however, the theoretical values of $f_n(K)$ depend on the target wave functions used. Any comparison of the PWBA and experiment, therefore, must be done with a realistic estimate of the wave function effect, even when the incident energy is high enough for the PWBA to be valid

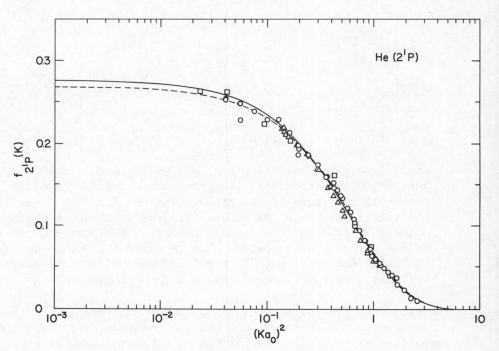

Fig. 2. Generalized oscillator strength for the $1^1S \rightarrow 2^1P$ excitation of He. The solid curve was computed from highly correlated wave functions (Ref. 6) and the broken curve from less correlated wave functions (Ref. 7). The circles are electron-impact experimental data by Lassettre et al. (Refs. 7,8), the squares are those by J. Geiger, (Rev. 9), and the triangles are those by Vriens et al. (Ref. 10). Experimental incident energies vary from 200 eV to 25 keV. (From Ref. 6).

In Figs. 2 and 3, we compare theoretical GOS[6,7] and "experimental" ones[8-10] for the excitation of He to the 2^1P and 2^1S states,
respectively. The solid curves[6] were computed from highly accurate
wave functions, uncertainties from which are expected to be less than
the thickness of the curves drawn. Figure 2 shows an excellent agreement between theory and experiment for an allowed transition, whereas
Fig. 3 indicates that the PWBA is not reliable at the incident energies of 300-500 eV for a forbidden transition.

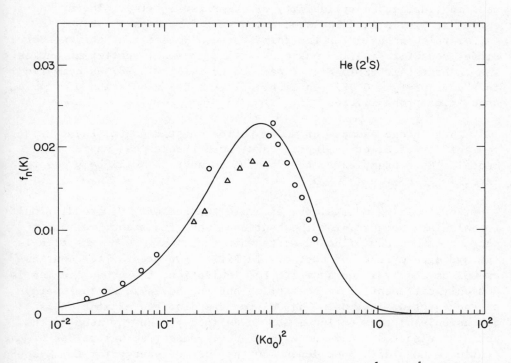

Fig. 3. Generalized oscillator strength for the $1^1S \to 2^1S$ excitation
of He. The solid curve was computed from highly correlated
wave functions (Ref. 6). The circles are electron-impact
experimental data by Lassettre et al. (Ref. 7) and the triangles are those by Vriens et al. (Ref. 10). Experimental
incident energies vary from 400 eV to 600 eV. (From Ref.
6)

Frequently, the ratio of incident energy to excitation energy, $u = T/E_n$, is used as a sole criterion for the validity of the PWBA. Such a practice is unjustified because the values of u for the experimental data presented in Figs 2 and 3 are similar (u = 15-25) but the figures clearly indicate different degrees of validity of the PWBA. For the resonance transition of Na, the experimental data[11] in Fig. 4 were measured at u = 50-70, but agreement with theory[12] is not as good as in the resonance transition of He presented in Fig. 2. Although the wave functions used for Na (Hartree-Fock) are inferior to those for He (50-term Hylleraas), the theoretical uncertainty in Fig. 4 is ∿1%, far smaller than the scatter in the experimental data.

The fact that $f_n(K)$ reduces to the dipole f_n as $K \to 0$ can be used in judging the reliability of theoretical $f_n(K)$. The broken curve[7] in Fig. 2 is calculated using wave functions which are inferior to those used for the solid curve. We see that the two curves differ most for small values of K, i.e., at small scattering angles. The theoretical GOS in Fig. 4 reduced to $f_n(0) = 0.987$ as compared to the value $f_n = 0.982$ recommended in the NBS tabulation of atomic transition probabilities.[13]

This is one example of using photoabsorption data (f_n) to assess reliability of electron-impact data, both theoretical and experimental. More examples of the "photon connection" will be presented later.

For ionizing collisions, accurate theoretical GOS are difficult to calculate because, for the continuum wave functions, we cannot match the accuracy of bound-state wave functions. One exception is the hydrogenic case, of course. In Fig. 5, we present the continuum GOS of the hydrogen atom.[14] For ionization, the cross section is not only differential in scattering angles, but also in the energy of the ejected electron. The GOS for the discrete excitation is dimensionless; the continuum GOS is defined per energy loss of the incident electron, as is done in Fig. 5. Hence, actual values of the continuum GOS, $df(K)/dE$, depend on the unit of energy for E. Rydberg energy is used in Fig. 5.

The continuum GOS near the ionization threshold resembles those for discrete excitations. However, a sharp peak at $(Ka_0)^2 \approx E/R$ emerges as E becomes much larger than the binding energy of the ejected electron. If the target electron were free and at rest, the momentum-energy conservation laws would have required the GOS to be a delta function with its peak at $(Ka_0)^2 = E/R$. The peak in the

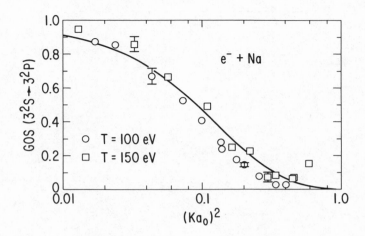

Fig. 4.   Generalized oscillator strength for the $3^2S \to 3^2P$ excitation
          of Na. The circles and squares are electron-impact data by
          Shuttleworth et al. (Ref. 11). The solid curve was computed
          from the Hartree-Fock wave function (Ref. 12). (From Ref. 12)

continuum GOS is referred to as the Bethe ridge or the binary peak.

       Another point to remember is that df(K)/dE describes the angu-
lar distribution of scattered electron and not the ejected electron.
In fact, df(K)/dE is obtained by integrating over the angles of the
ejected electron.

B.   Integrated Cross Sections

       Often experimental arrangement is such that it is easier to mea-
sure integrated cross sections rather than angular distributions.
Theoretically, one simply integrates over the angular distribution
of the scattered electron. Bethe developed[3] a simple but powerful
method for integrating the PWBA angular distribution.   The Bethe
approximation for the integrated cross section not only retains
essential physics contained in the PWBA, but it also provides a
compact form for the cross section—in terms of two to three constants
constants that are independent of the incident energy. This form is
particularly useful for applications in which the cross sections for
a wide range of incident energies are needed.

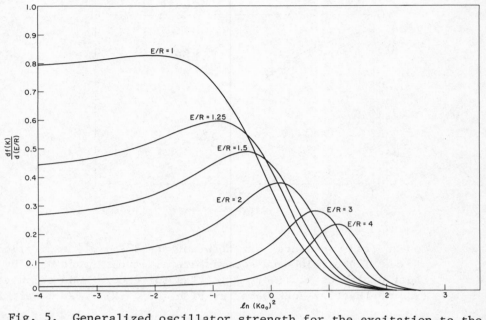

Fig. 5.   Generalized oscillator strength for the excitation to the
          continuum states of H.   The excitation energy (in rydbergs)
          from the ground state is denoted by E/R.   (From Ref. 14.)

1. Integrated Cross Section in the PWBA

Integration over the angles in the PWBA is equivalent to integration over $(Ka_0)^2$ in Eq. (27), or essentially integration of the GOS between the minimum and maximum momentum transfer. Let

$$Q \equiv (Ka_0)^2 \ . \tag{31}$$

Then, Eq. (27) can be rewritten as

$$d\sigma_n = \frac{4\pi a_0^2}{T/R} \frac{R}{E_n} \frac{f_n(Q)}{Q} \, dQ \ , \tag{32}$$

and the integrated cross section is given by

$$\sigma_n \equiv \int d\sigma_n = \frac{4\pi a_0^2 R^2}{T E_n} \int_{Q_{min}}^{Q_{max}} \frac{f_n(Q)}{Q} \, dQ \ , \tag{33}$$

where the limits of integration [cf. Eqs. (14a) and (15)]

$$Q_{min} = (k_i - k_f)^2 a_0^2$$

$$= \frac{E_n^2}{4TR} \left[1 + \frac{E_n}{2T} + O(\frac{E_n^2}{T^2}) \right] \quad , \tag{34}$$

and

$$Q_{max} = (k_i + k_f)^2 a_0^2 \quad ,$$

$$= \frac{4T}{R} \left[1 - \frac{E_n}{2T} + O(\frac{E_n^2}{T^2}) \right] \quad . \tag{35}$$

Equation (33) can easily be visualized by plotting $f_n(Q)$ vs. $\ln Q$ as is done in Fig. 2. Then, the area under the GOS curve bounded by Q_{min} and Q_{max} is the integral on the RHS of Eq. (33), i.e.,

$$\sigma_n = \frac{4\pi a_0^2 R^2}{TE_n} \int_{\ln Q_{min}}^{\ln Q_{max}} f_n(Q) \, d(\ln Q)$$

$$\equiv (4\pi a_0^2 R^2 / TE_n) G \quad . \tag{36}$$

The integral G depends, as it should, on E_n and T through Q_{min} and Q_{max}. A straightforward procedure to obtain σ_n is to evaluate G by numerical integration for each value of T of interest.

The Bethe approximation concerns the evaluation of G for high T in such a way that its T dependence is factored out and numerical integration is used only once. For this purpose, we note that for $E_n/T \ll 1$,

$$Q_{min} \cong E_n^2 / 4TR \quad , \tag{34a}$$

and

$$Q_{max} \cong 4T/R \quad . \tag{35a}$$

When T is large, $Q_{min} \ll 1$ and for a dipole-allowed transition, $f_n(Q) \cong f_n$ in the vicinity of Q_{min}. Also, Q_{max} becomes sufficiently large that it can be replaced by ∞. The first step in the Bethe approximation is to express G in terms of these simpler but approximate limits of integration:

$$G \cong \int_{\ln(E_n^2/4TR)}^{\infty} f_n(Q)d(\ln Q) \quad . \tag{37}$$

Then, we choose $Q_0 > Q_{min}$ such that the area of a rectangle bounded between $Q = E_n^2/4TR$ and Q_0 with height f_n becomes the same as G. This procedure is illustrated in Fig. 6. The correct value of G is the sum of areas I and II, excluding the shaded parts. The value Q_0 is chosen such that areas II and III, including the shaded upper left corner, are the same. Both shaded areas contribute to higher order corrections, from which the leading dependence on E_n and T can also be factored out. One numerical integration of $f_n(Q)$ is necessary to determine Q_0 accurately. Once Q_0 is known,

$$G \cong f_n \int_{\ln(E_n^2/4TR)}^{\ln Q_0} d(\ln Q) = f_n[\ln(T/R) + \ln(4R^2Q_0/E_n^2)]$$

and

$$\sigma_n = \frac{4\pi a_0^2}{T/R} [A_n \ln(T/R) + b_n] \quad , \tag{38}$$

where A_n and B_n are defined by

$$A_n \equiv f_n R/E_n \quad , \tag{39}$$

$$B_n \equiv A_n \ln(4R^2 Q_0/E_n^2) \quad . \tag{40}$$

Bethe parameters A_n and B_n depend only on the target property and not on T nor on the type of incident particle as long as it is structureless. For a bare ion of charge z and speed v_i, corresponding σ_n is given by

$$\sigma_n \cong \frac{4\pi a_0^2 z^2}{T/R} [A_n \ln(T/R) + B_n] \quad , \tag{38a}$$

where T should be evaluated with v_i as specified in Eq. (16), and not by the kinetic energy of the projectile. For example, proton-impact data can be scaled to give electron-impact cross sections and vice versa, by using Eq. (38a) in the asymptotic region.

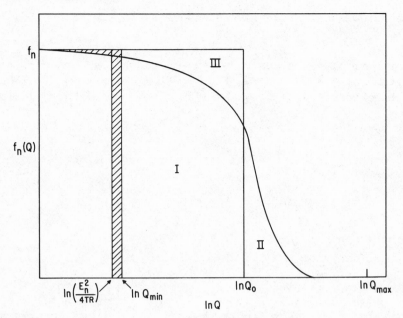

Fig. 6. Illustration of the Bethe approximation for integrated cross
sections. The sum of areas I and II is the integral G in
Eq. (36), corresponding to the exact PWBA integrated cross
section. The Bethe approximation replaces the integral by
a rectangle by choosing Q_0 such that, areas II and III are
equal. The shaded areas contribute to a higher order term
denoted by C_n in Eq. (41).

The Bethe approximation is equivalent to an expansion of the Born
cross section, Eq. (33), in powers of $1/T$. We can extend the power
series by one more term,[15] i.e.,

$$\sigma_n \cong \frac{4\pi a_0^2 z^2}{T/R} \; [A_n \ln(T/R) + B_n + C_n R/T] \; . \qquad (41)$$

Again, C_n is a constant that is independent of T, but now it depends
not only on the target property but also on the type of projectile.
The shaded area in Fig. 6 contributes to C_n.

Furthermore, for electron-impact excitations, the electron-
exchange effect and distortion of the projectile wave functions in
the field of the target (i.e., departure from the plane-wave description) compete with the C_n term in Eq. (41). In this sense, the
original work by Bethe,[3] in which he evaluated A_n and B_n for the

hydrogen atom, extracted all the essential physics contained in the PWBA.

For a forbidden transition, $A_n = 0$ and B_n is obtained from the integration of the GOS (e.g., Fig. 3) from $Q = 0$ to ∞. Values of the Bethe parameters for some Li-like ions are presented in Table I.

2. The Fano Plot and Line Strength

Asymptotic behavior of an integrated cross section can be seen clearly by plotting

$$Y = \sigma_n T/4\pi a_0^2 z^2 R \tag{42}$$

as a function of

$$X = \ln(T/R) \quad . \tag{43}$$

Then, at high T, Y will approach a straight line with slope A_n and the Y intercept B_n. For low T, the straight line will curve as the C_n term in Eq. (41) and other effects not included in the PWBA, become important.

The X-Y plot is known as the Fano plot.[17] This is a very power-ful tool in investigating systematics of both theoretical and experi-mental data. For instance, electron-impact data on a discrete transi-tion can be put into a Fano plot (even if they are relative cross sections!) and be examined for consistency by checking:

a) how smoothly do the data points approach a straight line as T increases; and

b) the slope against the value of $A_n = f_n R/E_n$ derived from photo-absorption data, if they are available. This is another example of using photon data for a consistency check on electron impact data. For a forbidden transition, the slope vanishes; i.e., the Fano plot approaches a horizontal line at high T.

An example of the Fano plot is presented in Fig. 7 for the 2^1P excita-tion of He.

Sometimes the collision strength, a quantity proportional to Y, is used:

Table I. Bethe parameters for the excitation of Li-like ions by
          electron impact. Theoretical data are evaluated from the
          Hartree-Fock wave functions.   Entries in parentheses in
          columns 3 and 4 are based on experimental energies and f_n
          values recommended in Ref. 16.

Ion	Excited state	E_n/R	A_n	B_n	C_n
Li	2^2P	0.1353 (0.1359)	5.658 (5.54)	17.330	−0.226
	3^2S	0.2490	—	0.578	−0.0797
	3^2P	0.2791 (0.2819)	0.0121 (0.020)	0.220	0.0223
	3^2D	0.2815	—	0.923	−0.172
C^{3+}	2^2P	0.5924 (0.5883)	0.493 (0.486)	1.271	−0.124
	3^2S	2.752	—	0.0604	−0.103
	3^2P	2.910 (2.917)	0.0684 (0.0675)	− 0.135	0.179
	3^2D	2.951	—	0.158	−0.266
N^{4+}	2^2P	0.7417 (0.7348)	0.323 (0.320)	0.820	−0.105
	3^2S	4.150	—	0.0408	−0.105
	3^2P	4.349 (4.354)	0.0540 (0.0535)	− 0.126	0.192
	3^2D	4.406	—	0.110	−0.273
Ne^{7+}	2^2P	1.187 (1.178)	0.131 (0.133)	0.324	−0.071
	3^2S	10.02	—	0.0174	−0.110
	3^2P	10.34 (10.34)	0.0288 (0.0283)	− 0.0903	0.212
	3^2D	10.45	—	0.0498	−0.283
Ar^{15+}	$2^2P_{1/2}$	2.363	0.0112	0.0272	−0.0126
	$2^2P_{3/2}$	2.594	0.0225	0.0505	−0.0278
Fe^{23+}	$2^2P_{1/2}$	36.24	0.0495	0.0118	−0.0087
	$2^2P_{3/2}$	47.89	0.101	0.0183	−0.0231

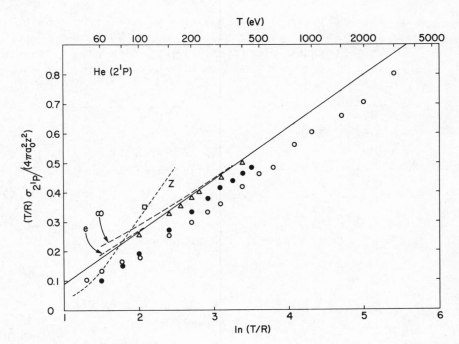

Fig. 7. Integrated cross section for the $1^1S \rightarrow 2^1P$ excitation of
         He. The solid straight line is the Bethe asymptotic cross
         section [Eq. (38)], and the broken curves marked e and ∞
         are the Bethe cross section with the projectile dependent
         term [Eq. (41)], the former for electrons and the latter
         for heavy particles such as protons and α particles. The
         Bethe parameters were computed from highly correlated wave
         functions (Ref. 6). The triangles are experimental data by
         Vriens et al. (Ref. 10), and other symbols represent electron
         impact experimental data by other investigators. See Ref.
         6 for details. (From Ref. 6.)

$$S_n \equiv \sigma_n T\omega_i / \pi a_0^2 R = 4\omega_i z^2 Y \ , \qquad\qquad (44)$$

where ω_i is the multiplicity of the initial state of the target.

3. Comparison of the PWBA, CBA, and DWBA on Discrete Excitations

     At high incident energies, the PWBA is expected to be valid,
and hence the CBA and DWBA results are expected to merge toward those
from PWBA. In this type of comparison, however, it is important that
all theoretical data be evaluated using identical wave functions;

otherwise differences in wave functions may mask the effects of dif-
rerences in theoretical models.

Major departures from the PWBA results occur for low T. The
leading causes for the departures are:

(a)   for target ions, difference in the threshold behavior of the
       cross sections;

(b)   distortion of the projectile wave function;

(c)   electron-exchange effect between the incident and bound
       electrons in the target atom; and

(d)   polarization of the target charge distribution during the
       collision due to the presence of the incident electron.

The first two causes can be corrected by the use of a Coulomb
or distorted wave for the projectile, i.e., by the CBA or DWBA.

The third cause is usually corrected by requiring antisymmetrization of the projectile with the bound electrons. The same procedure
sometimes is used in the PWBA—known as the Born-Oppenheimer approximation—but the results have been unsatisfactory. It is rare that
the Born-Oppenheimer approximation brings genuine improvements to
the PWBA results.

The exchange effect is intrinsically a strong interaction
between the projectile and the affected bound electrons, which is
likely to modify both the projectile and the bound-state wave functions.
Any treatment of the exchange effect that leaves the bound-state
description unchanged is likely to be incomplete.

The fourth cause cannot be handled within the framework of the
first order perturbation theory at all. In a simplified picture,
the target must be considered as a linear combination of many of its
excited states during the collision, which is the basic premise of the
strong coupling methods such as the close-coupling and R-matrix
theories.

In Fig. 8 we compare the PWBA, CBA, and DWBA (with exchange)
cross sections[18] with experimental data[19] for the 2^2P excitation of
C^{3+} by electron impact. As expected, the PWBA data agree better with
other theories at high incident energies, and the DWBA results agree

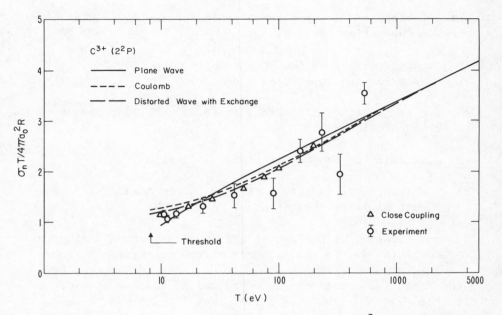

Fig. 8. Integrated cross section for the $2^2S - 2^2P$ excitation of
C^{3+}. The solid curve is the PWBA result calculated by the
present author from the Hartree-Fock wave functions, the
broken and dashed curves are the CBA and DWBA results calculated by Mann (Ref. 18), also from the Hartree-Fock wave
functions. The triangles are the results of a 5-state
close-coupling calculation by Gau and Henry (Ref. 20). The
circles with error bars are the experimental data by Taylor
et al. (Ref. 19). The PWBA cross section vanishes at the
threshold, whereas all other theoretical results are finite
as they should be.

slightly better with the experiment near the threshold than those
from the CBA. Because of the finite energy resolution of experiments,
theoretical data near the threshold should be folded with experimental
resolution when comparing with the experiment.

The DWBA results without exchange (not shown in Fig. 8)— i.e.,
without antisymmetrization of the total wave function Ψ_0—are very
close to those of the CBA in Fig. 8. The antisymmetrization reduces
cross sections for all incident energies. The scatter in the experimental data is too large to draw any conclusion on asymptotic behavior

For ions with complicated electronic structure, use of correlated wave functions may be at least as important as, if not more
than, the use of advanced collision theory. Ions of the Be, Mg, and

Zn isoelectronic sequences belong to this category.   Also, highly charged ions exhibit relativistic effects through changes in energy levels, coupling schemes and orbital sizes.   Correlation and relativistic effects in collision cross sections will be discussed in more detail later.

III.   IONIZATION BY ELECTRON-IMPACT

Although the ionization process has been studied experimentally from the early days of modern physics—e.g., discharge in gases, production of x rays, Geiger counters—we have only a limited success in predicting ionization cross sections from theory.

An obvious departure from discrete excitations is that two electrons that emerge after an ionizing collision (scattered and ejected) are indistinguishable. The faster one of the two is referred to as the "primary" electron and the slower one as the "secondary" electron. This is strictly an operational definition. Such a distinction becomes ambiguous when many electrons are ejected simultaneously during or immediately after a collision.   For instance, multiple ionization makes substantial contributions (through the decay of inner-shell holes) to the total ionization cross section of alkali-like ions, for which inner-shell electrons outnumber valence electrons by a large margin.

A comprehensive theory of ionization can provide detailed information for both primary and secondary electrons, i.e., ionization cross sections differential in

(a)   incident energy T,

(b)   energy loss, E, of the primary electron,

(c)   scattering angles of the primary electron Θ_p and ϕ_p,

(d)   kinetic energy of the secondary electron W, and

(e)   scattering angles of the secondary electron Θ_s and ϕ_s.

If only one electron is ejected, E and W uniquely identify the shell from which the electron was ejected.   Also, one can choose the reference plane to be that defined by the initial and final momenta of the primary electron, i.e., set $\phi_p = 0$.

In the literature, ionization cross sections which specify T, E, Ω_p (Θ_p and Φ_p), W, and Ω_s (Θ_s and ϕ_s) are known as the "triple" differential cross sections, denoted by $d^3\sigma/d\Omega_p d\Omega_s dW$. Those which specify only T, W, and Ω_s are called "double" differential cross sections, $d^2\sigma/d\Omega_s dW$. A double differential cross section is obtained from corresponding triple differential cross section by integrating over Ω_p and summing over all E that produce secondary electrons of kinetic energy W, e.g., secondaries from different shells. The "single" differential cross section, $d\sigma/dW$, is derived from the double differential cross section by integrating over Ω_s. Finally, the total ionization cross section, σ_{ion}, is obtained by integrating $d\sigma/dW$ over W. One can also define cross sections which are double and single differential in the primary electron variables by integrating the triple differential cross section over Ω_s and Ω_p, respectively. For a theoretical treatment, integration over Ω_s often leads to simpler expressions. In this section, we shall concentrate on the systematics of the secondary electrons and the total ionization cross section, after a brief discussion of triple differential cross section.

A. Triple Differential Cross Sections

In principle, triple differential cross sections provide the most complete and stringent comparison of theory and experiment. However, possible combinations of variables are so numerous that meaningful and feasible combinations must be selected judiciously.

1. General Comments

For fast primary electrons, the angular distribution is sharply peaked in the forward direction. Most of them emerge in the forward cone of $\Theta_p \lesssim 20°$. For this reason, it is impractical to measure triple differential cross sections for large scattering angles of fast primary electrons.[21,22] Even under favorable experimental conditions, a coincidence counting rate of one per second is common. Unfortunately theorists have failed so far to provide guidance to experimentalists identifying most significant combinations of variables and targets. This is one of the challenging tasks left for theorists. Some experimental features observed in triple differential cross sections are:

(a) The angular distribution of secondary electrons shows two peaks, one in the forward hemisphere referred to as the binary peak, and the other in the backward direction, which is called the recoil peak;

(b) the binary peak occurs at an angle slightly larger than the momentum transfer direction, presumably from mutual repulsion of the two outgoing electrons;

(c) the binary peak sometimes splits into two lobes, suggesting the existence of a node; and

(d) the recoil peak is small compared to the binary peak in most cases, particularly for fast secondaries, but they may become comparable in magnitude for slow secondaries.

No theory has yet accounted for *all* the features mentioned above, even qualitatively. For instance, the second feature mentioned above could be explained properly only if the theory includes interaction between the two outgoing electrons in the computation of wave function for the secondary electron. The PWBA predicts both binary and recoil peaks to be symmetric around the momentum transfer direction. Symmetry of the collision process is such that the secondary-electron angular distribution must be symmetric with respect to the collision plane ($\Phi_p = 0$), but not necessarily with respect to $\vec{p}_i$, $\vec{p}_f$, or $\vec{K}$. Also, correct shape of the recoil peak would be difficult to predict from any theory that does not incorporate electron-exchange properly.

2. Zero-Angle Scattering of Fast Primary Electrons—Poor Man's Synchrotron Light Source

An ingeneous method of simulating a high-energy (VUV to soft x ray) photon source, similar to those available from electron synchrotrons, has been developed and extensively used by van der Wiel and co-workers.[23,24] The method is based on the "photon connection" mentioned earlier for the inelastic scattering of fast electrons.

In the PWBA, df(Q)/dE reduces to continuum dipole oscillator strength df/dE at Q = 0. When primary electron energy is much higher than energy loss, i.e., E/T << 1, Q_{min} at $\Theta_p = 0$ becomes sufficiently small so that the GOS can be replaced by the dipole limit, df/dE. Then, using the PWBA formula, Eq. (32), one can deduce

(a) df/dE, where E corresponds to photon energy $h\nu$, by measuring the primary-electron energy loss cross section at $\Theta_p = 0$; or

(b) $d^2f(\Theta_s)/dEd\Omega_s$, angular distribution of photoelectrons, by measuring the angular distribution of secondary electrons in coincidence with the primary electrons that lost its kinetic energy by E and emerge at $\Theta_p = 0$.

Apparatus for this type of experiment is cheaper, simpler, and easier to operate than an electron synchrotron. This is why it is called a "poor man's synchrotron." Such an apparatus provides a continuous range of $E = h\nu$ as long as E/T remains small. Also, it can, in principle, provide information on the ionization yield; i.e., the ratio between photoionization ($\int [d^2f(\Theta_s)/dEd\Omega_s]d\Omega_s$) and photoabsorption (df/dE).

Disadvantages of this method are in the energy resolution and its dependence on the Born theory to relate electron-impact data to photon data. Good energy resolution, however, is important only when the cross section depends sharply on the photon energy such as near a resonance. For a routine survey of photoionization cross sections, this method offers many advantages. Primary energy must be kept high to insure the validity of the PWBA. Moreover, when $Q_{min} \gtrsim 0.01$, the GOS may not be close to its dipole limit yet. (See Fig. 4, for example.) In such a case, the angular distribution ($\Theta_p \lesssim 5°$) of the primary electron should be measured and converted into the GOS to assure proper convergence to the dipole limit.

B. Secondary Electrons

Detailed information on angular energy distributions of secondary electrons is needed not only to verify various theories but also as vital input to plasma modeling in fusion research, energy deposition modeling in radiation research, and in the study of upper atmosphere. More experimental data on secondary electrons are available,[25-28] though not abundant, than those on primary electrons in ionizing collisions.[29] Theoretical studies have not progressed beyond the PWBA,[30,31] although many qualitative aspects have been identified to analyze experimental data and extrapolate them to variable ranges not covered by the experiments.[32,33]

1. Angular Distribution of Secondary Electrons

From energy-momentum conservation one can easily show that, when an electron of kinetic energy T collides with another at rest, the struck electron will be ejected with kinetic energy E in the direction given by

$$\cos\Theta_b = \sqrt{E/T} \,, \tag{45}$$

where Θ_b is measured from the direction of the incident momentum.

Although Eq. (45) was derived for a free target electron (without exchange effect), we can still use it for bound electrons *as a qualitative guide* by interpreting E as the energy loss of the primary electron; i.e., the kinetic energy W of the secondary electron plus its binding energy, B:

$$E = W + B \quad . \tag{46}$$

Of course, for a multi-shell target, it is difficult to determine E when only W is measured—a common situation for most experiments on secondary electrons. In this case, we must sum over all energy losses consistent with the energy conservation law; i.e., sum over all inner shells that could have produced a secondary electron of kinetic energy W. For example, for a multi-shell atom,

$$\frac{d^2\sigma}{dWd\Omega_s} = \Sigma_j \ \frac{d^2\sigma}{dE_j d\Omega_s} \quad , \tag{47}$$

with

$$E_j = W + B_j \quad , \tag{48}$$

where B_j is the binding energy of the jth orbital.

For a free electron, the secondary electron peak will be sharp; it can be ejected only in the direction of Θ_b. A bound electron, however, has its own momentum distribution while in a bound orbit, and when ejected, its angular distribution is broadened around the "binary" peak at Θ_b. One can simply use addition of the ejected electron momentum with its average orbital momentum to provide the width of the binary peak:

$$\tan \Delta\Theta_b = \sqrt{E/U} \quad , \tag{49}$$

where U is the orbital *kinetic* energy of the ejected electron. In Table II, we list average orbital kinetic energies and binding energies of rare gases and some diatomic molecules. The orbital kinetic energies are larger in magnitude than the binding energies for nonhydrogenic atoms, particularly for outer orbitals with many radial nodes. Note that the virial theorem holds only for the *total* potential and kinetic energies, but not for each orbital. Radial nodes reduce the probability of finding an electron near the nodes; this implies that the electron is moving fast near the nodes, raising its average kinetic energy.

Table II. Binding (B) and kinetic (U) energies of rare gases and diatomic molecules. Binding energies are experimental data, and the kinetic energies are those calculated from nonrelativistic Hartree-Fock wave functions. (Unit: eV)

Shell	He		Ne		Ar		Kr		Xe	
	B	U	B	U	B	U	B	U	B	U
1s	24.59	39.51	866.9	1259.1	3203.0	4192.9	14325.6	17146.1	3456.4	38899.6
2s			48.47	141.88	320.0	683.1	1921.0	3406.9	5452.8	9240.0
2p			21.60	116.02	245.9	651.4	1692.3	3375.0	4889.4	8229.7
3s					29.24	103.5	295.2	829.8	1093.2	2481.8
3p					15.82	78.07	216.8	773.7	957.7	2412.9
3d							93.0	650.3	672.3	2283.8
4s							27.5	115.8	213.8	696.9
4p							14.22	82.72	146.7	635.3
4d									68.21	495.7
5s									23.3	110.4
5p									12.56	79.74

	H_2		N_2		O_2	
	B	U	B	U	B	U
$1\sigma_g$	15.43	31.96	409.9	601.78	543.5	794.84
$1\sigma_u$			409.9	602.68	543.5	795.06
$2\sigma_g$			37.3	69.53	40.3	78.19
$2\sigma_u$			18.78	62.45	25.69	90.40
$1\pi_u$			16.96	55.21	18.88	72.24
$3\sigma_g$			15.59	44.27	16.42	60.08
$1\pi_g$					12.07	82.14

Although there are sizable differences in detail among the measured angular distributions of secondary electrons in the literature[25-28] all of them exhibit binary peaks, whenever they are discernible, in the vicinity of Θ_b given by Eq. (45).

Another important point is the asymptotic behavior of the angular distribution at high incident energy, T.   For a given energy W
and angle Θ_s of a secondary electron, the same asymptotic behavior
(as T increases) as that discussed for the integrated cross section
[Eq. (38)] applies to the doubly differential cross section.   It can
be shown, analogous to Eq. (38), that[32]

$$\frac{d^2\sigma}{dWd\Omega_s} = \frac{4\pi a_0^2}{T} [A(W,\Theta_s)\ln(T/R) + B(W,\Theta_s) + \ldots] , \tag{50}$$

where

$$A(W,\Theta_s) = \Sigma_j \frac{R}{E_j} \frac{1}{4\pi} \frac{df}{dE_j} [1 - \beta(E_j)P_2(\cos \Theta_s)/2] \tag{51}$$

with the summation over all binding energies B_j of the target (provided that $T > B_j$), and E_j is defined by Eq. (48).   The RHS of Eq.
(51) is simply a summation over the photoelectron angular distribution divided by the photon energy, E_j, where β is known as the asymmetry parameter and P_2 is the Legendre polynomial of the second order.
Again, this is one of the photon connections.   A large body of data
exists on the photoelectron angular distribution in the literature.[34]

Moreover, since $P_2(\cos \Theta_s)$ is an even function of $\cos \Theta_s$, $A(W,\Theta_s)$
should be symmetric with respect to 90°, i.e.,

$$A(W,90°-\theta) = A(W,90°+\theta) \tag{52}$$

for all angles $\theta \leq 90°$.   This symmetry can be used in checking consistency of experimental angular distribution if data exist for high
values of T so that one can plot the Fano plots for a given W but at
two supplementary angles.   Then, at high T, the data points for the
two angles should become parallel; i.e., they exhibit same slope,
although their heights—the $B(W,\Theta_s)$ term in Eq. (50)—are different
for the two angles.   Electron-impact data[25] on N_2 in Fig. 9 clearly
show expected symmetry [Eq. (52)] in slope.

Some investigators[26-28] reported a sharp forward peak ($\Theta_s \lesssim 15°$)
in the angular distribution.   Although it is tempting to associate

the peak with exchange effects,[26,35] it is more likely that the peak
is an experimental artifact.[36]

2. Energy Distribution of Secondary Electrons

Although the extreme forward and backward angles pose severe
difficulties in measuring the angular distribution of secondary
electrons, the energy distribution $d\sigma/dW$ (obtained by integrating
$d^2\sigma/dWd\Omega_s$ over Ω_s) is insensitive to these angles because of the
$\sin\Theta_s$ in $d\Omega_p = 2\pi \sin\Theta_s d\Theta_s$. On the other hand, experimental results
are sensitive to the secondary electron energy, and most data in the
literature show signs of difficulties for slow secondaries, as we shall
see presently.

A powerful method of analyzing experimental data on energy distributions is to use the Platzman plot[33] in which the ratio of measured cross section to the Rutherford cross section is plotted as a
function of the energy loss of the primary electron.   For a multi-
shell atom, one can still use the lowest binding energy to determine the energy loss; this amounts to an assumption that all secondaries come from the valence shell.   In atomic ions, this assumption
will not cause any problem because the cross sections are much larger
for valence electrons than inner ones. (For molecules, some refine-
ments are necessary[38] because, in general, there are several valence
molecular orbitals and corresponding ionization potentials.)

The basic idea of the Platzman plot is that the ratio

$$Y = \frac{(d\sigma/dW)_{observed}}{(d\sigma/dW)_{Rutherford}} \tag{53}$$

is equivalent to the effective number of free electrons participating
in the ionizing collision, because the Rutherford cross section

$$\left(\frac{d\sigma}{dW}\right)_{Ruth.} = \frac{4\pi a_0^2}{T}\left(\frac{R}{W}\right)^2 \tag{54}$$

is exact if the target electron is unbound and at rest.   (We dis-
regard the indistinguishability of the two electrons for a moment.)
In reality, Eq. (54) will diverge for a secondary electron with $W = 0$.
To avoid the divergence, we use [see Eq. (46)]

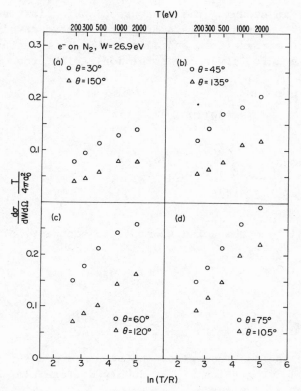

Fig. 9.   Fano plot for the angular distribution of secondary electrons
          from N$_2$ by electron impact. The circles and triangles are
          electron-impact data by Opal et al. (Ref. 25). Note the sym-
          metry in the slopes of the data for supplementary angles.
          (From Ref. 32.)

$$E \equiv E_1 = W + B_1 \quad , \tag{55}$$

where B_1 is the lowest ionization potential.   With this approximation,

$$Y = \frac{d\sigma}{dW} \frac{T}{4\pi a_0^{\,2}} \left(\frac{E}{R}\right)^2 \tag{56}$$

is the (dimensionless) ordinate of the Platzman plot.   As the abscissa, one can use either E/R or R/E.   A plot with E/R as the abscissa

is convenient in studying the cross section for fast secondaries. A plot with R/E as the abscissa is better suited to elucidate details for slow secondaries, and also it has an advantage that the area under the curve gives the total ionization cross section:

$$\sigma_{ion} \equiv \int_{0}^{W_{max}} (d\sigma/dW)dW \tag{56a}$$

$$= \frac{4\pi a_0^2}{T/R} \int_{R/E_{max}}^{R/B_1} Yd(R/E) \quad , \tag{56b}$$

where, according to the definition of the secondary electrons,

$$W_{max} = (T - B_2)/2 \quad , \tag{57a}$$

and

$$E_{max} = W_{max} + B_1 \quad . \tag{57b}$$

For high incident energies, again we can use the photon connection. Upon integration of the doubly differential cross section, Eq. (50), over Ω_s, we get

$$\frac{d\sigma}{dW} = \frac{4\pi a_0^2}{T} \left[\frac{R^2}{E} \frac{df}{dE} \ln(T/R) + B(W) + \ldots \right] \quad . \tag{58}$$

and accordingly,

$$Y = E(df/dE)\ln(T/R) + (E/R)^2 B(W) + \ldots \quad . \tag{59}$$

We expect to see the following trends in the Platzman plot when the incident energy is high:

(a) When photoionization cross section is significant—for slow secondary electrons in general—the shape of the Platzman plot should resemble that of the photoionization, or the first term on the RHS of Eq. (59);

(b) For fast secondaries such that binding energy can be ignored, $W \gg B_1$, Y should approach effective number of free electrons participating in the ionization, i.e., the total number of valence electrons;

(c) Near W_{max} [Eq. (57a)], where the exchange effect is strong,
     the shape of the Platzman plot should resemble that of the
     Mott cross section instead of the Rutherford cross section;
     and

(d) Auger electrons appear as additional peaks superposed on the
     above features.

   In Fig. 10, we present the Platzman plot for the energy distribution of secondary electrons from Ar produced by 500 eV incident
electrons.[25] The shape of measured secondary electron spectra clearly
resembles that of photoelectrons except for the peak near W = 200 eV.
The peak actually consists of several closely spaced LMM Auger peaks.
Steps in the solid curves in Fig. 10 represent a new threshold for
ionizing 2p electrons. The curve marked M is based on the Mott cross
section[39]—collision of two unbound electrons with exchange. As in
the case of the Rutherford cross section, we introduced the binding
energy into the Mott cross section to

(e) avoid divergence for W = 0 and

(f) preserve the symmetry in cross section between the two out-
     going electrons.

   The modified Mott cross section is:

$$\left(\frac{d\sigma}{dW}\right)_{Mott} = \frac{4\pi a_0^2 R^2}{T} \; \Sigma_j N_j \left[\frac{1}{E_j^2} + \frac{1}{(T-W)^2} - \frac{1}{E_j(T-W)} \right], \qquad (60)$$

where N_j is the occupation number of the jth orbital, and the summation is over all orbitals that satisfy $B_j \leqq T$. The curve marked BM
in Fig. 10 is based on the binary encounter theory with exchange corrections. It is clear that the binary encounter theory cannot reproduce the delicate shape of the secondary electron spectra.

   The dip in Fig. 10 at E/R $\sim$ 3.5 is the Cooper minimum, where a
leading term in the dipole matrix element for ionizatin vanishes.
Note that the secondary electron cross section reduces to the Mott
cross section at the Cooper minimum; the Mott cross section does not
include dipole interaction.

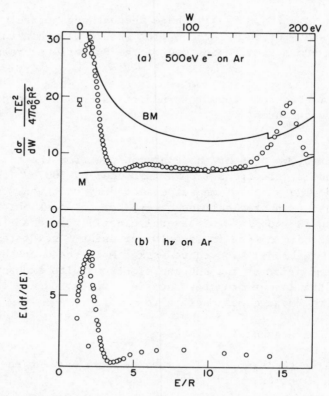

Fig. 10. Energy distribution of electrons ejected from Ar: (a) by
electron impact, and (b) by photoionization. In (a), the
circles are experimental data by Opal et al. (Ref. 25), the
square is by Grissom et al. (Ref. 40), and the triangle is
the PWBA cross section extrapolated from discrete excitations (Ref. 41). The curves marked BM and M are Binary-
Mott and Mott cross sections, respectively. The circles in
(b) are photoionizatin cross sections compiled by Berkowitz
(Ref. 34). The abscissa in (a) is the sum of secondary-
electron energy W and the lowest ionization potential, and
that in (b) is the photon energy. (From Ref. 33.)

An independent measurement of the cross section for the production
of zero-kinetic energy secondary electrons by Grissom et al.[40] and a
theoretical estimate of the same cross section by extrapolation of
bound-state excitation cross sections[41] all support the close relationship between photoionization and secondary electron spectra.

An example of the Platzman plot with R/E abscissa is given in
Fig. 11. Again, there is a striking resemblance between the shape

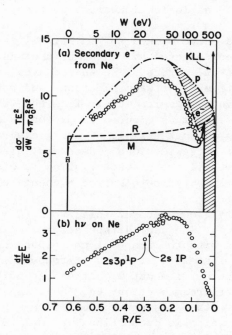

Fig. 11.   Platzman plot of secondary electrons ejected from Ne:
           (a) by electron and photon impact, and (b) by photoioni-
           zation. In (a), the circles are the experimental data by
           Opal et al. (Ref. 25).The curve marked R is the Rutherford
           cross section [Eq. (60) *without* the last two terms in the
           brackets], M is the Mott cross section [Eq. (60)], e⁻ is
           the theoretical cross section for electron impact (T =500
           eV) that is consistent with known σ_{ion}, and p is the cor-
           responding one for proton impact. The square is the ex-
           perimental value by Grissom et al. (Ref. 40) and the tri-
           angle is the PWBA cross section extrapolated from discrete
           excitation (Ref. 41). The arrow marked KLL is the KLL
           Auger peak. The shaded area represents the difference
           between cross sections for electron and proton impact.
           In (b), the circles are photoionization data compiled by
           Berkowitz (Ref. 34).   (From Ref. 33.)

of the secondary electron spectra and that of the photoelectrons.
In addition, the electron impact data[25] should be renormalized to
the dot-dash curve marked "e⁻" so that the area under the curve
matches the known total ionization cross section.

The dot-dashed curve marked "p" in Fig. 11 is an estimated cross
section for electrons ejected by 0.918 MeV protons, which has the
same speed as 500 eV electrons. The shaded area in Fig. 11, which
represents the difference between electron- and proton-impact ionization cross sections, results mainly from the fact that the maximum energy of the electrons ejected by proton is not restricted by
Eq. (57a). This is another example of using electron-impact data to
infer proton-impact cross sections. Secondary electrons ejected by
fast protons exhibit all the characteristics of electron-impact data
except for those attributed to electron exchange effects.[42,43]

Spectra of very slow secondary electrons ($W \overset{<}{\sim} 5$ eV) are often
difficult to measure. One can extrapolate the missing part, however, by comparison with the shape of corresponding photoelectron
spectra, which are easier to measure. Extrapolation to fast secondaries can also be carried out with the help of the photoionization
spectra and the Mott cross section, Eq. (60), as is done in Figs.
10 and 11. Clearly, the area under the Platzman plot with R/E
abscissa (Fig. 11) which is proportional to σ_{ion}, is more sensitive
to the extrapolation for slow secondaries than that for fast secondaries. In fact, 50-70% of the total ionization cross section comes
from the production of secondary electrons whose kinetic energies
are less than the lowest ionization potential of the target.

For a fixed value of W, $d\sigma/dW$ is expected to show the familiar
asymptotic behavior [Eq. (41)] as T increases. In fact, one can combine the Platzman plot and the Fano plot as a three-dimensional plot
by using Y defined by Eq. (56) as the z axis, $\ln(T/R)$ as the y axis,
and R/E as the x axis. Then the curve on a plane parallel to the y-z
plane is a Fano plot exhibiting the T dependence of $d\sigma/dW$ for a given
W or E, approaching a straight line with the slope E(df/dE). On the
other hand, the curve on a plane parallel to the x-z plane is a
Platzman plot with the area under the curve equal to $\sigma_{ion}T/4\pi a_0^2 R$.
When σ_{ion} and df/dE are well known, details of the three-dimensional
plot can be determined to a remarkable degree to satisfy contraints
imposed on $d\sigma/dW$ by the values of σ_{ion} and df/dE. The energy distribution of secondary electrons ejected from He by electron impact
determined by the three-dimensional plot is presented in Fig. 12.
An advantage of the cross sections determined in this way is its

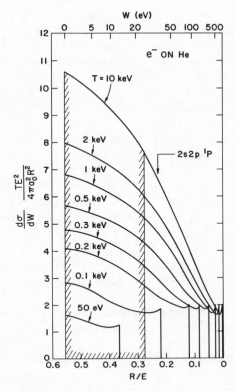

Fig. 12.   Recommended energy distribution for secondary electrons from
            He produced by electron impact. The shaded area corresponds
            to the fraction of secondary electrons too slow to ionize
            other He atoms. A small peak resulting from the decay of
            the doubly-excited 2s2p ^{1}P state should appear at the posi-
            tion indicated by the arrow for each curve with T $\overset{>}{\sim}$ 0.2 keV.
            (From Ref. 42.)

versatility in covering continuous ranges of T and W, particularly for
high T. The shaded area in Fig. 12 represents the contribution to
σ_{ion} by secondary electrons with W < 24.6 eV, the first ionization
potential of He.

For highly stripped ions, $d\sigma/dW$ should resemble at high T the
shape of the hydrogenic df/dE, calculated from appropriate initial
state, not necessarily 1s, interlaced with many autoionizing peaks.
Well isolated autoionizing peaks appear as narrow peaks in the
Platzman plots such as in Figs. 11 and 12, and hence contribute little

to total ionization. For instance, experimental data on $d\sigma/dW$ of He show a small peak for the $2s2p^1P$ excitation at the place marked in Fig. 12, but the area of the autoionizing peak is insignificant. If there are a large number of autoionizing peaks concentrated near the threshold and all of them produce slow secondary electrons ($W \stackrel{<}{\sim} 10$ eV), then it is obvious from the properties of the Platzman plot that such peaks will add significantly to the total ionization cross section. When an unusually large number of autoionization peaks occurs right above the ionization threshold, the Platzman plot is an effective way to estimate the contribution of such peaks to σ_{ion}.

In summary, we note that the Platzman plot is not tied to any particular theory, and it can be used with experimental data (absolute or relative) for any T. The plot is especially useful:

(g) in extrapolating $d\sigma/dW$ to slow secondaries by comparing the shape with that of photoionization;

(h) in identifying Auger and autoionization peaks when their energies are known;

(i) in extrapolating $d\sigma/dW$ to fast secondaries to match the shape and *order of magnitude* to those of the Mott cross section;

(j) in identifying similarities and dissimilarities between electron- and proton-impact $d\sigma/dW$; and finally,

(k) in checking consistency with σ_{ion} if it is known, or in integrating $d\sigma/dW$ to obtain σ_{ion}.

Comparisons (g) and (j) are meaningful only when T is sufficiently high ($\stackrel{>}{\sim} 500$ eV for ionization of neutral atoms and molecules). However, even at lower T, many systematic problems in experiments—i.e., poor transmission of slow secondaries—can be detected simply by putting experimental data into the Fano and/or Platzman plot. As an illustration, in Fig. 13, $d\sigma/dW$ of He with 200 eV incident electrons measured by Opal et al.,[25] DuBois and Rudd,[27], and by Shyn and Sharp[28] are compared with the theoretical cross section determined by the three-dimensional analysis.[33] We find that the shape of the Opal data is in best agreement with theoretical shape, but their data require renormalization by $\sim 20\%$. The DuBois data show a clear sign of missing some slow secondaries; the Shyn data also show a similar trend for $W \stackrel{<}{\sim} 5$ eV. The small peak near $W = 32$ eV in the data by

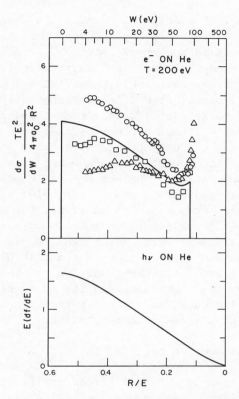

Fig. 13.   Comparison of energy distributions for secondary electrons
           from He: (a) by electron impact, and (b) by photoioniza-
           tion.   In (a), the solid curve is the recommended cross
           section from Fig. 12, the circles are the experimental data
           by Opal et al. (Ref. 25), the squares are those by Shyn
           and Sharp (Ref. 28), and the triangles are those by DuBois
           and Rudd (Ref. 27).   In (b), the solid curve is the photo-
           ionization cross section compiled by Berkowitz (Ref. 34).
           Note the small peak of electrons from the 2s2p ^{1}P state in
           the data by Opal et al. near W = 35 eV.

Opal et al. represents the autoionization of the 2s2p ^{1}P state.

C.   Total Ionization Cross Section

     Reliability of theoretical ionization cross sections depends
mainly on the collision theory and the continuum wave function for

the ejected electron. Preceding discussions on the energy distribution of secondary electrons clearly indicate that slow secondaries
contribute most to ionization cross sections. Hence, it is important
to describe slow secondaries well in computing ionization cross
sections. Wave functions for slow secondaries are more sensitive to
effective screening, correlation effects, interaction with the primary electron, excited ion states, etc. than those for fast secondaries. At present, perturbation theories mentioned in Section II
and their variants are the only practical methods for calculating
ionization cross sections. Among these methods, as usual, the PWBA
has been the most popular one. Since the two electrons that emerge
after a collision are indistinguishable, the electron exchange effect should be treated properly. In addition, for a nonhydrogenic
atom or ion, choice of wave functions introduces more complexity.
As a result, there are many versions of PWBA, e.g., with or without
exchange, with or without correlation in the initial and/or final
state of the target, etc. All these variants of the PWBA, however,
have one goal in common: How to reduce the theoretical ionization
cross section near its peak, and attain better agreement with experiment for intermediate and low primary energies.

1. Ionization Cross Section for Fast Incident Electrons

Most modifications to the PWBA affect ionization cross sections
only slightly for high T. As we have seen already, the dipole term
becomes the leading part of the PWBA cross sections at high T, and
the Fano plot is effective in investigating the asymptotic behavior
of ionization cross sections. Again, σ_{ion} in the PWBA can be expressed in the Bethe form for high T:

$$\sigma_{ion} = \frac{4\pi a_0^2 z^2}{T/R} [A_{ion} \ln(T/R) + B_{ion} + C_{ion} R/T] \quad , \tag{61}$$

where

$$A_{ion} \equiv \int_{B_1}^{\infty} (R/E)(df/dE)dE \quad , \tag{62}$$

and B_{ion} and C_{ion} are constants independent of T.

It is possible to evaluate A_{ion} from photoionization data in principle, but direct computation of B_{ion} and C_{ion} is impractical. Instead, one can use sum rules to sum Bethe parameters for all inelastic scattering and subtract those for all discrete transitions. Then the remainder corresponds to the Bethe parameters for total ionization plus decay of inner-shell excited states through nonionizing channels (e.g., through fluorescence). For atoms of low nuclear charge, the Auger process is the dominant mode for the decay of inner-shell excited states, and hence the difference between the cross section for total inelastic scattering, σ_{tot}, and that for the total discrete excitation, σ_{exc}, is practically the total ionization cross section.

The sum rule method is preferable because:

(a) The Bethe parameters for the total inelastic scattering depend on the ground-state wave function of the target and a complete information on its continuum dipole oscillator strength, which can be obtained either from photoionization experiments or by direct computation—an easier task than a direct computation of the ionization cross section itself;

(b) Reliable Bethe parameters for discrete excitations can be calculated for low-lying states by using accurate bound-state wave functions; and

(c) Those for the Rydberg states can be extrapolated by the quantum defect method.

In many cases, the continuum oscillator strengths are known in sufficient detail and accuracy to determine A_{ion} through Eq. (62). Hence, by using the sum rule method, we can avoid all difficulties and uncertainties associated with continuum wave functions in direct computation of σ_{ion}.

One simple but effective method of calculating exchange corrections to σ_{ion} is to use the Mott cross section, Eq. (60). In Eq. (60), the first term in the brackets arises from direct (Coulomb) interaction, the second term from the exchange interaction, and the third term represents the interference between the direct and exchange interactions. Hence, the exchange correction to σ_{ion} is obtained by integrating the second and third terms in Eq. (60) over the allowed ranges of E_j:

$$
\sigma_{exch} = \frac{4\pi a_0^2 R^2}{T} \Sigma_j N_j \int_{B_j}^{(T+B_j)/2} \left[\frac{1}{(T-E_j+B_j)^2} - \frac{1}{E_j(T-E_j+B_j)} \right] dE_j
$$

$$
= \frac{4\pi a_0^2 R^2}{T} \Sigma_j \frac{N_j}{T+B_j} [1 + \ln(B_j/T) - B_j/T] . \tag{63}
$$

For the asymptotic region where $B_j \ll T$, Eq. (60) is simplified further

$$
\sigma_{exch} \cong \frac{4\pi a_0^2}{(T/R)^2} \Sigma_j N_j [1 + \ln(B_j/T)] . \tag{63a}
$$

Details of the sum-rule method is given elsewhere.[44] We simply note that in the sum-rule method, the first Bethe parameter, A_{ion} in Eq. (61), is more reliable than the second one, B_{ion}. The value of A_{ion} is calculated directly from bound-state wave functions alone whereas B_{ion} requires a weighted sum of all continuum dipole oscillator strengths, which is sensitive to the reliability of the dipole data, either experimental or theoretical. Also, *since the Bethe cross section is an asymptotic formula, it should not be used at low incident energies.*

We present typical results (H$^-$, He, and Li$^+$) in Figs. 14-16. The utility of the Fano plot is clearly demonstrated in Fig. 14, where the data by Tisone and Branscomb[45] exhibit irregular asymptotic behavior. The Bethe parameters for ionization for light atoms and ions are presented in Table III.

When the target has inner-shell structure, a direct calculation requires shell-by-shell summation of the ionization cross section. Furthermore, as the incident energy is increased, more partial waves are required escalating numerical complexity.   The sum-rule method does avoid this problem and includes all inner-shell events automatically by definition:

$$
\sigma_{ion} = \sigma_{tot} - \sigma_{exc} . \tag{64}
$$

On the other hand, the sum-rule method does not provide any information on individual inner-shell events. A combination of direct calculation for low and intermediate T and the sum-rule cross section for high T is the most effective way to determine an ionization

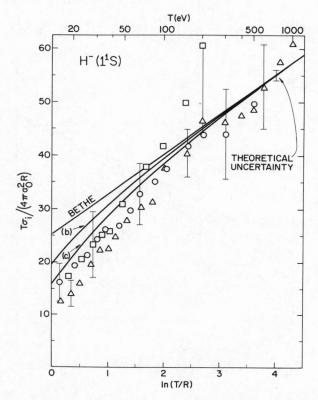

Fig. 14.   Fano plot for the total ionization of H⁻ by electron impact. The straight line marked BETHE is the Bethe asymptotic cross section [Eq. (60) *without* the C_{ion} term], and
the curves marked (b) and (c) are the Bethe cross sections
without exchange [Eq. (61)] and with the Mott exchange [Eq.
(61) plus Eq. (63)]. The squares are experimental data by
Tisone and Branscomb (Ref. 45), the circles are those by
Dance et al. (Ref. 46), and the triangles are those by
Peart et al. (Ref. 47). (From Ref. 44.)

cross section for a wide range of T.   As will be shown later, the
Bethe cross section can also be extended to relativistic range of
T ($\gtrsim$ 10 keV) in a trivial manner.

2.   Ionization cross sections at intermediate and low incident energies

To describe ionizing events, it is necessary to determine continuum wave functions for three electrons; incident, scattered, and

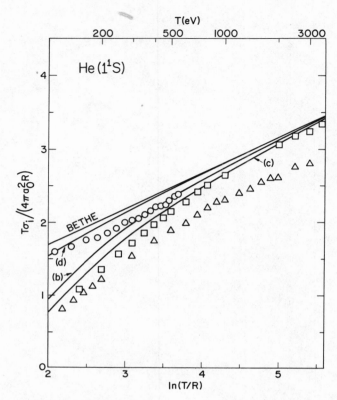

Fig. 15. Fano plot for the total ionization of He. The curve labeled (d) is the Bethe cross section [Eq. (61)] with C_{ion}
for heavy projectiles, and the circles are proton-impact
experimental data by Hooper et al. (Ref. 48). Other curves
are for electron impact, and their labels are the same as
those used in Fig. 14. The squares are electron-impact
data measured by Smith (Ref. 49), and the triangles are
those by Schram et al. (Ref. 50). (Form Ref. 44.)

ejected. Even within the PWBA, in which plane waves are used for
the first two wave functions, different options for the third wave
function, e.g., Coulomb wave, distorted wave, etc. exist. For highly
stripped ions, it is obvious that plane waves for slow incident
electrons would be a poor choice. On the other hand, the wave function for the electron ejected from a highly stripped ion may not
differ much from a Coulomb wave. One of the requirements that should
be satisfied is the orthogonality between the continuum wave function
for the ejected electron and its bound-state wave function. For this

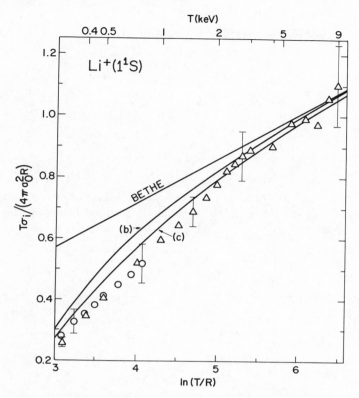

Fig. 16.   Fano plot for the total ionization of Li$^+$ by electron impact. The circles are experimental data by Lineberger
et al. (Ref. 51), and the triangles are those by Peart et
al. (Ref. 52). The labels for the theoretical curves are
the same as those used in Fig. 14. (From Ref. 44.)

purpose, it is preferable to calculate all distroted waves in the
potential of the initial state of the target; or for Coulomb waves,
use the same screened charge throughout the calculation.

Coulomb functions have been used often in the computation of
ionization cross sections of ions, but the use of distorted waves
for all three continuum wave functions is relatively new.[53] An urgent need for the ionization cross sections of impurity ions in
Tokamak plasma has stimulated the atomic physics community and some
theoretical[53-57] and experimental[58] data on multiply charged ions
are available now. We compare theory and experiment on the ionization of C^{3+}, N^{4+}, and O^{5+} in Figs. 17-19, respectively. Recent

Table III. Bethe cross sections for ionization.[a] The Bethe formula
is given by

$$\sigma_{ion} = \frac{4\pi a_0^2 z^2}{T/R} \left[A_{ion} \ln(T/R) + B_{ion} + C_{ion} R/T \right] + \sigma_{exch} \ ,$$

$$\sigma_{exch} = \frac{4\pi a_0^2 z^2}{(T/R)^2} \ \Sigma_j N_j \left[1 + \ln(B_j/T) \right] \ ,$$

where N_j and B_j are the number of electrons and binding
energies of the jth orbital, respectively. The Bethe
parameters, A_{ion}, B_{ion}, and C_{ion} are dimensionless.

Atom	A_{ion}	B_{ion}	C_{ion}	B_j (eV)[b]	N_j
H	0.2834	1.2566	−2.6294	R =13.606	1
H-like ions	$0.2834/z^2$	$84.241/z^2$	−2.6294	$z^2 R$	1
H^-	7.484	25.11	−5.545	0.7552	2
He	0.489	0.714	−5.519	24.59	2
Li^+	0.1445	0.137	−5.439	75.64	2
Li	0.536	2.783	−7.728	5.392 (2s) 67.42 (1s)	1 2
C^{3+}	0.107	0.0498	−7.927	64.49 (2s) 366.8 (1s)	1 2
N^{4+}	0.0750	−0.0077	−7.951	97.89 (2s) 521.2 (1s)	1 2
O^{5+}	0.0554	−0.0385	−7.968	138.1 (2s) 702.9 (1s)	1 2

a. The Bethe formula should be used only for high incident electron
energies, beyond the peak in σ_{ion}

b. Binding energies of the valence orbitals are experimental ones,
and those for the inner orbitals are from the Hartree-Fock
orbital energies.

DWBA results by Younger[54] agree closely with CBA results[55,56] from
the threshold to 2.25 times the threshold. The CBA results in
Figs. 17-19 do not include the ionization of 1s electrons, and hence
their high T results do not agree with experiment. The sum-rule
results agree well with experiment in the slopes of the asymptotic
cross sections, but the heights disagree. As was mentioned, the sum-
rule method requires accurate df/dE to determine the height of the
Fano plot. For the Li-like ions, df/dE based on the Hartree-Slater
potential was used in the sum-rule cross section.

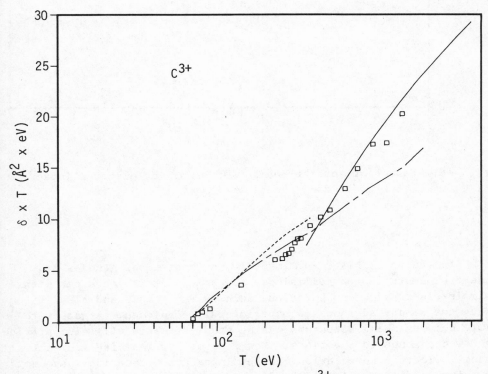

Fig. 17. Fano plot for the ionization of C^{3+} by electron impact.
The squares are the experimental data by Crandall et al.
(Ref. 58). The solid curve is the Bethe sum rule cross
section (Ref. 57), the broken curve is the CBA cross section
by Moores (Ref. 56), and the dot-dash curve is the scaled
hydrogenic CBA by Golden and Sampson (Ref. 55).

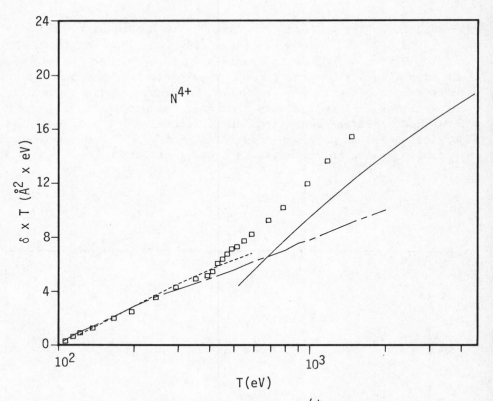

Fig. 18. Fano plot for the ionization of N^{4+} by electron impact.
          See Fig. 17 for legend.

     In the past, both experimental and theoretical studies of
electron-neutral atom collisions have been centered on atoms with
simple electronic configurations such as rare gases and alkali-atoms.
With the recent advances in experimental techniques, targets with
more complex shell structure can be studied. This will also provide
a new challenge to theorists. For ions of low ionicity and neutral
atoms, electron correlation becomes important. Theorists have not
yet found any convenient and effective method to account for the
final-state correlation in ionizing collisions. For highly stripped
ions, the decay of inner-shell hole states must be studied carefully,
because most of them decay through fluorescence as the nuclear charge
increases. Finally, the subject of multiple ionization has not been
studied well. Again, experimental data on multiple ionization will
stimulate the development of comprehensive theory.

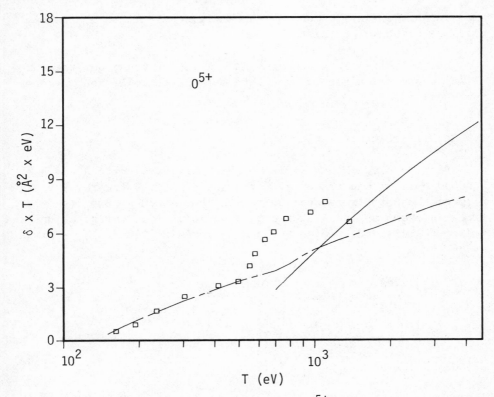

Fig. 19. Fano plot for the ionization of O^{5+} by electron impact.
          See Fig. 17 for legend.

IV. RELATIVISTIC AND CORRELATION EFFECTS

     So far, we have used nonrelativistic collision theory, but rela-
tivistic kinematics begin to affect collision cross sections for
$T \gtrsim 10$ keV. In fact, most electron-impact cross sections begin to
rise beyond $T \sim 1.5$ MeV, known as the relativistic rise. In addition,
inner-shell electrons of medium to heavy atoms are affected by relativistic structure effects that change binding energies, orbital
sizes, and coupling schemes.

     For light atoms, the electron correlation effect can be studied
independent of the relativistic structure effect. For heavy atoms,
however, the correlation and relativistic effects affect each other

because changes in the coupling scheme also alter configurations that are strongly correlated.

The relativistic kinematics is well understood, but the correlation and relativistic structure effects in collision are new subjects.

1. Relativistic kinematics

Fortunately, the collision of high energy electrons can be handled well within the PWBA.[59],[60] For instance, the relativistic form of the Mott cross section for the collision of two free electrons is known as the Møller cross section.[61] In our notation, the Møller formula that replaces the Mott formula [Eq. (60)] is given by[33]

$$\frac{d\sigma}{dW} = \frac{4\pi a_0^2 \alpha^2 R}{\beta^2} \; \Sigma_j N_j \left[\frac{1}{E_j^2} + \frac{1}{(T-W)^2} + \frac{\alpha^4}{4(R+\alpha^2 T/2)^2} \right.$$

$$\left. - \frac{1}{E_j(T-W)} \frac{(1+\alpha^2 T/R)}{(1+\alpha^2 T/2R)^2} \right] \quad , \tag{65}$$

where α is the fine structure constant, and $\beta = v_i/c$, with the speed of light c. At the maximum secondary electron energies [Eq. (57a)] the Møller cross sections are 3% and 54% larger than the Mott cross sections for incident electrons of 100 keV and 1 MeV, respectively.

Bethe[59] and Fano[60] have shown that the relativistic kinematics raises the angular distribution of primary electrons in the extreme forward direction. Consequently, only dipole-allowed transitions are affected by the relativistic kinetmatics. The net change in the integrated cross sections (σ_n and σ_{ion}) is to rewrite the Bethe format [Eqs. (38) and (61)] to

$$\sigma_n = \frac{4\pi a_0^2 \alpha^2 z^2}{\beta^2} \left\{ A_n \left[\ln\left(\frac{\beta^2}{1-\beta^2}\right) - \beta^2 \right] + B'_n \right\} \tag{66}$$

and

$$\sigma_{ion} = \frac{4\pi a_0^2 \alpha^2 z^2}{\beta^2} \left\{ A_{ion} \left[\ln\left(\frac{\beta^2}{1-\beta^2}\right) - \beta^2 \right] + B'_{ion} \right\} \quad , \tag{67}$$

where

$$B' = A \ln(2mc^2/R) + B \quad . \tag{68}$$

The term in the square brackets in Eqs. (66) and (67) result from the combination of the relativistic expression for Q_{min}, Eq. (34), and the contribution from the small component of the relativistic wave functions of the target.

One can readily see from Eqs. (66) and (67) that the extension of the nonrelativistic Bethe cross sections to relativistic incident energies is a trivial task once the values of the Bethe parameters are known. Also, the Fano plot should use $\sigma\beta^2/4\pi a_0^2$ $\alpha^2 z^2$ as the ordinate and $\ln[\beta^2/(1-\beta^2)] - \beta^2$ as the abscissa; then the asymptotic cross section should follow a straight line with the slope, A. **An** example of the relativistic Fano plot for the ionization of He by relativistic electrons and positrons is presented in Fig. 20. As expected from the PWBA, the sign of the charge of the incident particle does not matter at high incident energies. The experiment[62] was done with a small amount of H_2 mixed with He so that all He atoms in discrete excited states ionized H_2 molecules. All ions, both of He and H_2, were detected by a Geiger counter. Hence, the experiment measured the total inelastic scattering cross section, σ_{tot}, and it is in excellent agreement with the Bethe cross section[63] calculated from correlated wave functions.

2. Electron correlation

The correlation effect between the incident and bound electrons should be treated in the framework of a collision theory. We consider here only the correlation among the bound electrons themselves. Qualitatively, we know what class of atoms and ions are subject to strongly correlated structure effects. For instance, the Be-like ions are known to have strong correlation effects,[64] and all group II atoms (Be, Mg, Ca, Sr, Ba, Rn, Zn, Cd, and Hg) and ions of the same electronic configurations are also affected seriously by the correlation effect.[65,66] In addition, atoms and ions with d shells (e.g., transition metals) show strong mixing of configurations. In fact, only alkali-like configurations (Li, Na, etc.) can be treated without elaborate configuration mixing to describe the target.

Details such as angular distributions are more susceptible to the correlation effect because they are likely to be the results of strong interference among various partial waves or interaction

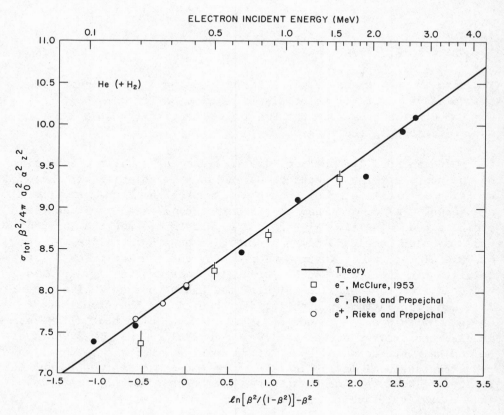

Fig. 20.   Fano plot for the total inelastic scattering of He.   The
solid line is the relativistic Bethe cross section [Eq.
(67)] from Ref. 63, the solid and open circles are experimental data by Rieke and Prepejchal (Ref. 62), and the
squares are those by G. W. McClure [Phys. Rev. 90, 796
(1953)].   (From Ref. 63.)

channels. Optically forbidden transitions could be coupled to allowed transitions through configuration mixing (e.g., 1s2p $^3P_1$ level
mixed with 1s2p $^1P_1$), and consequently have unexpectedly large cross
sections at high incident energies.   For allowed transitiions, cross
sections decrease in most cases when correlated wave functions are
used. The correlation effect in the dipole oscillator strength of
an allowed transition serves as a reliable indicator of the correlation effect in electron-impact cross sections.

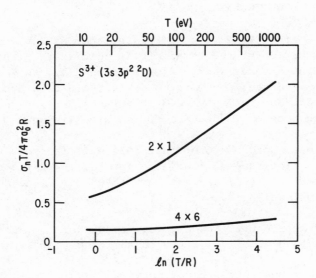

Fig. 21. Fano plot for the $3s^2 3p\ ^2P \rightarrow 3s3p^2\ ^2D$ excitation of S^{3+}
         by electron impact. The labels n × m denote n configur-
         ations for the initial target state, and m configurations
         for the final target state used in the DWBA calculations
         by Mann (Ref. 67).

         In Fig. 21, two DBWA calculations by Mann[67] on the $3s^2 3p\ ^2P \rightarrow$
$3s3p^2\ ^2D$ excitation of S^{3+} (Al-like) are presented. The two calculations use wave functions of different configuration composition
both for the initial and final states. The admixture of the $3s^2 3d$
in the final state is the most important correlation effect in this
case. Since the correlation effect could affect either or both of
the initial and final states, it is difficult to anticipate the net
result of the correlation effect without using reliable wave functions.
For heavy atoms, it is necessary to use relativistic correlated wave

functions even for those transitions that involve valence electrons, particularly if they are from s or p shells.[68]

V.   THEORY OF SLOW COLLISIONS

For slow projectiles, the collision time is long enough for the charge distribution of the target to be polarized by the projectile, and in turn the altered charge distribution distorts the projectile wave function. Hence, the first-order perturbation treatment of the interaction Hamiltonian as we did in various versions of the Born approximation becomes inadequate for slow collisions.

An obvious approach to correct this deficiency is to try higher-order perturbation theory, i.e., the interaction Hamiltonian appears in the scattering matrix element more than once. The second Born and Glauber approximations belong to this category.

Another approach is to describe the projectile and the target electrons on equal footing, i.e., abandon the simple product-type wave function, Eq. (7), and use a linear combination of product-type wave functions that include many real and/or virtual excited states of the target and corresponding continuum functions for the projectile. The close-coupling and R-matrix theories are typical examples.

Each theoretical method mentioned above merits full exposirion, and for details the reader should consult excellent and extensive reviews available in the literature.[69-73]   In this lecture, we shall concentrate

 (a) on basic ideas of these methods without mathematical details
     and

 (b) on the advantages and disadvantages of the methods for slow
     collisions as compared to the simpler methods discussed so far.

1.   Second Born Approximation

The second Born approximation extends the first Born approximation by summing over all intermediate excitations as in any second-order perturbation theory. The transition matrix element in Eq. (11) is replaced by an expression that involves summation over *all* excited states (including continuum) of the target as well as the integration of corresponding intermediate momenta of the scattered electron, with an energy dependent denominator as usual.[69] A simplified picture

of this process is that the target is first excited to one of the
infinitely many intermediate states and then returns to the final
state specified. For instance, the 1s → 2s excitation of H by electron impact is, in the second Born approximation, treated as the
sum of all two-step processes of the type 1s → nℓ → 2s. Of these,
1s → 2p → 2s is the most important correction to the direct, first
order process 1s → 2s. One can easily see that this is no simple
task to perform, even with the help of high-speed computers.
Furthermore, the perturbation treatment can be extended further to
higher orders, and there is no proof that the perturbation series
will converge. Some prototype calculations exist on the scattering
by H, usually with additional shortcuts. The examples are insufficient to draw any conclusions on the utility of the method for
targets of more complex electronic structure, and also on ionizing
collisions. Because of its perturbative approach, the second Born
approximation is not expected to be reliable for very slow collisions (e.g., threshold region). Also, the incident electron is
always treated differently from the bound electrons in such a way
that it is difficult to treat resonances which are the most prominent
features in the scattering of slow electrons.

2. Glauber Approximation

     The Glauber approximation[69,70] is equivalent to an infinite
order perturbation with the perturbation series summed over in a
particular way—H_{int} appears in an exponential form—so that the
scattered electron paths close to the incident momentum, i.e.,
small-angle scattering, are emphasized. This method reduces to the
PWBA in the limit of high incident energy. One drawback of this
method is that the interaction matrix element contains five-dimen-
sional integration.   For small atoms and ions, analytic Roothaan-
type wave functions[74,75] can be used, and the five-fold integration
can be reduced to one-dimensional integration which can be carried
out numerically. The Glauber results are available for the electron-
impact excitation and ionization of several light atoms and ions.[70]
The theoretical results are in better agreement with experiment than
the PWBA results for intermediate projectile energies ($T \lesssim 500$ eV).
The difficulties associated with the five-fold integration, however,
force the theory to depend on analytic target wave functions only—
a severe, if not fatal, restriction for ionizing collisions.   Also,
the Glauber theory is inherently a high-energy approximation, and
as is the case of the second Born approximation, the incident electron is treated differently from the bound electrons.   This makes
the study of near-threshold resonances difficult.   Because small⁻

angle scattering is emphasized in the Glauber approximation, the theoretical results on dipole-allowed transitions are more reliable than those on forbidden transitions because the former peak sharply in the forward direction.

3. Close-Coupling Approximation

In this method the incident electron is treated on the same footing as the bound electrons, i.e., the total wave function for the colliding system is given by, instead of Eq. (7),

$$\Psi(1,2,\ldots N, \ N+1) = \tilde{A} \sum_n \psi_n(1,2,\ldots N)\phi_n(N + 1) \ , \tag{69}$$

where 1,2, etc. stand for the set of variables and quantum numbers to identify bound and incident electrons, $\tilde{A}$ is the antisymmetrization operation, ψ_n is a target wave function representing one of the initial and excited states, and ϕ_n is the corresponding wave function for the scattered electron.   The target wave functions ψ_n are considered known, and ϕ_n is determined by substituting Eq.(69) into the Schrödinger equation.   The asymptotic ($r \rightarrow \infty$) form of ϕ_n provides the cross section for exciting the target to state n. Well-designed computer programs based on this method are available.[71]   The close-coupling approximation is most appropriate for very slow collisions, particularly for detailed study of threshold behavior and resonances.   Since the polarization of the target is described by ψ_n in Eq. (69) and they remain fixed during the solution of the Schrödinger equation, the effectiveness of the method depends critically on the number of target wave functions, ψ_n, included in Eq. (69).   In particular, the results are sensitive to ψ_n whose excitation energies are higher than the incident electron energy.   On the other hand, the distortion of the projectile wave function is represented very well by the close-coupling method.   In reality, one must truncate the expansion, Eq. (69), to a modest length, and this restricts the application of the close-coupling method to low and modest incident energies.   Sometimes pseudo-state wave functions—those with no *real* corresponding states of the target—are used in the expansion to reduce its length, but there are no general principles to guide the choice of such pseudo-state wave functions.   Also, pseudo-state wave functions could produce fictitious resonances.   The calculation by Gau and Henry (Ref. 20), which included five states of C^{3+} (Li-like), is compared to other theoretical results and experimental data in Fig. 7.   It is difficult at present to handle ionization in the framework of the close-coupling approximation.   For ionizing collisions, two

continuum orbitals must be included in Eq. (69), and the joint
asymptotic behavior of the two unbound electrons must be determined
to obtain ionization cross sections.   Such asymptotic behavior is
not well understood theoretically.   On the other hand, the exchange
effect—between bound and incident electrons—is built in the formulation by the antisymmetrization operator in Eq. (69).

To simplify computaional procedure, one can reduce the number
of product functions in Eq. (69) and substitute some (N+1)-electron
bound-state functions, $\Phi(1,2,\ldots,N,N+1)$, to represent the correlation between the projectile and the target electrons:

$$\Psi(1,2,\ldots N+1) = \sum_n a_n \psi_n (1,2,\ldots,N)\phi_n (N+1) + \sum_m b_m \Phi_m (1,2,\ldots,N+1). \tag{70}$$

One can select ϕ_n and Φ_m from simpler, uncoupled calculations, and
then determine coupling constants a_n and b_m by applying the Hulthén-
Kohn variational principle.   This process leads to a set of coupled
linear equations for a_n and b_m, an eminently simpler computational
problem than solving Eq. (69) directly for ϕ_n.   This procedure is
called linear algebraic method, analogous to the configuration
interaction method for bound states.   A more sophisticated approach
keeps only b_m as a variational constant (i.e., all $a_n = 1$) and solves
resulting variational equations for ϕ_n and b_m.   Here, the computational procedure becomes more involved than just solving for a_n and
b_m, but better results are expected because more flexibility is
allowed for ϕ_n, from which actual cross sections are derived by taking their asymptotic forms.   Note that both in Eqs. (69) and (70),
functions ψ_n, ϕ_n, and Φ_m are solved in coupled equations for all
electrons in all the configuration space, particularly for r = 0 to ∞
with full electron exchange effect, which may not be significant at
large r.

4.   R-Matrix Method

In the R-matrix method, the configuration space for the projectile is divided into two regions: an internal region near the target
where it is difficult to distinguish the incident electron from
those bound to the target, and an external region where the projectile electron (before and after the collision) interacts with the
target only through long-range Coulomb interaction and is distinguishable from the bound electrons, i.e., exchange interaction is negligible.   In the internal region, the total wave function for the
colliding system has the same form as in the close-coupling

approximation, Eq. (69).   In the external region, however, the total
wave function is characterized only by the asymptotic behavior of the
projectile to simplify the task of solving the Schrödinger equation.
The solution in the two regions are then joined together at the
boundary—usually a sphere enclosing most of the target charge distribution. The boundary conditions that the external solutions must
satisfy on the sphere are the elements of the R matrix.

    Burke and co-workers[72,74] have been the moving force in develop-
ing and applying the R-matrix method to ions, atoms, and molecules.
In principle, the R-matrix method can pack more target wave functions
in the expansion, Eq. (69), in the internal region than the close-
coupling method. As in the case of the close-coupling method, the
R-matrix method is well suited to the study of slow collisions near
the threshold. The R-matrix calculations on the collision strengths
averaged over the Maxwellian distribution of plasma electrons could
sometimes show drastic (up to a factor of 6) departure from those
calculated by the DWBA or even by the R-matrix method with only a few
bound-state functions in Eq. (69).[76] As in the case of the close-
coupling method, computer programs are available for the R-matrix
calculations, but both methods are still too cumbersome to calculate
cross sections to highly-excited states, including ionization.

VI.   CONCLUSIONS

    It is clear from the discussions presented so far that there is
no single theory that describes electron-atom and electron-ion collisions satisfactorily for all incident energies.   We must therefore
make compromises between accuracy and expediency. For instance, a
brute force extension of the DWBA calculations to high incident
energies (T $\gtrsim$ 1 keV) would be costly and even wasteful; the PWBA
with appropriate high-energy exchange corrections (Mott, Ochkur)
would be sufficient and quicker. On the other hand, perturbative
calculations near the threshold should be verified by strong-coupling
methods such as the close-coupling and R-matrix methods, in particular
on the effects of resonances near the threshold.

    Unfortunately, we do not have much choice for ionization cross
sections. There are no strong-coupling theories that can handle
ionizing collisions at present, and we must bear with the perturbative theories. For many applications, cross sections must be known
for all incident energies. In such cases, particularly those involving dipole-allowed excitations for which their cross sections
remain high even at high incident energies, accurate asymptotic

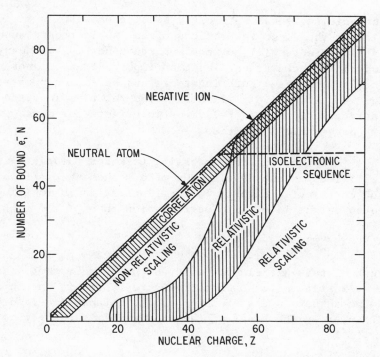

Fig. 22   Qualitative illustration of importance of correlation,
           scaling, and relativistic effects in atomic and collision
           properties. Scaling of collision properties should be
           studied along isoelectronic sequences. Properties of ions
           with low ionicity, Z - N, will be dominated by the cor-
           relation effect for light ions and both by the correlation
           and relativistic effects for medium and heavy ions.

values in the form of the Bethe parameters simplify the task of
presenting the cross sections. Also, the Bethe cross section
format serves as a reliable anchor with which the results of other
theories at lower incident energies can be fitted into compact,
analytic forms for simplicity and convenience.

     Furthermore, for practical applications, knowledge on the scal-
ing of collisional properties as functions of nuclear charge along
isoelectronic sequences will be of great help. Most of the existing
knowledge on the scaling is based on hydrogenic properties, but
they are certainly altered by screening and the correlation effects
in many-electron ions.[12] Scaling of collision properties are basically determined by the scaling of target ionic properties along

isoelectronic sequences. No clear scaling properties will be discernible for the first few members of an isoelectronic sequence; correlation effects will be more important here. For heavy atoms, relativistic effects are at least as important as, if not more than, the correlation effect both in atomic and collision properties. In Fig. 22, we qualitatively illustrate the relative importance of correlation, scaling, and relativistic effects for ions with various nuclear charges and ionicities.

In my opinion, the most pressing task for theorists in electron-atom collisions is the development of a method that allows for electron correlation effects in the final state in ionizing collisions--even on the level of a perturbation theory.

ACKNOWLEDGEMENTS

Much of this article was written while the author was visiting the Joint Institute for Laboratory Astrophysics in the summer of 1981. The author is grateful for the assistance provided by JILA, in particular by the staff of the Atomic Collisions Information Center.

REFERENCES

1. L. I. Schiff, Quantum Mechanics (McGraw Hill, New York, 1955) 2nd Ed., p. 207.
2. E. Fermi, Notes on Quantum Mechanics (University of Chicago Press, Chicago, 1961) p. 100.
3. H. A. Bethe, Ann. Physik, 5, 325 (1930).
4. H. A. Bethe and R. W. Jackiw, Intermediate Quantum Mechanics (Benjamin, New York, 1968), 2nd Ed., p. 305.
5. L. D. Landau and E. M. Lifshitz, Quantum Mechanics. Non-Relativistic Theory (Pergamon, London, 1965), 2nd Ed., p. 581.
6. Y.-K. Kim and M. Inokuti, Phys. Rev. 175, 176 (1968).
7. E. N. Lassettre and E. A. Jones, J. Chem. Phys. 40, 1218 (1964); S. M. Silverman and E. N. Lassettre, ibid. 40, 1265 (1964)
8. E. N. Lassettre, M. E. Krasnow, and S. Silverman, J. Chem. Phys. 40, 1242 (1964); A. Skerbele and E. N. Lassettre, ibid. 45, 1077 (1965); E. N. Lassettre, ibid. 45, 3214 (1966).
9. J. Geiger, Z. Physik 175, 530 (1963).
10. L. Vriens, J. A. Simpson, and S. R. Mielczarek, Phys. Rev. 165, 7 (1968).
11. J. Shuttleworth, W. R. Newell, and A. C. Smith, J. Phys. B 10, 1641 (1977).

12. Y.-K. Kim and K. T. Cheng, Phys. Rev. 18, 36 (1978).

13. W. L. Wiese, M. W. Smith, and B. M. Miles, Atomic Transition Probabilities, Vol. II, NSRDS-NBS 22 (National Bureau of Standards, 1969).

14. M. Inokuti, Rev. Mod. Phys. 43, 297 (1971).

15. Y.-K. Kim and M. Inokuti, Phys. Rev. A 3, 665 (1971).

16. G. A. Martin and W. L. Wiese, Phys. Rev. A 13, 699 (1976); Phys. Chem. Ref. Data 5, 537 (1976).

17. U. Fano, Phys. Rev. 95, 1198 (1954).

18. J. Mann, in Los Alamos Scientific Laboratory Report LA-8267-MS (1980).

19. P. O. Taylor, D. Gregory, G. H. Dunn, R. A. Phaneuf, and D. H. Crandall, Phys. Rev. Lett. 26, 1256 (1977); D. Gregory, G. H. Dunn, P. A. Phaneuf, and D. H. Crandall, Phys. Rev. A 20, 410 (1979).

20. J. N. Gau and R. J. W. Henry, Phys. Rev. A 16, 986 (1977).

21. K. Jung, E. Schubert, H. Ehrhardt, and D.A.L. Paul, J. Phys. B 9, 75 (1976) and references therein.

22. S. P. Hong and E. C. Beaty, Phys. Rev. A 17, 1829 (1978) and references therein.

23. G. R. Wright, M. J. van der Wiel, and C. E. Brion, J. Phys. B 10, 1863 (1977) and references therein.

24. C. E. Brion, Radiat. Res. 64, 34 (1975); A. P. Hitchcock, C. E. Brion, and M. J. van der Wiel, Chem. Phys. 45, 461 (1980).

25. C. B. Opal, E. C. Beaty, and W. K. Peterson, Atom. Data 4, 209 (1972); E. C. Beaty, Radiat. Res. 64, 70 (1975).

26. N. Oda, Radiat. Res. 64, 80 (1975) and references therein.

27. M. E. Rudd and R. D. DuBois, Phys. Rev. A 16, 26 (1977); R. D. DuBuois and M. E. Rudd, ibid. A 17, 843 (1978).

28. T. W. Shyn and W. E. Sharp, Phys. Rev. A 19, 557 (1979).

29. E. N. Lassettre, E. M. Krasnow, and S. Silverman, J. Chem. Phys. 40, 1242 (1964); S. M. Silverman and E. N. Lassettre, ibid. 40, 1265 (1964).

30. S. T. Manson, Phys. Rev. A 4, 1260 (1971); A 5, 668 (1972).

31. K. L. Bell and A. E. Kingston, J. Phys. B 8, 2666 (1975).

32. Y.-K. Kim, Phys. Rev. A 6, 666 (1972).

33. Y.-K. Kim, Radiat. Res. 61, 21 (1975); 64, 205 (1975); in the Physics of Electronic and Atomic Collisions, Eds., J. S. Risley, and R. Geballe (University of Washington Press, Seattle, 1975), p. 741.

34. J. Berkowitz, Photoabsorption, Photoionization, and Photo-electron Spectroscopy (Academic Press, New York, 1979).

35. S. Tahira and N. Oda, J. Phys. Soc. Japan 35, 582 (1973).

36. N. Oda and F. Nishimura, in Abstracts of the Xth International
     Conference on the Physics of Electronic and Atomic Collisions
     (Commissariat a L'Energie Atomique, Paris, 1977), p. 362.
37. See footnote on p. 575, Ref. 5.
38. H. C. Tuckwell and Y.-K. Kim, J. Chem. Phys. 64, 333 (1976).
39. See Eq. (145.17) on p. 575, Ref. 5.
40. J. T. Grissom, R. N. Compton, and W. R. Garrett, Phys. Rev.
     A 6, 977 (1972).
41. Y.-K. Kim, M. Inokuti, and R. P. Saxon, in Abstracts of Papers
     of the VIIIth International Conference on the Physics of
     Electronic and Atomic Collisions, Eds., B. C. Cobic and M.
     V. Kurepa (Institute of Physics, Beograd, 1973), p. 688.
42. Y.-K. Kim, Radiat. Res. 64, 96 (1975); L. H. Toburen, S. T.
     Manson, and Y.-K. Kim, Phys. Rev. A 17, 148 (1978).
43. V. I. Ochkur, Soviet Phys. JETP 18, 503 (1964); 20, 1175 (1965).
44. Y.-K. Kim and M. Inokuti, Phys. Rev. A 3, 665 (1971).
45. G. Tisone and L. M. Branscomb, Phys. Rev. Lett. 17, 236 (1966);
     Phys. Rev. 170, 169 (1968).
46. D. F. Dance, M.F.A. Harrison, and R. D. Rundel, Proc. Roy. Soc.
     (London) A229, 525 (1967).
47. B. Peart, D. S. Walton, and K. T. Dolder, J. Phys. B 3, 1346
     (1970).
48. J. W. Hooper, D. S. Harmer, D. W. Martin, and W. E. McDaniel,
     Phys. Rev. 125, 2000 (1962).
49. P. T. Smith, Phys. Rev. 36, 1293 (1930).
50. B. L. Schram, F. J. de Heer, M. J. van der Wiel,, and J. Kisten-
     maker, Physica 31, 94 (1965); B. L. Schram, A.J.H. Boerboom,
     and J. Kistemaker, ibid. 32, 185 (1966); B. L. Schram, H. R.
     Moustafa, J. Schutten, and F. J. de Heer, ibid. 32, 734 (1966)
51. W. C. Lineberger, J. W. Hooper, and E. W. McDaniel, Phys. Rev.
     141, 151 (1966).
52. B. Peart, D. S. Walton, and K. T. Dolder, J. Phys. B 2, 1347
     (1969).
53. S. Younger, Phys. Rev. A 22, 1425 (1980); A 23, 1138 (1981).
54. S. Younger, Phys. Rev. A 22, 111 (1980).
55. L. B. Golden and D. H. Sampson, J. Phys. B 10, 2229 (1977).
56. D. L. Moores, J. Phys. B 11, L1 (1978).
57. Y.-K. Kim and K. T. Cheng, in Abstracts of the XIth Internation-
     al Conference on the Physics of Electronic and Atomic
     Collisions, Eds., K. Takayanagi and N. Oda (The Society for
     Atomic Collision Research, Japan, 1979) p. 218.
58. D. H. Crandall, R. A. Phaneuf, and P. O. Taylor, Phys. Rev.
     A 18, 1911 (1978); D. H. Crandall, R. A. Phaneuf, B. E.
     Hasselquist, and D. C. Gregory, J. Phys. B 12, L249 (1979).

59. H. A. Bethe, Z. Physik 76, 293 (1932); in Handbuch der Physik,
       Eds., H. Geiger and K. Scheel (Springer, Berlin, 1933),
       Vol. 24/1, p. 273.
60. U. Fano, Phys. Rev. 102, 385 (1956); Ann. Rev. Nucl. Sci. 13,
       1 (1963).
61. C. Møller, Ann. Physik 14, 531 (1932).
62. F. F. Rieke and W. Prepejchal, Phys. Rev. A 6, 1507 (1972).
63. M. Inokuti, Y.-K. Kim, and R. L. Platzman, Phys. Rev. 164,
       55 (1967).
64. A. W. Weiss, Phys. Rev. 122, 1826 (1961).
65. Y.-K. Kim and P. S. Bagus, Phys. Rev. A 8, 1739 (1973).
66. W. D. Robb, J. Phys. B 7, 1006 (1974).
67. J. Mann, personal communication, 1981. See also Ref. 18.
68. J. P. Desclaux and Y.-K. Kim, J. Phys. B8, 1977 (1975).
69. For a unified view of the second Born, eikonal, and the Glauber
       approximations, see C. J. Joachain, Quantum Collision Theory
       (North Holland, Amsterdam, 1975).
70. For the Glauber approximation, see F. T. Chan. M. Lieber, G.
       Foster, and W. Williamson, Jr., Adv. Electronic Electron
       Phys. 49, 133 (1979) and references therein.
71. For the application of the close-coupling theory to electron-ion
       collisions, see R.J.W. Henry, Phys. Rep. 68, 1 (1981); M.J.
       Seaton, Adv. At. Mol. Phys. 11, 83 (1975) and references
       therein.
72. For a review of the R-matrix theory, see P. G. Burke in
       Atomic Physics 5, Eds., R. Marrus, M. Prior, and H. Schugart
       (Plenum, New York, 1977) p. 293.
73. For an extensive review of the theoretical methods for slow
       collisions, see B. H. Bransden, and M.R.C. McDowell, Phys.
       Rep. 30, 207 (1977).
74. C.C.J. Roothaan, Rev. Mod. Phys. 23, 69 (1951); 32
       179 (1960).
75. E. Clemente and C. Roetti, Atom. Data Nucl. Data Tables 14,
       177 (1974).
76. K. L. Baluja, P. G. Burke, and A. E. Kingston, J. Phys. B 14,
       1333 (1981) and references therein.

POTENTIAL ENERGY CURVES

FOR DISSOCIATIVE RECOMBINATION

Steven L. Guberman

Boston College
885 Centre St.
Newton, MA 02159

INTRODUCTION

The direct dissociative recombination (DR) of a diatomic molecular ion, AB^+, with an electron is described by

$$AB^+ + e^- \rightarrow AB^* \rightarrow A + B \qquad (1)$$

where AB^* is a repulsive state of the neutral molecule which dissociates directly to A and B. The dissociation prevents the emission of an electron by AB^* and accounts for the high rate of electron recombination with molecular ions compared to atomic ions. This paper presents a discussion of ab initio calculations of potential energy curves for the AB^* and AB^+ states in (1). Results for recent large scale calculations for O_2 are presented along with additional results for H_2 and He_2. These calculations provide answers to the following questions: What are the identities of the molecular states responsible for DR? What are the translational energies and states of the resulting atoms? How do the atomic state quantum yields vary with ion vibrational excitation and electron temperature?

DR is an important process in both atmospheric and laboratory plasmas. In the Earth's upper atmosphere DR of O_2^+ is an important electron sink and a major source of $O(^1S)$ (Frederick et al., 1976; Kopp et al., 1977) and $O(^1D)$ (Sharp et al., 1975; Cogger et al., 1977; Link et al., 1981; Torr et al., 1981). DR of O_2^+ may be the major contributor to a corona of hot O atoms surrounding the earth's thermosphere (Yee et al., 1980). DR is a source of hot O atoms on Venus (Nagy, et al., 1981) and may impart enough kinetic energy to

167

allow for the escape of hot O atoms from the atmosphere of Mars
(McElroy, 1972; Nier et al., 1976). Numerous laboratory studies of
DR of O_2^+, H_2^+ and He_2^+ have been reported and the most recent of
these will be compared to the calculated results. The AB* states
discussed here are also significant in indirect DR where the electron is initially captured into a vibrationally or rotationally
excited Rydberg state followed by dissociation along AB*. This process is thought to be important above 1000K (Mul et al., 1979) and
will not be treated here. For reviews of much of the previous
theoretical and experimental work on DR the reader is referred to
the articles by Bardsley, Dalgarno, and Mitchell and McGowan in this
volume and to previous reviews by Bardsley (1968a, 1968b), Bardsley
and Biondi (1970), Bates (1974, 1979a), Dalgarno (1979), Dolder and
Peart (1976) and Massey and Gilbody (1974).

ATOMIC AND MOLECULAR RECOMBINATION

        In order to provide some perspective concerning the magnitude
of molecular ion recombination rates a brief comparison to atomic
ion recombination is presented here. At low electron densities
where collision with a second electron is unlikely, atomic recombination can proceed by radiative or dielectronic recombination.
In the former process the electron is captured into a vacant orbital
of the ion and is stabilized by emission of a photon. About
10^{-8} seconds are needed for a dipole allowed transition compared to
roughly 10^{-14} seconds needed for a 1 eV electron to traverse an
atom. Radiative recombination is therefore expected to be slow.
As an example, the rate constant for radiative recombination of O^+
is about 10^{-12} cm^3/sec at 1000°K (Bates, 1974). For molecular ions
we can expect the radiative recombination rate to be about the same
magnitude as that found for atomic ions. However, for molecular
ions the DR rate overwhelms the radiative rate. For O_2^+ at 1000°K
the DR rate is 10^{-7} cm^3/sec (Mul et al., 1979; Walls et al., 1974).
In dielectronic recombination the electron is captured into an
excited orbital and simultaneously a target electron is excited to a
vacant orbital. The formation of the neutral doubly excited atom is
similar to the formation of AB* in (1). However for atoms the energy
needed to reach the lowest accessible doubly excited state can be
large. For He the energy needed is 35.2 eV while for O, about 0.5
eV is needed. For DR, as we will see below, electrons with
near zero energy can often lead to the formation of "doubly excited"
states of AB* in (1). In order to prevent autoionization, the
doubly excited atom can only stabilize by the relatively slow photon emission. Dielectronic recombination will only be important at
very high electron temperatures. Calculations (Burgess, 1964) for

He^+ show that at $10^6 °K$ dielectronic recombination is two orders of
magnitude faster than the radiative rate but near $10^5 °K$ it is equal
to the radiative rate. For O^+, recent calculations (Beigman and
Chichkov, 1980) show that near $5000°K$ the rate for dielectronic
recombination is about 3×10^{-13} cm^3/sec exceeding the rate for radiative recombination at this temperature.

DR MECHANISM

    DR of O_2^+ was first discussed by Kaplan (1931), Massey (1937)
and Bates and Massey (1947) and for He_2^+ by Bates (1950a). A mechanism for DR was first proposed by Bates (1950b).

    DR of a diatomic ion with an electron will proceed with a high
rate if there is a favorable crossing between the potential energy
curves for AB^* and AB^+ in (1). Fig. 1 shows a favorable crossing
for DR from the lowest vibrational state of AB^+. For an electron
energy of ε the cross section for DR is a maximum if the potential
energy curves cross (as in Fig. 1) so as to give a maximum Franck
Condon overlap for the vibrational wavefunctions of the bound ion
and neutral continuum states. Once the dissociating atoms separate
to distances greater than R_C, autoionization back to the molecular
ion and a free electron is impossible and recombination is assured.
As mentioned in the Introduction we see that near zero energy electrons can participate in DR if the potential curve crossings are
favorable. A high DR cross section from v=0 of the molecular ion
at low electron energies requires the intersection of AB^* potential
curves near the midpoint of the v=0 vibrational level. Molecular
ions in the upper atmosphere and in laboratory plasmas can often be
generated with substantial populations in excited vibrational levels.
Vibrationally excited ground electronic state homonuclear diatomic
ions generated under near collision free conditions will have long
lifetimes (10^5 sec) before radiative decay to the lowest vibrational
level. Because of the importance of the Franck Condon factor between bound and continuum nuclear motion levels, vibrational excitation of the ion will have a large effect on DR cross sections.
We expect maximal cross sections for DR at low electron energies
for AB^* curves which intersect excited ion vibrational levels near
the turning points where vibrational wavefunctions have maximum
amplitudes. For electron capture into AB^* potential curves which
intersect the small R turning points of the bound vibrational levels,
autoionization can occur while the atoms are dissociating for
separations less than the large R crossing point. As a result cross
sections for DR will be highest for capture at large R turning point
intersections.

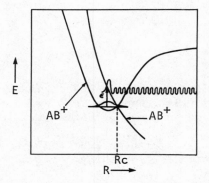

Fig. 1. Potential energy curves of a diatomic ion, AB^+, and a
         neutral dissociative state, AB^*, leading to DR of the v=0
         level by an electron of energy ε.

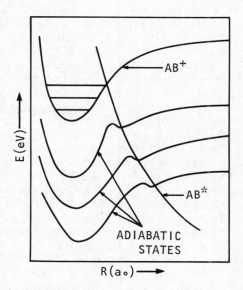

Fig. 2. Potential energy curves for adiabatic bound states showing
         perturbations due to the repulsive diabatic AB^* state.

NATURE OF THE DISSOCIATING STATES

At first glance the calculation of the Born-Oppenheimer states AB* for DR appears to present a difficult problem. Any AB* state arising from neutral atomic asymptotes must pass through an infinite number of states of the same symmetry before crossing above the ground state of the ion. According to the non-crossing rule, states of the same symmetry cannot cross. Therefore, it would appear impossible for a single molecular state arising from neutral separated atoms to cross the molecular ion curve. However there is ample experimental evidence for such states. AB* states have been populated by photon and electron impact and dissociate to neutral atoms or autoionize to atomic ions or molecular ions. Indeed these states can be calculated by restricting the character of the wavefunction. For example if a repulsive AB* state arises from valence states of the atoms, at intermediate internuclear distances this state can cross through the ion state if Rydberg character is not included in the wavefunction. Avoided crossings with Rydberg states that have the molecular ion as their series limit are eliminated by excluding Rydberg character from the wavefunction. However avoided crossings between tight valence molecular states will still be evident. The curves with only valence character are termed diabatic whereas the curves allowed to have the optimum mixture of Rydberg and valence character are termed adiabatic. A comparison of the adiabatic and diabatic states is shown schematically in Fig. 2. The diabatic AB* states cross through the molecular ion and can account for DR and dissociative ionization by photon and electron impact. The wiggle in the adiabatic state is due to the character of the diabatic repulsive state mixing into the adiabatic state. The simplest example of such a state is the diabatic repulsive $^1\Sigma_g^+$ state of H_2 which has primarily $(1\sigma_u)^2$ character at intermediate R. The state rises through a number of bound states of H_2 and is responsible for perturbations in the E, F and G, K and higher states (Wolniewicz and Dressler, 1977). The $^1\Sigma_g^+$ diabatic state is the lowest accessible route for direct DR of H_2^+. Once the dissociating atoms arising from DR pass the internuclear separation at which the ion and neutral curves intersect the dissociating probability will partition itself among the diabatic Rydberg states that cross the dissociating diabatic AB* state. Thus it is possible to populate many atomic and molecular Rydberg states during dissociation.

In H_2 all the excited states of interest are Rydberg states. However it is easy to obtain an exact description of the one electron states of the molecular ion. In order to obtain the diabatic AB* states of H_2, we simply project out of the wavefunction the relevant one electron H_2^+ states (see discussion below). For O_2, however, it is much more difficult to obtain the exact description of the ground state of the molecular ion. The diabatic

AB* states are valence states and they are obtained by eliminating Rydberg character from the O_2 wavefunction. For He_2 the example discussed below uses frozen atomic orbitals to obtain the diabatic AB* states. The construction of the wavefunction and the details of these techniques is discussed further in the following sections.

WAVEFUNCTIONS

A brief outline is presented here of some of the important features of large scale calculations. The purpose of this section is to convey some feeling for the current techniques used and the magnitude of the effort needed in calculations which yield quantitative accuracy for molecular spectroscopic properties. This section reviews the highlights of techniques used for the O_2 and H_2 DR calculations reported here.

Configurations and Orbitals

The many electron spatial wavefunction, Ψ_α, is expressed as a superposition of terms or configurations, Φ_i,

$$\Psi_\alpha = \sum_{i=1}^{M} C_i^\alpha \, \Phi_i(\alpha) \tag{1}$$

The C_i^α are coefficients to be determined. The Φ_i consist of sums of products of one electron orbitals. Each Φ_i is restricted to one particular set of orbitals. The sum of products arises from the necessity to permute the electrons among the one electron spatial orbitals in order to satisfy Pauli's principle and in order to have a pure spin symmetry. For Φ_i we can write

$$\Phi_i = Op(\phi_1(1)\phi_1(2)\phi_2(3)\phi_3(4)\ldots\ldots) \tag{2}$$

where Op is a general many electron operator which generates the required sum of orbital products. Each product has the electrons permuted in a different way among the orbitals in Φ_i. Also Op puts the correct coefficient in front of each permuted product so as to have the proper spin and spatial symmetry and satisfy Pauli's principle. For example, for the $^1\Sigma_g^+$ ground state of H_2 the most important term (largest C_i) is

$$\Phi_1\left(^1\Sigma_g^+\right) = 1\sigma_g(1) \, 1\sigma_g(2) \tag{3}$$

and the second most important term is

$$\Phi_2\left(^1\Sigma_g^+\right) = 1\sigma_u(1) \; 1\sigma_u(2) \quad . \tag{4}$$

For the first excited state of H_2 the most important term is

$$\Phi_1\left(^3\Sigma_u^+\right) = \frac{1}{\sqrt{2}} \left[1\sigma_g(1) \; 1\sigma_u(2) - 1\sigma_u(1) \; 1\sigma_g(2)\right] \quad . \tag{5}$$

where we have already operated with Op. For the ground state of O_2 the most important or Hartree-Fock (HF) configuration is given by

$$\Phi_1\left(^3\Sigma_g^-\right) = \text{Op} \left(1\sigma_g^2 \; 1\sigma_u^2 \; 2\sigma_g^2 \; 2\sigma_u^2 \; 3\sigma_g^2 \; 1\pi_u^4 \; 1\pi_g^2\right) \tag{6}$$

whereby $1\sigma_g^2$ we mean $1\sigma_g(1) \; 1\sigma_g(2)$.

Basis Functions

The orbitals shown above, ϕ_i, are in turn expanded over basis functions, χ_i, centered on the nuclei:

$$\phi_i = \sum_{j=1}^{M'} c_j^i \; \chi_j \quad . \tag{7}$$

Since we cannot take $M'=\infty$ in (7) we must be prudent in our choice of basis functions. A natural choice for basis functions could be the exact one electron solutions for the hydrogen atom. These solutions are linear combinations of the functions

$$\chi_{n\ell m\zeta} = Nr^{n-1} \; e^{-\zeta r} \; Y_{\ell m} \tag{8}$$

where $Y_{\ell,m}$ is an angular factor, N is a normalization constant, and r is distance of the electron from the nuclear center. The functions in (8) are called Slater functions. A serious problem with Slater functions is that some of the integrals that must be calculated in order to solve for c_j^i in (7) and c_i^α in (1) cannot be done in closed form. An example is the two electron, two center integral,

$$\iint \chi_i^a(1) \; \chi_j^b(1) \; \frac{1}{r_{12}} \; \chi_k^a(2) \; \chi_\ell^b(2) \; d\tau_1 d\tau_2 \quad , \tag{9}$$

where the integration is over all space and a and b refer to the atomic centers. These integrals can be time consuming to compute. The problem becomes especially difficult for three and four center

integrals. In order to circumvent the integral evaluation problem
Boys (1950) suggested the use of Gaussian functions which have the
form

$$\chi_{stv\zeta} = N_{s,t,v,\zeta} \; x^s y^t z^v \; e^{-\zeta r^2} \tag{10}$$

where x, y, and z are the cartesian coordinates and s, t and v are
positive integers. $N_{s,t,v,\zeta}$ is a normalization constant. The advantage of Gaussian functions is that all the integrals can be done
in closed form and evaluated quite rapidly by computer. However an
advantage of Slater functions is that it can be shown that they have
the proper limiting behavior at both small and large r. Nevertheless, by taking combinations of several Gaussians, each with a
different exponent, we can correctly describe the electron distribution in the important regions of space. Because the computational time needed to solve for the C_j^i in (7) rises rapidly with the
number of basis functions, it is most convenient to combine individual (primitive) Gaussian functions into groups with fixed coefficients. The coefficient before each group is optimized in molecular
calculations while the coefficients within a group are often optimized in atomic calculations. Each group of Gaussians is referred
to as a contracted set. Dunning (1970,1971,1977) has reported
extensive studies of the construction of Gaussian basis sets in
molecular and atomic calculations. Huzinaga (1965,1971,1977,1979)
has reported optimized exponents for Gaussian basis sets for atoms
extending in size to the fourth row of the periodic table.
Raffenetti (1973) has devised an efficient method of contracting
Gaussian functions and has written an efficient computer program
BIGGMOLI (Raffenetti, 1976), for evaluating these integrals. A
slightly revised version of this program has been used in the O_2
calculations reported here.

    As an example of the effectiveness of Gaussian basis sets,
consider first the lowest state of the H atom. A single Gaussian
with an optimized exponent gives E = -.42441 hartrees (h)
(Huzinaga, 1965) which is in error by 15% compared to the exact
value given by a single Slater of -0.5h. Addition of a second
Gaussian reduces the error to 2.8%. A four term Gaussian expansion gives E = -.49927 and is in error by only 0.15%. For H_2^+
(for which combinations of Slater functions are not the exact basis
functions) a basis set on each center of nine 1s Gaussians contracted to five Gaussians and two $2p_z$ Gaussians (s=t=0 and v=1 in
(10)) contracted to a single Gaussian gives an electronic energy of
E = -1.10184h at R=2.0 Bohr (a_o) (Guberman, 1982) compared to the
exact energy of E = -1.10263h (Bates et al., 1953) for an error of
only .07%. For the ground state of the hydrogen molecule (Guberman,
1982) the basis set used for H_2^+ was supplemented on each atom with
eight $2p_x$ and eight $2p_y$ Gaussians contracted to four $2p_x$ and four

$2p_y$ Gaussians plus four $3dx^2$ (s=2, t=v=0 in (10)), four $3dy^2$ and four $3dz^2$ functions contracted to four $3d\sigma$ functions. The C_j^i in (7) were optimized for the one electron states of H_2^+ and the final wavefunction took the form shown in (1) with 82 terms. The calculated total energy was $E = -1.168045h$ at $R = 1.4$ a_o compared to the highly accurate energy of $E = -1.1744699h$ of Kolos and Wolniewicz (1964) for an error of only 0.54%. Using Gaussian basis sets similar in size to that for the ground state in a study of ten low lying excited states of H_2 gives energies which differ on the average by only .0029h from exact energies (Guberman, 1982). For the O_2 calculations reported here we have used a Gaussian basis set of nine 1s functions, five 2p functions and two 3d functions taken from Huzinaga (1965) and contracted to three 1s, two 2p and one 3d function. The s and p contractions are from Dunning and Hay (1977) and the d contraction is from Dunning (1971). This basis set is shown in Table 1. The ground state O atom energy in this basis is -74.798844h (Guberman, 1977) and is only .01054h above the HF limit energy (H. F. Schaefer III et al., 1969). The HF limit energy is the lowest energy obtainable with a one configuration wavefunction, i.e., it is the energy found in the limit of a very large basis set. For O_2 at $R = 2.2819$ a_o the HF energy in this basis is -149.63642h which is .0295h above the HF limit energy at 2.282 a_o (Cade and Wahl, 1974). The HF limit Slater basis set consisted of two 1s, two 2s, one 3s, three 2p, two 3d and one 4f function on each center (Cade and Wahl, 1974). If we assume that the error consists of twice the atomic error plus a molecular extra error, the molecular extra error is only .0084h. Indeed the Gaussian basis set used here accounts for 82% of the molecular HF limit binding energy. Thus the small Gaussian basis shown in Table 1 is able to account for most of the HF binding energy found in the much larger limiting basis set. It is energy differences such as binding energies and excitation energies rather than absolute energies that are of interest to chemists and are the real tests of basis set accuracy. Table 3 of Dunning and Hay (1977) shows excellent agreement between atomic excitation energies calculated in the Gaussian basis of Table 1 and HF limit energies. Further calculations by the author on the excited states of O_2 in a Gaussian basis show that general agreement is obtained for binding energies and excitation energies with similar wavefunctions calculated in a Slater basis (Saxon and Liu, 1977). Indeed the agreement is such that it is difficult to justify the use of Slater orbitals unless one has a highly efficient and rapid Slater integrals computer program.

Orbital Optimization

In this section a brief description is given of how we determined the optimum in C_j^i in (7). We can in theory skip this step and instead write the configurations in (2) in terms of basis functions instead of orbitals and go ahead to the next step of determining the

Table 1.   Contracted Gaussian Basis Set for the O Atom[a].

Exponents	Type[b]	Contraction Coefficients
7817.	s	.001176
1176.	s	.008968
273.2	s	.042868
81.17	s	.143930
27.18	s	.355630
9.532	s	.461248
3.414	s	<u>.140206</u>
9.532	s	-.154153
.9398	s	<u>1.056914</u>
.2846	s	1.0
35.18	p	.019580
7.904	p	.124200
2.305	p	.394714
.7171	p	<u>.627376</u>
.2137	p	1.0
1.8847	d	.357851
.55826	d	.759563

[a]The exponents listed are from Huzinaga (1965) and
the contraction coefficients are from Dunning and
Hay (1977) and Dunning (1971).
[b]The Gaussian types are defined in Eq. (10) with
$s+t+v=\omega$. For s, p and d type Gaussians $\omega=0,1$ and
2 respectively. For the p and d Gaussians all possibilities for the positive integers s, t and v are
included except the combination $s=t=1$, $v=0$ for d.

C_i^α in Eq. (1). However M in Eq. (1) would then be extraordinarily
large making the calculation of Ψ impossible in practice. The calculation of molecular orbitals represents, in a sense, a further
contraction of the basis set. These molecular orbitals are
often pseudoeigenfunctions of one electron Schroedinger equations.
As a result we can interpret an orbital as containing an electron
which sees an averaged field due to all the other electrons. This
independent particle interpretation allows one to gain considerable
qualitative understanding of molecular binding.

We consider here two approaches to solving for molecular orbitals: the single configuration HF method and the Multiconfiguration
Self Consistent Field (MCSCF) method. In order to determine optimum orbitals we must consider the energy expression for the wavefunction in (1):

$$E = \frac{< \Psi | \mathcal{H} | \Psi >}{<\Psi | \Psi >} \tag{11}$$

where $\mathcal{H}$ is the Hamiltonian operator for N electrons and N´ nuclei,

$$\mathcal{H} = \sum_{i=1}^{N} h_i + \sum_{i>j}^{N} \frac{1}{r_{ij}} + \sum_{a>b}^{N´} \frac{z_a z_b}{R_{ab}} \tag{12}$$

where i is an electron index, h_i contains the one electron kinetic
energy and electron-nuclear interaction terms, a and b are indices
of the nuclei, and z_a is the nuclear charge. In order to determine
the optimum orbitals, i.e., the orbitals that lead to the minimum
energy in (11), the energy is required to be stationary ($\delta E = 0$)
for small changes in the C_i^j. Often it is also required that the
orbitals be orthonormal, $<\phi_i | \phi_j> = \delta_{ij}$. The orthogonality restric-
tion leads to a considerable simplification in the resulting equations. This condition is relaxed in the calculations on He_2 described below. From these requirements pseudoeigenvalue equations
arise for the orbitals. Applying the basis set expansion of the
orbitals shown in (7) one then solves for the C_j^i by standard matrix
techniques. The potential that an orbital "sees" is dependent upon
all the other orbitals which are in turn dependent upon the first
orbital. As a result an iterative procedure (which hopefully con-
verges in a few iterations) is needed to determine all the orbitals.
These orbitals are generally denoted HF orbitals if they are determined from a wavefunction which has the most important single configuration, i.e. M=1, in Eq. (1). If the orbitals are determined
from a multiconfiguration wavefunction we refer to the procedure as
MCSCF. The resulting pseudoeigenvalue equations are then more
complicated since in MCSCF both C_i^α in (1) and C_i^i in (7) are optimized. The reader is referred to the articles by Wahl and Das
(1977), Bobrowicz and Goddard (1977), and Hinze (1973) for the

details of these procedures. The O_2 MCSCF calculations reported
here were done with a revised version of a program originally written
by Das and Wahl (1972).

It has been recognized for several years that there are
many problems with the HF description of molecules. A serious problem is the incorrect dissociation behavior of HF wavefunctions, i.e.,
as the intermolecular distance increases, single configuration HF
wavefunctions often become a superposition of ionic and neutral atoms
rather than dissociating to the proper states of the separated atoms.
Closed shell molecules which dissociate to closed shell atoms can
dissociate correctly in HF. However many chemically interesting
systems do not involve closed shell species. A simple example of
incorrect HF dissociation is the ground state of H_2. The $1\sigma_g^2$ single
configuration HF wavefunction with $1\sigma_g = 1/\sqrt{2}\,(1s_a+1s_b)$ dissociates

to ionic and neutral terms. However the $\dfrac{1}{\sqrt{2}}\,(1\sigma_g^2-1\sigma_u^2)$ two config-

uration wavefunction, where $1\sigma_u = \dfrac{1}{\sqrt{2}}\,(1s_a-1s_b)$, dissociates correctly

to two ground state H atoms. For larger systems MCSCF provides the
solution to the incorrect dissociation problem. For example, for
the ground state of O_2 six configurations are needed to properly
dissociate to two ground state O atoms (Guberman, 1977). These
configurations are referred to as the configurations for proper dissociation (CPD).

Configuration Interaction

The orbitals determined from the MCSCF procedure are determined in a static averaged potential due to all the other electrons.
However this is only an approximation since each of the electrons
in fact "sees" a dynamic instantaneous potential, i.e., for any
arbitrary movement of one electron, all the other electrons will
adjust. The additional energy is called correlation energy and is
found by considering configuration interaction (CI). In CI we take
the orbitals determined in the MCSCF procedure plus additional orbitals which can be formed from the basis set and construct a wavefunction as in (1). Now, however, the CI wavefunction in (1) extends
over many more configurations and orbitals than the MCSCF wavefunction. If Ψ is constructed so that the configurations are orthonormal, the C_i^α in (1) can be determined by solving the matrix equation

$$HC = CE \qquad\qquad\qquad\qquad\qquad\qquad\qquad (13)$$

where C is a matrix in which each column is the αth linearly independent set of C_i^α coefficients in (1). Each element of the H matrix
is an integral over the Hamiltonian operator between configurations
shown in (1),

$$H_{ij} = <\Phi_i | \mathcal{H} | \Phi_j> \quad .$$ (14)

The eigenvectors and the eigenvalues of H are then found by diagonalizing the H matrix. In general the CI H matrix can often have
several thousand terms and standard techniques of matrix diagonalization are not very useful. A discussion of iterative techniques
for diagonalizing large CI matrices and further details of CI calculations can be found in the article by Shavitt (1977). The program
used in the CI calculations on O_2 and H_2 was written by the author
and coworkers (R. C. Ladner, P. J. Hay, W. J. Hunt, F. W. Bobrowicz,
N. W. Winter and T. H. Dunning) at the California Institute of
Technology. The technique used to construct the H matrix is described in Bobrowicz (1974) while the diagonalization technique
was devised by Shavitt (1977) and Shavitt et al (1973).

     In the case of H_2 the determination of the types of configura-
tions to include in (1) is quite straightforward. Namely, for two
electrons and a particular total spatial symmetry one can generate
all possible ways of distributing the electrons among all the orbitals formed from an adequate basis set and still have a relatively
small wavefunction by modern standards. As discussed above, 82
configurations resulted when this full CI is generated for the
ground state of H_2. For systems with more electrons such as O_2, a
full CI in a basis set such as that shown in Table 1 is out of the
realm of practical computations. However we can construct a good
approximation to a full CI wavefunction from the following consideration. If we take Φ_i in Eq. (14) to be the HF configuration, H_{ij}
will vanish for all Φ_j which differ by more than two orbitals (a
double excitation) from Φ_i if the orbitals are orthogonal. Because
of the matrix elements of the electron-electron repulsion operator
in the Hamiltonian, configurations differing by a triple excitation
or higher from the HF configuration can be ignored to first order.
Indeed it is now known that most (i.e. 90%) of the correlation
energy can often be obtained with single and double excitations.
However, because of the incorrect dissociation of the HF wavefunction it is often inadequate to generate the CI wavefunction from
only the HF configuration. Since the HF wavefunction treats the
equilibrium separation, R_e, much better than it treats larger internuclear distances (except in closed shell systems which dissociate
correctly in HF) a CI generated only from the HF configuration will
also be biased to give a better description at R_e. Such calculations can give dissociation energies and fundamental frequencies
that are too large compared to experiment (Guberman, 1977). The
situation is remedied if the configurations which dissociate the
molecule correctly (CPD) are used as reference configurations.
However all single and double excitations from the CPD can lead to
too many configurations to be handled with current programs. As a
result an approximation is used in which all double excitations are

generated from the CPD; however in each configuration only a single
electron is allowed in the orbitals not occupied in the CPD. These
orbitals make up the virtual space. This approximation is based on
the first order wavefunction of Schaefer and Harris (1968) originally applied to a single reference configuration. Calculations on
the excited states of O_2 using this approach give adiabatic excitation energies and dissociation energies that differ by about 0.15
eV from experiment. Calculated fundamental frequencies differ by
only 50 cm^{-1} from experiment while equilibrium separations are
about .06 a_o larger than experiment (Guberman, 1982).

Practical Considerations

     The calculations for O_2 and H_2 were performed on a CDC 6600
computer which has an available memory of about 97000 words.
The programs and the matrix elements to be manipulated must be
squeezed into this memory. The first step in the determination
of molecular wavefunctions and potential energy curves is a
calculation of integrals over the contracted Gaussian basis functions. These integrals are used to construct the MCSCF equations and the H matrix needed for determination of the CI wavefunction. There are overlap integrals and integrals over the one electron operator, h_i, in Eq. (12) consisting of kinetic energy and
nuclear attraction integrals. For N total basis functions there
are only $N(N+1)/2$ of each of the types of one electron integrals.
As a result these integrals do not become large in number and their
computer storage presents no problem. However for the two electron
integrals there are $\frac{1}{8} (N^4+2N^3+3N^2+2N)$ integrals. This expression
is arrived at if one considers the interchange symmetry of some of
the indices in the two electron integrals. For example, for the
integral shown in (9) switching index i with index j or index k
with index ℓ or indices k and ℓ with indices i and j does not change
the value of the integral so that we need only store in the computer
a single representative value. For the basis set shown in Table 1
there are a total of 28 contracted Gaussians, 14 on each O atom.
From the above formula one would expect a total of 82621 two electron integrals leaving very little room in the computer for the
integral or MCSCF programs. In fact the situation is not quite so
hopeless. If one considers the spatial symmetry of the orbitals it
turns out that for O_2 most of the two electron integrals are zero
by symmetry and calculation shows that there are at R = 2.2819 a_o
only 14706 non zero integrals which can easily fit in the machine
memory. Usage of only the non-zero integrals requires efficient
algorithms which indicate whether a required integral is zero or
non-zero and, if non-zero, where its value can be found in memory.
We can see from the above expression that the number of two electron
integrals needed in a particular calculation can get rapidly out of

hand for molecules with more than three first row atoms or molecules
with little symmetry.

In order to obtain the wavefunction for $^3\Sigma_g^-$ states of O_2 a
first order wavefunction was generated from the CPD of each of the
$^3\Sigma_g^-$ states giving a total of 2496 terms. Determination of the
eigenvectors or eigenvalues requires diagonalization of a 2496×2496
H matrix. Obviously storage of the full matrix in memory is out of
the question. It would not even fit into the fast memory of the
largest supercomputers. Fortunately, however, these matrices are
characteristically quite sparse. In the case of $X^3\Sigma_g^-$ only 7.5% of
the matrix elements are non-zero. However this still leaves almost
half a million matrix elements which must be considered. Iterative
techniques have been developed which require that only a single row
of the lower triangular H matrix need be in memory at any instant
(Shavitt et al., 1973 and Shavitt, 1977). The remainder of the H
matrix can be kept on disk or tape. Typically only a few of the
lowest roots of the matrix are determined. For the case considered
here the maximum number of non-zero elements in a lower triangular
row of the H matrix is 280 and the diagonalization is quite manage-
able. For the $X^3\Sigma_g^-$ state of O_2 the integrals take about 80 seconds
and the determination of the lowest six states of this symmetry from
the 2496×2496 CI matrix took about 29 minutes on a CDC 6600 computer. About 65% of the 29 minutes was spent in the diagonalization
while the remainder involved calculating the H matrix.

DISSOCIATIVE RECOMBINATION OF O_2^+

Calculations were performed on all 62 valence states of O_2
which arise from the $O(^3P)$, $O(^1S)$ and $O(^1D)$ separated atom limits
and the $X^2\Pi_g$ and a $^4\Pi_u$ states of O_2^+ (Guberman, 1982). DR of a $^4\Pi_u$
will be discussed separately (Guberman, 1982). The number of terms
in the CI wavefunctions ranged from 672 for the $^5\Delta_g$ states to 3359
for $X^2\Pi_g$. In each case the CI wavefunction consisted of a first
order wavefunction generated from each term of the CPD reference
state to the full virtual space. Since we obtained all the valence
states for each symmetry this was equivalent to generating a full
CI in the space of the occupied valence orbitals (all the orbitals
in (6) plus $3\sigma_u$) followed by a first order CI generated from each
of the terms in the valence full CI. In all states the core $1\sigma_g$
and $1\sigma_u$ orbitals, which do not contribute much to molecular binding
(corresponding mostly to atomic 1s orbitals), were kept fully occupied. The resulting valence state adiabatic excitation energies
are shown in Table 2 where they are compared to the results of
Saxon and Liu (1977) and experiment (Krupenie, 1972). The results
differ on the average by about 0.16 eV from the experimentally known
adiabatic excitation energies and are in excellent agreement with
the results of Saxon and Liu (1977).

Table 2. O_2 Adiabatic Excitation Energies (eV)

	This Work	Saxon–Liu[a]	Experiment[b]
a $^1\Delta_g$	1.1870	1.098	0.9817
b $^1\Sigma_g{}^+$	1.6954	1.776	1.6360
c $^1\Sigma_u{}^-$	3.9297	3.888	4.0987
A´ $^3\Delta_u$	4.1766	4.130	4.3066
A $^3\Sigma_u{}^+$	4.2452	4.206	4.3889
B $^3\Sigma_u{}^-$	5.9435	6.079	6.1737
ΔE^c	.1564	.153	
$^5\Pi_g$	4.8817	4.746	
$(1\ ^3\Pi_g)^d$	6.5691	–	
$2\ ^3\Sigma_g{}^-$	6.5977	6.699	
$2\ ^3\Pi_g$	6.7055	6.884	
$3\ ^3\Pi_g$	7.1782	7.072	
$(1\ ^1\Pi_g)^d$	8.1720	–	
$2\ ^1\Pi_g$	8.2344	8.411	
$1\ ^1\Delta_u$	8.6235	8.570	
$^1\Phi_g$	9.6050	9.282	
$^1\Sigma_u{}^+$	10.058	10.430	
$4\ ^1\Pi_g$	11.5336	11.667	

a. Saxon and Liu (1977).
b. Krupenie (1972).
c. ΔE is the average absolute value of the difference
    between the calculated and experimental energies.
d. These states have shallow wells and may not support
    a vibrational level.

In order to accurately describe DR it is necessary to accurately reproduce both the slopes of the dissociating states and the energy and internuclear distance of the intersection of the repulsive states with the molecular ion states. However it is very difficult to calculate an adiabatic ionization energy with the same accuracy as adiabatic excitation energies. Because O_2^+ has one less electron it has less correlation energy to account for than the neutral states. As a result comparable calculations performed on the neutral and ionic state will lead to ionization potentials that are too small compared to experiment. The calculated adiabatic electronic ionization potential of O_2 was 10.66 eV compared to the experimental value of 12.052 eV (Krupenie, 1972). This is mostly an atomic effect. In the figures presented below we have used the experimental (Krupenie, 1972) potential curve for $X^2\Pi_g$ for R<2.8 a_0. For R>2.8 a_0 the experimental $X^2\Pi_g$ curve has been joined to the calculated curve. $X^2\Pi_g$ is plotted so as to reproduce the experimental ionization potential relative to $X^3\Sigma_g^-$. However the curve has been shifted on the abscissa so that the R_e value for the ion state is the same as the calculated R_e value for this state. For $X^2\Pi_g$ this requires a shift to larger R by .0518 a_0. By plotting the ion curves at the calculated R_e we compensate for most of the error in the calculated R_e of the neutral states. For six low lying neutral states the average difference between experiment and theory for R_e is .0650 a_0. The results described here use in most cases a larger wavefunction than the previously reported results (Guberman, 1979) and are in good qualitative agreement with the earlier study.

About half of the 62 states arising from the valence asymptotes of O_2 can be eliminated as unimportant for DR of $X^2\Pi_g$ by examination of the configurational structure of the dissociating states. The cross section for DR is directly proportional to Γ, an electronic coupling matrix element,

$$\Gamma = 2\Pi |<\Psi_\alpha(k)| \mathcal{H} | \Psi_{Res}>|^2 \qquad (15)$$

where Ψ_α is a multiconfiguration wavefunction of the ion plus a free electron with momentum $\hbar k_\alpha$. Because of the two electron operator, $\dfrac{1}{r_{ij}}$ in $\mathcal{H}$, states with important configurations which differ by more than two orbitals from the primary configuration of the ground state of the ion plus a free electron will have small width matrix elements. These considerations are similar to those discussed earlier in regard to the role of triple and higher excitations in CI wavefunctions. For example, the wavefunction of the ground state of O_2^+ plus a free electron is given by

$$\Psi\left({}^2\Pi_g, \phi_k\right) = \sum_{i=1}^{3359} C_i^{\,{}^2\Pi_g} \, \Phi_i\left({}^2\Pi_g, \phi_k\right) \tag{16}$$

where ϕ_k is the wavefunction of the free electron. The most important configuration has $C_1^\alpha = 0.95$ and is given by

$$\Phi_1\left({}^2\Pi_g, \phi_k\right) = \mathrm{Op}\left(\dots 3\sigma_g^2 \, \pi_{ux}^2 \, \pi_{gx} \, \pi_{uy}^2\right)\phi_k \; . \tag{17}$$

where $\dots$ denotes $1\sigma_g^2 \, 1\sigma_u^2 \, 2\sigma_g^2 \, 2\sigma_u^2$. For the ${}^1\Sigma_u^+$ dissociating
neutral state the wavefunction is given by

$$\Psi\left({}^1\Sigma_u^+\right) = \sum_{i=1}^{1618} C_i^{\,{}^1\Sigma_u^+} \, \Phi_i\left({}^1\Sigma_u^+\right) \; . \tag{18}$$

The dominant terms in (18) are the equivalent configurations,

$$\Phi_1 \; {}^1\Sigma_u^+ = \mathrm{Op}\left(\dots 3\sigma_g^2 \, \pi_{ux}^2 \, \pi_{gx}^2 \; \pi_{uy} \, \pi_{gy}\right) \tag{19}$$

and

$$\Phi_2\left({}^1\Sigma_u^+\right) = \mathrm{Op}\left(\dots 3\sigma_g^2 \, \pi_{ux} \, \pi_{gx} \, \pi_{uy}^2 \, \pi_{gy}^2\right) \; , \tag{20}$$

The configurations in (19) and (20) each have coefficients of 0.6
in the CI wavefunction. Configuration (17) differs from (19) and
(20) by two orbitals so that one contribution to the width matrix
element will consist of

$$\langle \pi_{uy}\phi_{kx} | \frac{1}{r_{12}} | \pi_{gy}\pi_{gx} \rangle \tag{21}$$

multiplied by an overlap matrix element of the remaining orbitals.
The overlap matrix element is equal to unity in this case since all
orbitals are orthonormal.

On the other hand consider the seventh state of ${}^1\Sigma_g^+$ symmetry
which is dominated by the following two equivalent configurations:

$$\Phi_1\left(7\,^1\Sigma_g^+\right) = Op\ \left(\ldots 3\sigma_g^2\ 3\sigma_u^2\ \pi_{ux}^2\ \pi_{gx}^2\right)\quad \text{and} \tag{22}$$

$$\Phi_2\left(7\,^1\Sigma_g^+\right) = Op\ \left(\ldots 3\sigma_g^2\ 3\sigma_u^2\ \pi_{uy}^2\ \pi_{gy}^2\right)\ . \tag{23}$$

These configurations each have CI coefficients of .51 and differ by three orbitals from (17). These configurations will have zero matrix elements with the configuration in (17) and the width matrix element between the total wavefunctions will be small. A more quantitative statement concerning the relative importance of the dissociating states requires calculation of the width matrix element for states expected to have non-zero matrix elements between primary configurations. Nevertheless the above qualitative considerations allow for the identification of the states which are expected to be unimportant.

Dissociation to $O(^1S)$

There are 11 states which give rise to DR to $O(^1S)$. Of these states the width matrix elements with $X^2\Pi_g$ are expected to be small for 6 of the states leaving $^1\Sigma_u^+$, $2\,^1\Delta_g$, $4\,^1\Pi_g$, $5\,^3\Pi_g$ and $3\,^3\Sigma_u^-$ as likely routes for DR to $O(^1S)$. These states are shown in Fig. 3 with the states of the dissociating atoms shown in parentheses after the molecular state symbol.

Fig. 3 shows that the lowest accessible state for production of $O(^1S)$ from $X^2\Pi_g$ is $^1\Sigma_u^+$ which also dissociates to $O(^1D)$. The state crosses $X^2\Pi_g$ between v=1 and v=2. At low electron energies DR from v=1 will give atoms with 0.4 eV kinetic energy. The next accessible state, $5\,^3\Pi_g$, crosses the ion curve at v=10. Dissociation along this route will produce 2.5 eV atoms. Therefore we can expect that the rate constant for production of $O(^1S)$ by DR of v=0 will be very small.

Zipf (1979b;1980) has reported experimental evidence of the strong dependence of $O(^1S)$ production on the vibrational excitation of the ion. In order to bring measurements of an auroral rocket experiment (O'Neil et al., 1979) into agreement with laboratory experiments, Zipf proposed that the DR source of $O(^1S)$ has a quantum yield which is five times smaller than the previous laboratory measurements of O_2^+ DR (Zipf, 1970). The laboratory measurements (Zipf, 1970) which found an $O(^1S)$ quantum yield of 10% were believed to be vibrationally cold; however more recent experiments on N_2^+ (Zipf, 1979a) formed in similar conditions indicate that the ions are vibrationally excited. The earlier results (Guberman, 1979) calculated with a smaller wavefunction (in good agreement

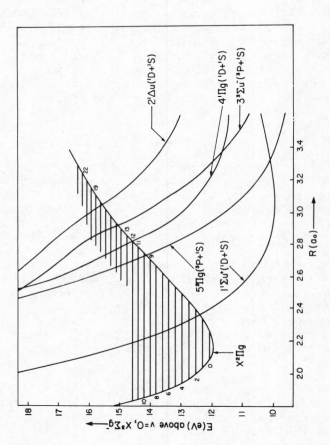

Fig. 3. Potential energy curves which are predicted to have non-negligible electronic matrix elements for DR of the ground state of O_2^+ leading to $O(^1S)$.

with the results reported here) and the current results indicate
that vibrationally excited O_2^+ will have a larger DR quantum yield
for $O(^1S)$ than O_2^+ which is mostly in v=0. This led Zipf to conclude that the auroral rocket experiment encountered vibrationally
cold O_2^+ at D region altitudes where quenching by O_2 was rapid.
However other experiments at higher F region altitudes find $O(^1S)$
quantum yields of 9.4% (Kopp et al., 1977) in agreement with Zipf's
earlier measurements. These results indicate that at F region
altitudes quenching of excited O_2^+ may be less effective than at D
region altitudes and that the F region $O(^1S)$ arises from DR of
vibrationally excited O_2^+ (Zipf, 1979b) as in the earlier experiments (Zipf, 1970).

A plasma spectroscopy experiment has been reported by Zipf
(1980) in which $X^2\Pi_g$ O_2^+ ions were produced with v<12 and v<3 by
charge exchange with Ar_2^+ and Kr_2^+ respectively. For O_2^+ with
v<12 an $O(^1S)$ quantum yield of about 10% was observed while for v<3
ions, a 2% quantum yield was observed. These results appear to
support the observations made above and may indicate that most of
the $O(^1S)$ arises from 4<v<12. From Fig. 3 and the above discussion
of the electronic width matrix elements the results of Zipf (1980)
indicate that the $5^3\Pi_g$ and $4^1\Pi_g$ states are important routes for DR
of the excited levels with v<12. The $5^3\Pi_g$ state may be dominant
because of its statistical weight. Future calculation of the electronic widths and DR cross sections by the author are expected to
identify the relative importance of these states.

A recent paper by Bates and Zipf (1980) presents a strong
argument that vibrationally excited O_2^+ (v>2) may be efficiently
relaxed by collision with O at altitudes where $O(^1S)$ is thought to
arise from vibrationally excited O_2^+ DR. It was proposed that $O(^1S)$
could be generated from a Rydberg $^3\Pi_u$ state (Borst and Zipf, 1971)
which is predissociated by $^3\Pi_u$ $(^1S+^3P)$. The former state appears
to be the bound Rydberg $3s\sigma_g$ $^3\Pi_u$ state (Katayama, 1981; Katayama
et al., 1981) while the latter state is the $5^3\Pi_u$ $(^1S+^3P)$ calculated
here. The inner wall of the Rydberg state crosses the large R
turning point of v=2 of $X^2\Pi_g$ while the $5^3\Pi_u$ state crosses the
Rydberg state at the large R turning point of v=3 and the large R
turning point of v=12 of $X^2\Pi_g$ (Guberman, 1982). Therefore these
states do not appear to offer an important mechanism for DR of v=0
of O_2^+, $X^2\Pi_g$.

Keto et al. (1981) have investigated the possibility of creating a laser based on an $O(^1S)$ – $O(^1D)$ population inversion arising
in electron beam excited mixtures of Ar doped with O_2. With O_2^+
created by charge exchange of Ar_2^+ with O_2 an 11% quantum yield of
$O(^1S)$ was found in agreement with the results of Zipf (1980).
However, attempts to increase the $O(^1S)$ quantum yield by using
Ne_2^+ charge exchange to access a higher dissociating state yielding

$O(^1S) + O(^1S)$ were not successful. As we have seen above, the
pertinent state is the $7^1\Sigma_g^+$ state. The configurational character
of this state allows for the prediction of a very small width matrix
element and therefore a relatively small cross section for DR, consistent with the experimental results.

Dissociation to $O(^1D)$

There are 21 molecular singlet states and 18 triplet states
that lead to $O(^1D)$ via DR. Ten singlet states are expected to have
large electronic matrix elements for DR of $X^2\Pi_g$ and are shown in
Fig. 4. The lowest accessible state is $^1\Delta_u$ which intersects the
large R turning point of $v=0$ and dissociates to two ^1D atoms. Note
that there are many routes giving rise to two ^1D atoms as opposed
to the case for two ^1S atoms which from symmetry considerations can
only arise from a single state, $7^1\Sigma_g^+$. $^1\Sigma_u^+$ and $1^1\Delta_u$ are the only
states accessible from the low vibrational levels. At higher
energies $2^1\Pi_g$ becomes accessible near $v=4,5$. The well in the $2^1\Pi_g$
state near $R = 3.0$ a_0 is due to an avoided crossing with $1^1\Pi_g$ (see
Fig. 6) which dissociates to ^3P atoms. Therefore dissociation
along $2^1\Pi_g$ can be expected to proceed partially in a diabatic manner through the curve crossing and will yield ^3P as well as ^1D
atoms. The change in slope for $R < 2.6$ a_0 in $3^1\Sigma_g^+$ and for $R <$
3.0 a_0 in $2^1\Delta_u$ is due to a mixing with states of the same symmetry
arising from $O^+ + O^-$ asymptotes.

Fig. 5 shows the important triplet states leading to $O(^1D)$.
Each of these states dissociates to $O(^1D) + O(^3P)$ and because of
spin are statistically more important than the singlets. The lowest state, $1^3\Sigma_u^-$, is the diabatic extension of the upper state of
the well known Schmann-Runge system. From Fig. 5 we see that this
state would be expected to have a large DR cross section from $v=1$
of $X^2\Pi_g$ and is the only accessible state for $v\leq2$. The next accessible state, $2^3\Pi_g$, comes at the large R turning point of $v=3$. $2^3\Pi_g$
has a small well near $R = 3.3$ a_0 due to an avoided crossing with
$1^3\Pi_g$ (see Fig. 6) which dissociates to two ^3P atoms. As a result
we can expect dissociation along $2^3\Pi_g$ to lead to additional ^3P
atoms by "jumping" the avoided crossing.

Dissociation to $O(^3P) + O(^3P)$

The most favorable route for DR of $v=0$ of $X^2\Pi_g$ O_2^+, $1^3\Pi_u$,
is shown in Fig. 6, which depicts the states expected to have large
widths. Of all 62 possible dissociating states of O_2 the $1^3\Pi_u$
state is expected to have the largest cross section for DR of $v=0$
and is probably the most important route in the many experiments
which have measured total recombination rates for O_2^+. The total
rate at room temperature is 2×10^{-7} cm^3/sec (Mul et al., 1979).
Clearly, vibrationally cold O_2^+, mostly in $v=0$, will lead primarily

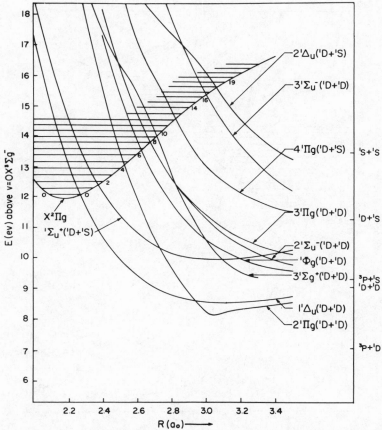

Fig. 4. Singlet states which are predicted to have non-negligible
         electronic matrix elements for DR of the ground state
         of O_2^+ leading to $O(^1D)$ atoms.

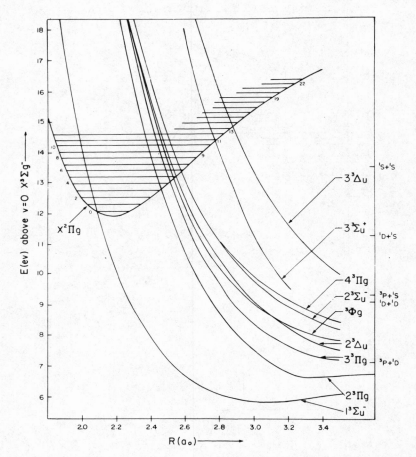

Fig. 5.   Same as Fig. 4 except that triplet states dissociating
          to O(^{1}D) atoms are shown.

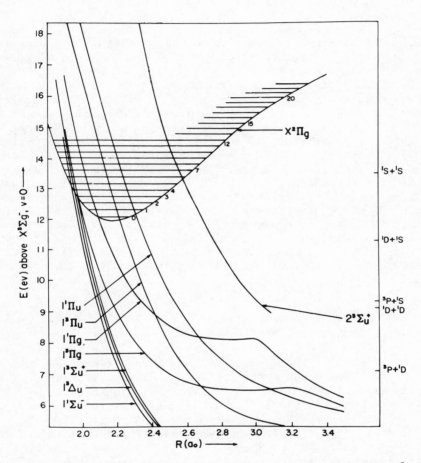

Fig. 6.   Same as Fig. 4 except that states leading to O(^{3}P) +
          O(^{3}P) are shown.

to $O(^3P)$ with the $1^1\Delta_u$ state of Fig. 4 and the $1^3\Sigma_u^-$ state of Fig. 5 providing some $O(^1D)$. Therefore these qualitative arguments lead to the prediction that the magnitude of atomic state quantum yields from v=0 will be in the order $^3P > {}^1D > {}^1S$. This is in agreement with the result found by Zipf (1970), which appears to be appropriate for vibrationally excited O_2^+ (Zipf, 1979b,1980). Dissociation on $1^3\Pi_u$ from v=0 will lead to hot ^3P atoms each with 3.5 eV kinetic energy.

It is clear from the above results that vibrational excitation will be of major importance in determining the DR yield of ground and excited atomic states and in determining the magnitude of the total DR cross section. A comparison of trapped ion (Walls et al., 1974) and merged electron ion (Mul et al., 1979) measurements of the total DR cross section of O_2^+ indicates reasonable agreement between the two studies for electron energies below 1 eV, even though the degree of ion vibrational and possibly electronic excitation is probably different for these two experiments. However the extent of this difference is not known. Above 1 eV the merged ion experiment does not reproduce some dramatic structure seen in the ion trap experiment. This structure may be due to indirect DR through the $3s\sigma_g$ $^3\Pi_u$ state (Katayama, 1981; Katayama et al., 1981). Calculations currently underway of the width matrix elements, nuclear vibrational wavefunctions, and DR cross sections in this laboratory are expected to clarify the nature of the variation of the DR cross section with electron energy and ion vibrational excitation.

DISSOCIATIVE RECOMBINATION OF H_2^+

Some aspects of the H_2 calculations have already been discussed above. The basis set for the dissociative states is of the same size as that discussed above for the H_2 ground state. The calculations on the dissociating states were originally done (Guberman, 1982) to describe dissociative ionization of H_2 by photon and electron impact. The repulsive states involved in dissociative ionization of H_2 above 23 eV are the same states which provide routes for DR of H_2^+ at lower energies.

The H_2 CI wavefunctions were expanded over optimized H_2^+ orbitals. As opposed to the case for O_2, all of the H_2 excited states of interest are Rydberg states. As a result, we cannot obtain the AB^* states by leaving out Rydberg character from the basis set. However, the Rydberg AB^* states of interest are the repulsive states having the repulsive $1\sigma_u$, $^2\Sigma_u^+$ state of H_2^+ as the core orbital. These states are obtained by projecting out of the wavefunction all states having the $1\sigma_g$ orbital, i.e., the attractive state of the ion. Thus the problems with the non-crossing rule mentioned

earlier are avoided. The projection is carried out by simply not
including any $1\sigma_g$ orbitals in the wavefunction in (1). The calculated state will be an upper bound to the exact autoionizing state
in the space having $1\sigma_g$ projected out. Therefore the lower the
energy given by the dissociating state wavefunction, the more
accurate the description of the dissociating state. For further
details the reader is referred to Guberman (1982).

The results for the potential curves relevant to DR are shown
in Fig. 7. Clearly DR of the lower vibrational levels will be
dominated by $^1\Sigma_g^+$ which intersects the large R turning point of v=1.
DR from H_2^+ mostly in v=0 is expected to have a low rate. Note that
DR of v=1 at low electron energies can lead to H(2ℓ) atoms since
the $^1\Sigma_g^+$ state will cross through bound Rydberg states having the
H(1s) + H(2ℓ) limit. DR from v=1 and $^1\Sigma_g^+$ will require electron
energies of about 0.8 eV to produce H(3ℓ) atoms. The next acces-
sible state is $^3\Pi_g$ which can lead to DR from v$\geq$5 and $^3\Sigma_u^+$, $^1\Sigma_u^+$
and $^1\Pi_g$ which intersect near the v=6 large R turning point. At
low electron energies direct DR from v=5 along $^3\Pi_g$ and from v=6
along all the upper states can lead to H(3ℓ) and more highly excited
atoms. In the region near the crossing with $^2\Sigma_g^+$, the $^1\Sigma_g^+$ state
calculated here is about 0.23 eV below the result calculated by
Bottcher and Docken (1974), 0.97 eV below that calculated by
O'Malley (1969) and 0.30 eV below the result of Dastidar et al.

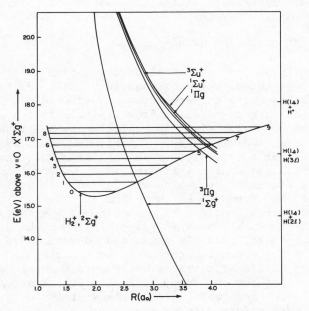

Fig. 7. States leading to DR of the lower vibrational levels of
         H_2^+.

(1979). It should be pointed out that O'Malley and Dastidar et al.
used considerably smaller wavefunctions than those reported here.
The results are therefore in good agreement with the results of
Bottcher and Docken and Dastidar, et al. There appears to be no
previous reported calculation for the $^3\Pi_g$ state. For $^3\Sigma_u^+$ the current result is only .07 eV below the previous result (Bottcher et
al., 1974) at R = 4.0 a_0 and .13 eV below results of Takagi and
Nakamura (1980) at R = 2.0 a_0. For $^1\Pi_g$ there are no previously
reported results. For $^1\Sigma_u^+$ the current result is only 0.22 eV below
the previous result of O'Malley (1969) near R = 4.0 a_0 and 0.11 eV
below the result of Takagi and Nakamura (1980) at R = 2.0 a_0.

     A review of the experimental and earlier theoretical work on
H_2^+ DR has been presented by Dolder and Peart (1976), Peart and
Dolder (1974), Auerbach et al. (1977) and Phaneuf, Crandall and Dunn
(1975). An important feature of these experiments is that they
involve vibrationally excited H_2^+ with most ions in v=1 and v=2 in
addition to considerable population in v=5 and 6. Therefore we can
expect the dissociative $^3\Pi_g$, $^3\Sigma_u^+$, $^1\Sigma_u^+$ and $^1\Pi_g$ states to play an
important role in these experiments. This appears to be confirmed
by the merged beam results of McGowan et al. (1976) in which the
removal of H_2^+ with v>2 from the beam led to a drop in the DR cross
section by a factor of more than 2. Direct DR appears to dominate
in these experiments. For energies greater than 0.07 eV structure
is seen in the cross section curves which is attributed to indirect
DR through bound H_2 Rydberg states (McGowan et al., 1979).

DISSOCIATIVE RECOMBINATION OF He_2^+

     In Fig. 8, repulsive potential energy curves for DR of He_2^+
are shown. These curves were obtained (Guberman, 1972) from wavefunctions that are quite different from those used for O_2 and H_2.
The wavefunctions are constructed from orbitals that are not required to be orthogonal. Also the orbitals are not required to
have gerade or ungerade symmetry, however, the total wavefunction
has the proper g or u symmetry. This lack of restrictions on the
orbitals provided for a highly interpretable wavefunction. Maxima
in the excited states of He_2 which were not due to avoided crossings
and which were considered to be anomolous at that time were explained in terms of the unfavorable overlap of a Rydberg orbital
on one He with the doubly occupied 1s orbitals on the other He
(Guberman and Goddard, 1972; Guberman and Goddard, 1975). Other
features of the He_2^* potential curves could be explained in terms
of the crossing of diabatic states which were constructed in the
same manner as that for the fully optimized states except that the
orbitals used were those which were optimum at R=∞ only. These
wavefunctions were referred to as frozen orbital (FO) wavefunctions.

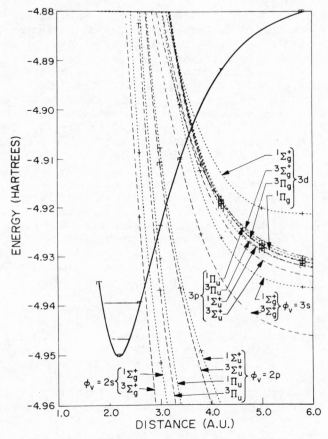

Fig. 8.   Potential energy curves for DR of He₂⁺ (Guberman, 1972).
The solid curve is the calculated ground state of He₂⁺.
The dashed curves are the FO diabatic states of He₂.

It was shown that a change in character of a fully optimized state corresponds to a crossing of two diabatic FO potential energy curves.

The repulsive or attractive nature of the FO curves for $R<5$ a_o is dependent upon the nature of the three electron ion core of the excited neutral states. These excited states consist of a diffuse Rydberg orbital surrounding a tight He_2^+ core. The tight He_2^+ core is a mixture of the attractive $^2\Sigma_u^+$ ground state of He_2^+ and the repulsive $^2\Sigma_g^+$ state. Consider the states dissociating to $He(1s^2)$ + $He(1s,2\ell)$. At intermediate internuclear distances where the diffuse Rydberg orbital surrounds both He atoms, states having primarily a Rydberg 2s orbital and a $^2\Sigma_g^+$ core will be repulsive. These are the $^3\Sigma_g^+$ and $^1\Sigma_g^+$ states and are the lowest states providing routes for DR as shown in Fig. 8. For a 2p Rydberg orbital the repulsive states must be of ungerade symmetry and correspond to the $^3\Pi_u$, $^1\Pi_u$, $^3\Sigma_u^+$, and $^1\Sigma_u^+$ states shown in Fig. 8. The Π states shown in Fig. 8 correspond to the $^2\Sigma_g^+$ He_2^+ state shifted to have the same $R=\infty$ limit as the four electron states.

The potential curves shown in Fig. 8 are intended to provide a qualitative picture of DR of He_2^+. Because the Σ states are obtained from frozen orbitals, these states can be expected to be more repulsive than the same states determined with the techniques used above for O_2 and H_2. The He_2 DR curves are not expected to be more than 0.5 eV above optimized DR curves. Nevertheless because the DR curves in Fig. 8 are very steep near the He_2^+ ground state, a lowering of 0.5 eV would not qualitatively change the conclusions reached here.

Potential curves for triplet states leading to DR of He_2^+ have been reported by Cohen (1976). His potential curves are more repulsive relative to He_2^+ than those reported here. The $^3\Sigma_g^+$ curve arising from $He(1s,2s)$ ^3S + $He(1s^2)$ ^1S crosses the ion between $v=2$ and $v=3$ compared to a crossing between $v=1$ and $v=2$ shown in Fig. 8. Both results are therefore in agreement concerning the slow rate expected for DR from $v=0$ with low energy electrons. Cohen's calculated dissociation energy was 2.424 eV compared to the accurate value of 2.469 eV (Liu, 1971). Since these calculations tend to give too low a value for the $He + He^+$ asymptote compared to the $He + He$ asymptote the good result for D_e calculated by Cohen may have the effect of placing the He_2^+ ground state at energies that are too low compared to the neutral states. This can result in neutral states crossing too high on the ion potential curve. The calculated dissociation energy obtained here was 1.97 eV (Guberman, 1972) and effectively cancels much of the error (0.7 eV) in the He ionization potential leading to potential curve crossings that may be more reliable in the results reported here.

Mulliken (1964) reported qualitative potential curves for excited states of He_2 which also led to the conclusion that DR from v=0 of He_2^+ is slow. Mulliken's curves were mostly adiabatic in nature and required jumping from one adiabatic curve to another in order to generate $He(1s,2s)$ ^3S from DR. The FO diabatic curves described here provide a single route for dissociation to the separated atoms but cross other diabatic curves of the same symmetry.

From Fig. 8 it appears that DR of v=0 of He_2^+ will be very slow while DR of v=1 may be appreciably faster along the $^3\Sigma_g^+$ dissociative state leading directly to $He(1s,2s)$ ^3S. Direct dissociation to $He(1s,2p)$ along the ungerade states shown in Fig. 8 will only be possible from higher vibrational levels (v$\geq$3). But it is possible to generate $He(1s,2p)$ ^3P from the crossing of the attractive $^3\Sigma_g^+$ state arising from $He(1s^2) + He(1s,2p)$ ^3P with the repulsive $^3\Sigma_g^+$ state arising $He(1s^2) + He(1s,2s)$ ^3S. An analogous situation applies to the production of $He(1s,2p)$ ^1P from dissociation along the $^1\Sigma_g^+$ state dissociating directly to $He(1s^2) + He(1s,2s)$ ^1S (Guberman, 1972).

Experimental measurements of DR of He_2^+ have taken a tortuous path since the first measurements over 30 years ago (Biondi and Brown, 1949). The experiments have been reviewed by Massey and Gilbody (1974), Deloche et al. (1976) and Bates (1979a;1979b). The measured rate constants differ by more than a factor of 100. The most recent results indicate that the rate constant is less than 5×10^{-10} cm^3/sec (Deloche et al., 1976) at room temperature. The experiments in microwave discharge afterglows are difficult because they involve He_2^+ which is mostly in v=0. Mulliken (1964) has remarked that the reported band spectra of He_2 taken from discharges at about 20 mm pressure involve almost entirely bands with v=0. Bates (1979b) argues that at 15 mm pressure He_2^+ is easily vibrationally relaxed by collision with He before recombination can occur. The potential curves shown in Fig. 8 in addition to those of Cohen (1976) and Mulliken (1964) indicate that DR from He_2^+ in v=0 will be very slow. Indeed the most recent experiments (Deloche et al., 1976; Boulmer et al., 1977) indicate that other recombination processes such as

$$He^+ + 2e^- \rightarrow He + e \; , \; He^+ + He + e^- \rightarrow He + He \; ,$$

$$He_2^+ + 2e^- \rightarrow He_2^* + e \; , \; \text{and} \; He_2^+ + He + e^- \rightarrow He_2^* + He$$

are significant in He afterglows.

Note that a comparison of Figs. 7 and 8 indicates that the lack of favorable routes for DR of H_2^+ could lead to rates that are not much larger than those for He_2^+. However, in order to avoid the formation of H_3^+ and heavier ions the H_2^+ experiments are performed in beams generated from ion sources that produce vibrationally excited H_2^+ (see the previous section). If H_2^+ ions were mostly in v=0 the DR rate would be expected to be only slightly faster than that for He_2^+.

ACKNOWLEDGEMENT

    The author gratefully acknowledges support from the Air Force Geophysics Laboratory (Contract No. F19628-79-C-0139) and N.A.S.A. (Grant No. NAG 2-89) for the research reported here.

REFERENCES

Auerbach, D., Cacak, R., Caudano, R., Gaily, T. D., Keyser, C. J., McGowan, J. Wm., Mitchell, J. B. A., and Wilk, S. F. J., 1977, J. Phys. B 10, 3797-820.
Bardsley, J. N., 1968a, J. Phys. B1, 349-64.
Bardsley, J. N., 1968b, ibid, 365-80.
Bardsley, J. N., and Biondi, M. A., 1970, Adv. Atom. Molec. Phys. eds. D. R. Bates and I. Estermann (New York: Academic Press), 6, 1-57.
Bates, D. R., 1950a, Phys. Rev. 77, 718L-9L.
Bates, D. R., 1950b, ibid, 78, 492-3.
Bates, D. R., 1974, Case Studies in Atomic Physics 4, 57-92.
Bates, D. R., 1979a, Adv. Atom. Molec. Phys., eds. D. R. Bates and B. Bederson (New York: Academic Press) 15, 235-62.
Bates, D. R., 1979b, J. Phys. B 12, L35-8.
Bates, D. R. and Massey, H. S. W., 1947, Proc. Roy. Soc. London A192, 1-16.
Bates, D. R., Ledsham, K., and Stewart, A. L., 1953, Phil. Trans. Roy. Soc. (London) A246, 215-40.
Bates, D. R., and Zipf, E. C., 1980, Planet. Space Sci. 28, 1081-6.
Beigman, I. L., and Chichkov, B. N., 1980, J. Phys. B 13, 565-9.
Biondi, M. A. and Brown, S. C., 1949, Phys. Rev. 75, 1700-5.
Bobrowicz, F., 1974, Ph.D. Thesis, California Institute of Technology (University Microfilms: Ann Arbor).
Bobrowicz, F. and Goddard III, W. A., 1977, The Self-Consistent Field Equations for Generalized Valence Bond and Open-Shell Hartree-Fock Wavefunctions in: "Modern Theoretical Chemistry. Methods of Electronic Structure Theory," H. F. Schaefer III, ed., Plenum Press, New York.
Borst, W. L., and Zipf, E. C., 1971, Phys. Rev. A 4, 153-61.
Bottcher, C. and Docken, K., 1974, J. Phys. B7, L5-8.
Boulmer, J., Devos, F., Stevefelt, J. and Delpech, J-F, 1977, Phys. Rev. 15, 1502-12.
Boys, S. F., 1950, Proc. Roy. Soc. London A200, 542-54.

Burgess, A., 1964, Astrophys. J. 139, 776-80.

Cade, P. E. and Wahl, A. C., 1974, At. Data Nucl. Data Tables 13, 339-89.

Cogger, L. L., Smith, L. S. and Harper, R. M., Planet. Space Sci. 25, 155-9.

Cohen, J. S., 1976, Phys. Rev. A 13, 86-98.

Dalgarno, A., 1979, Adv. Atom. Molec. Phys., eds. D. R. Bates and B. Bederson (New York: Academic Press) 15, 37-76.

Das, G. and Wahl, A. C., 1972, Argonne National Laboratory, Report No. ANL-7955.

Dastidar, K. R. and Dastidar, T. K. R., 1979, J. Phys. Soc. Japan, 46, 1288-94.

Deloche, R., Monchicourt, P., Cheret, M., and Lambert, F., 1976, Phys. Rev. A 13, 1140-76.

Dolder, K. T. and Peart, B., 1976, Rep. Prog. Phys. 39, 693-749.

Dunning, T. H., 1970, J. Chem. Phys. 53, 2823-33.

Dunning, T. H., 1971, Ibid, 55, 3958-66.

Dunning, T. H., 1977, Ibid, 66, 1382-3.

Dunning, T. H., and Hay, P. J., 1977, Gaussian Basis Sets for Molecular Calculations in: "Modern Theoretical Chemistry. Methods of Electronic Structure Theory," H. F. Schaefer III, ed., Plenum Press, New York.

Frederick, J. E., Rusch, D. W., Victor, G. A., Sharp, W. E., Hayes, P. B., and Brenton, H. C., 1976, J. Geophys. Res. 81, 3923-30.

Guberman, S. L., 1972, Ph.D. Thesis, California Institute of Technology, University Microfilms, Ann Arbor.

Guberman, S. L., 1977, J. Chem. Phys., 67, 1125-35.

Guberman, S. L., 1979, Int. J. Quant. Chem. S13, 531-40.

Guberman, S. L., 1982, J. Chem. Phys., submitted for publication.

Guberman, S. L., and Goddard, W. A., 1972, Chem. Phys. Letters 14, 460-5.

Guberman, S. L., and Goddard, W. A., 1975, Phys. Rev. A12, 1203-21.

Hinze, J., 1973, J. Chem. Phys. 59, 6424-32.

Huzinaga, S., 1965, J. Chem. Phys. 42, 1293-302.

Huzinaga, S., 1971, Approximate Atomic Functions II, unpublished.

Huzinaga, S., 1977, J. Chem. Phys. 66, 4245-5.

Huzinaga, S., 1979, J. Chem. Phys. 71, 1980-1.

Kaplan, J., 1931, Phys. Rev., 38, 1048-51.

Katayama, D. H., 1981, private communication.

Katayama, D. H. and Tanaka, Y., 1981, J. Mol. Spec. 88, 41-50.

Kato, J. W., Hart, C. F., Kuo, C.-Y., 1981, J. Chem. Phys. 74, 4433-44.

Kolos, W. and Wolniewicz, L., 1964, J. Chem. Phys. 41, 3663-73.

Kopp, J. P., Frederick, J. E., Rusch, D. W., and Victor, G. A., 1977, J. Geophys. Res., 82, 4715-9.

Krupenie, P. H., 1972, J. Phys. Chem. Ref. Data 1, 423-534.

Link, R., McConnell, J. C., and Shepherd, G. G., 1981, Planet. Space Sci. 29, 589-94.

Liu, B., 1971, Phys. Rev. Lett. 27, 1251-3.

Massey, H. S. W., 1937, Proc. Roy. Soc. (London) A163, 542-53.

Massey, H. S. W. and Gilbody, H. B., 1974, Electronic and Ionic
    Impact Phenomena (Oxford University Press: London) 4,
    2115-305.
McElroy, M. B., 1972, Science 175, 443-5.
Mul, P. M., and McGowan, J. Wm., 1979, J. Phys. B, 12, 1591-1601.
Mulliken, R. S., 1964, Phys. Rev. 136, A962-5.
Nagy, A. F., Cravens, T. E., Yee, J-H., Stewart, A. I. F., 1981,
    Geophys. Res. Letters 8, 629-32.
Nier, A. O., Hanson, W. B., Seiff, A., McElroy, M. B., Spencer,
    N. W., Duckett, R. J., Knight, T. C. D., Cook, W. S., 1976,
    Science 193, 786-8.
O'Malley, T., 1969, J. Chem. Phys. 51, 322-34.
O'Neill, R. R., Lee, E. T. P., Huppi, E. R., 1979, J. Geophys. Res.,
    84, 823-33.
Peart, B. and Dolder, K. T., 1974, J. Phys. B, 7, 236-43.
Phaneuf, R. A., Crandall, D. H., and Dunn, G. H., 1975, Phys. Rev.
    A, 11, 528-35.
Raffenetti, R. C., 1973, J. Chem. Phys. 58, 4452-8.
Raffenetti, R. C., 1976, BIGGMOLI, QCPE Program No. 328, Indiana
    University, Bloomington.
Saxon, R. and Liu, B., 1977, J. Chem. Phys. 67, 5432-41.
Schaefer III, H. F. and Harris, F. E., 1968, Phys. Rev. Lett. 21,
    1561-3.
Schaefer III, H. F., Klemm, R. A., and Harris, F. E., 1969, J. Chem.
    Phys. 51, 4643-50.
Sharp, W. E., Rusch, D. W. and Hays, P. B., 1975, J. Geophys. Res.
    80, 2876-8.
Shavitt, I., 1977, The Method of Configuration Interaction in
    "Modern Theoretical Chemistry. Methods of Electronic Struc-
    ture Theory," H. F. Schaefer III, ed. Plenum Press,
    New York.
Shavitt, I., Bender, C. F., Pipano, A., Hosteny, R. P., 1973, J.
    Comput. Physics 11, 90-108.
Takagi, H., and Nakamura, H., 1980, J. Phys. B. 13, 2619-32.
Torr, D. G., Richards, P. G., Torr, M. R., and Abreu, V. J., 1981,
    Planet. Space Sci 29, 595-600.
Wahl, A. C. and Das, G., 1977, The Multiconfiguration Self-Consistent
    Field Method in: "Modern Theoretical Chemistry. Methods of
    Electronic Structure Theory," H. F. Schaefer III, ed. Plenum
    Press, New York.
Walls, F. L., and Dunn, G. H., 1974, J. Geophys. Res. 79, 1911-5.
Wolniewicz, L. and Dressler, K., 1977, J. Mol. Spectrosc. 67, 416-39.
Yee, J. H., Meriwether, Jr., J. W., and Hays, P. B., 1980, J.
    Geophys. Res. 85, 3396-400.
Zipf, E. C., 1970, Bull. Amer. Phys. Soc. 15, 418.
Zipf, E. C., 1979a, Bull. Amer. Phys. Soc. 24, 129.
Zipf, E. C., 1979b, Geophys. Res. Lett. 6, 881-4.
Zipf, E. C., 1980, J. Geophys. Res. 85, 4232-6.

ELECTRON IMPACT EXCITATION OF IONS

D. H. Crandall

Physics Division
Oak Ridge National Laboratory*
Oak Ridge, Tennessee 37830 U.S.A.

1. INTRODUCTION

The ancient curiosity of "what produces the light" still
motivates research on electron impact excitation of ions. We have
developed extensive knowledge of the subject which relies on
detailed information about atomic structure and a conviction that
quantum theory, in principle, correctly describes the atomic
structure and the collision dynamics. Still, we are unable to
predict cross sections or rates for this process with sufficient
reliability to satisfy our needs.

In astronomy and astrophysics, observations from spacecraft
allow detection of light (from ions) without modification by the
Earth's atmosphere so that considerably more detailed data are
available. Correspondingly, a more accurate understanding of the
underlying basic processes is needed. In the development of a
controlled fusion energy source, the electron impact excitation of
ions provides unique information on the conditions in the plasma
and can be a decisive factor in the energy balance affecting
plasma evolution. A wide variety of investigations in spectroscopy, plasma properties, and basic collision studies requires
known cross sections for excitation of ions.

The electron impact excitation of ions has several unique
aspects. The light emitted during relaxation of excited ions is
generally of shorter wavelength than light from other processes.

*Work sponsored by the Office of Fusion Energy, U.S. Department of
Energy under contract W-7405-eng-26 with the Union Carbide
Corporation.

Ion excitation cross sections are finite and often largest at the
threshold energy. At the same time the calculations are most
difficult near threshold. The processes of excitation, ionization,
and recombination are so strongly interrelated in electron-ion
collisions that independent treatments of these problems may not
be sufficient.

In order to study electron impact excitation of ions under
sufficient control to determine specific cross sections as a
function of collision energy, it is generally necessary to produce
ion beams of specific species, charge, and atomic state and to
interact the ion beam with a controlled electron beam. Since
attainable ion beam densities are tenuous (typical beams of 10^5
particles/cm^3 compare to 10^{-11} torr pressure), few definitive
experiments have been accomplished. Attempts to study electron
impact excitation of ions without recourse to controlled ion beams
are typically unable to produce sufficiently detailed data to test
understanding of the basic process.

This paper will review electron-ion beams experiments primarily. The techniques, difficulties, and present trends in this
area will be discussed. Measured cross sections will be compared
with theoretical results and the current level of agreement
assessed.

2. THEORY

2.1 General

The theory of electron impact excitation of ions has been
recently reviewed by Henry;[1] it was discussed by Robb[2] at a preceeding NATO summer institute; and it has previously been reviewed
by several others.[3-5] There are a great number of individual
predictions of excitation for particular transitions of specific
ions. Henry[1] lists over 350 references, the Los Alamos compilation[6] gives comparative results from more than one calculation for
each of more than 100 transitions, and the current ORNL compilation
on excitation of iron ions[7] lists calculated cross sections for
roughly 800 transitions. This vigorous activity in excitation
theory has produced some concensus as to which approximations are
most reliable in spite of the paucity of experimental results
which are needed to serve as a standard to test the accuracy of
the various theoretical approaches. The discussion which follows
makes no attempt to provide equations which would allow actual
calculations but examines standard approaches for physical insight.

The accepted non-relativistic theoretical approaches all
begin with the Schroedinger equation for the system of N + 1
electrons:

$$H(Z,N+1)\Psi = E\Psi \tag{1}$$

where N and Z are the number of electrons and nuclear charge of
the ion, Ψ is the total wave function of the N + 1 electron system,
and E is the total energy, $E = E_i (Z,N) + k_i^2$ (which defines the
kinetic energy, k_i^2, of the incident electron in Rydbergs). The
Hamiltonian is

$$H(Z,N+1) = -\sum_{i=1}^{N+1} \left(\nabla_i^2 + \frac{2Z}{r_i} \right) + \sum_{j=i+1}^{N+1} \sum_{i=1}^{N} \frac{2}{r_{ij}} . \tag{2}$$

Having written these standard expressions we immediately
encounter difficulty because Eq. (1) is a many-body problem, never
less than three bodies. In principle there is no compromise in
expanding the many-body wave function in terms of products of
single-particle wave functions:

$$\Psi(N + 1) = A \sum_i \theta_i(N + 1) \ \chi_i(X_1 \ldots X_N) \tag{3}$$

where A antisymmetrizes the total function and X_n denotes coordinates of the nth electron. Substitution of Eq. (3) into Eq. (1)
results in a number of coupled differential equations. Generally,
expansion (3) is chosen so that the functions χ_i are the solutions
of the next simpler problem (i.e., they are the eigenvectors of
the N electron ion which is the target), and the θ_i (N + 1) are
coefficients as well as the wave function of the free or incident
electron. The wave functions for the bound-state system, χ_i,
are usually antisymmetrized products of one-electron functions
determined from the solution of a central field model such as the
Hartree-Fock method. It should be further noted that the index,
i, stands for all the system quantum numbers including parity,
angular momentum, and spin. Thus, Ψ is a sum over coupled, antisymmetrized, product-wave functions which leads to messy algebra.
One additional modification to Eq. (3) is usually specified:

$$\Psi(N + 1) = A \sum_i \chi_i \ \theta_i + \sum_j C_j \ \phi_j(X_1 \ldots X_{N + 1}) \tag{4}$$

where the C_j are numerical coefficients and ϕ_j are bound-state
wave functions of the ion or atom of next lower charge with N + 1
bound electrons. The addition of the second term in Eq. (4) is
needed to provide a more complete set mathematically,[1] but note
that the ϕ_j states are important physical quantities in this
problem as well. Some of these states are schematically represented for an arbitrary Li-like ion in Fig. 1. Of the eigenstates
represented in Fig. 1, only those below the $1s^2$ (^1S)k continuum
are truly bound states. The other discrete levels illustrated are
autoionizing resonances.

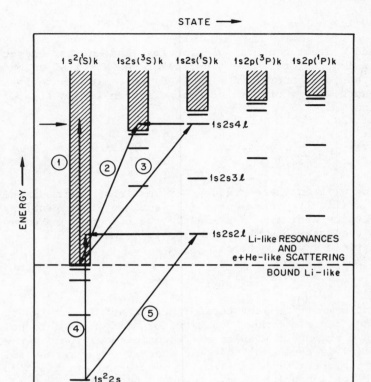

Fig. 1. Schematic representation of the energy levels of the
electron + He-like ion. The state of the He-like ions
is designated and the free electron is represented by k.
Discrete states of Li-like ions are illustrated below
each of the e + He-like continua. Transitions 1, 2, and
3 are all e + He-like scattering at the same energy (see
text). Transitions 4 and 5 schematically represent ionization of an initial Li-like ion — a fourth electron has
been added to the system for transitions 4 and 5.

        Figure 1 schematically illustrates some of the transitions
important in electron impact excitation studies. Transitions
labeled 1, 2, and 3 are collisions involving a free electron (k)
and the $(1s^2)$ ground state of an He-like ion. These three transitions represent: (1) elastic scattering, (2) direct excitation of
the $(1s2s)^3S$ excited state of the He-like ion, and (3) the excitation of the He-like ion via an intermediate recombination resonance
of an Li-like ion in which e + He$(1s^2)$ → Li$(1s2s4\ell)$ → e + He$(1s2s)^3S$.

Transition 3 is to a state of the type Φ_j in Eq. (4). Since the
incident electron in transition 3 becomes (temporarily) a bound
electron the transition is a true resonance which will only occur
when the energy of the free electron plus ($1s^2$) He-like ion is
matched to the particular ($1s2s4\ell$) Li-like bound state illustrated.
By contrast the transitions 1 and 2 are to final states where the
initial electron remains free and thus can occur at any value of
the free electron energy for 1 and any energy above the $(1s2s)^3S$
threshold for 2. Transitions like 3 produce structure in the
excitation cross sections (see Ref. 8 for recent calculations
including such resonance effects).

The inclusion of transition 4 and 5 on Fig. 1 is to illustrate
a different process which can be used to study excitation. Transition 5 represents inner-shell excitation of an initial Li-like ion
(another electron has been added to the system for transitions 4
and 5). Both 4 and 5 are ionization events, in which e + Li-like →
He-like + 2e. Transition 5 can occur for any energy above the
inner-shell excitation ($1s2s2\ell$) threshold (the continuum of the
fourth electron is not illustrated). However, this inner-shell
excitation-autoionization will have the abrupt threshold onset,
characteristic of an excitation cross section, and should be easily
distinguishable in the measured Li-like ionization cross section.
Looking ahead, Fig. 8 demonstrates the case cited here for qualitative discussion. To the extent that transition 5 just adds to
4 — no interference, the inner-shell excitation cross section can
be measured through the structure in the ionization cross section.

2.2 Coupled State Theory

Qualitatively, if expansion (3) is substituted into Eq. (1)
and projected onto the angular momentum components of a particular
channel, i, we obtain the close-coupling equations which are of the
form:

$$-\left(\nabla^2 + k_i^2\right)\theta_i(X) + \sum_f V_{if}(X)\theta_f(X) = 0 \qquad (5)$$

where X represents the space and spin coordinates and

$$V_{if}(X_1) = \frac{-2Z}{r_1}\delta_{if} + \int \Psi_i(X_2\ldots X_{N+1})\sum_{n=2}^{N+1}\frac{2}{r_{1,n}} \cdot$$

$$\Psi_f(X_2\ldots X_{N+1})\,dX_2\ldots dX_{N+1} \quad . \qquad (6)$$

The asymptotic form of this potential is

$$V_{if}(X)\underset{r\to\infty}{\sim}\frac{-2q}{r}\delta_{if} \qquad (6a)$$

where q = Z − N is the ionic charge, δ is the usual "delta function" operator, and note that f is an index here which runs over all of the states included in the expansion (4). In seeking solutions to the set of equations (5), various approximations are made, and suitable boundary conditions are imposed.

Assuming that the continuum electron moves in a central field, the form of θ_i can be:

$$\theta_i(X) = Y_{\ell m}(r)\delta(\text{spin})\frac{1}{r} F_i(r) \tag{7}$$

where the $Y_{\ell m}$ are spherical harmonics and the radial components $F_i(r)$ which satisfy Eq. (5) will have *asymptotic* form:

$$F_i(r) \underset{r \to \infty}{\propto} \frac{1}{\sqrt{k_i}} \{\delta_{if} \sin(\zeta_i) + \rho_{if} \cos(\zeta_i)\} \tag{8}$$

where $\zeta_i = k_i r - \frac{1}{2}\ell_1\pi + \frac{q}{k_i} \ell n(2k_i r) + \text{arg}\Gamma(\ell + 1 - \frac{iq}{k_i})$ is the asymptotic phase of the regular Coulomb function. Expression (8) has included the quantity we ultimately want. The term $\delta_{if} \sin(\zeta_i)$ represents elastic scattering while the term $\rho_{if} \cos(\zeta_i)$ represents inelastic scattering state i to state f. The ρ_{if} is an element of a matrix, related to the reactance or R-matrix, which has properties analogous to the more familiar scattering amplitude $f(\theta)$ used in the partial wave approach to simple elastic scattering problems (see Ref. 9 for example).

The matrix element ρ_{if} is obtained by solving the coupled-differential equations. Recall that f is still an index which runs over all states included so therefore, the number of equations must be limited, that is the number of functions (3) or (4) is truncated at I so that the index i (or f) runs from 1 to I. An I-state close-coupling approximation then results. The cross section for excitation of a channel i to channel f is ultimately represented, qualitatively, by

$$\sigma_{if} \propto |\rho_{if}|^2 . \tag{9}$$

Actual solutions by computer (for instance using the code RMATRX[10]) involves computation of a set of one-electron basis functions, considerable vector-coupling algebra, and many matrix inversions. Solutions for simple ions and a few states I are tractable, but complex ions and/or many states (I) are generally not attempted.

2.3 Distorted-Wave and Coulomb-Born Approximations

The cause of numerical difficulty with Eq. (5) is the coupled nature of the V_{if} ($i \neq f$) part of the potential in Eq. (6). A substantial simplification results from uncoupling the equation which can be accomplished by setting $V_{if} = \begin{cases} 0 & \text{if } i \neq f \\ V_{ii} & \text{if } i = f \end{cases}$ where V_{ii} is given by Eq. (6) with f = i. The differential Eqs. (5) now have only one term in what was the summation, and each differential equation contains only one index, i, and thus stands alone. In the matrix of ρ_{if}, the state f is one particular, selected, final state, and calculation of excitation to any selected final state is now a two-state problem. Again qualitatively, this is the distorted-wave approximation. Coupling of states by the incident electron has been lost but the technique is sufficiently simple to apply to quite complex ions. Of course the actual approximation of Eqs. (5) and (6) can be done many ways. Henry[1] distinguishes about ten variations of distorted wave in published work.

The Coulomb-Born approximation is a modification of the potential (6) which provides additional simplification. The asymptotic form of Eq. (6), $V_{if} = \frac{-2q}{r}\delta_{if}$, is used for the problem at all r. Note that the first term of Eq. (6) is not the same as this in that the nuclear charge Z has been replaced by the ionic charge, q. The screening of the bound electrons which was determined by the integral part of Eq. (6) in a flexible (r-dependent) manner even in distorted wave is now fixed to be simply full screening at any r. The plane-wave Born approximation takes $V_{if} \equiv 0$ which would correspond to q = 0 in the Coulomb-Born approach and is not generally appropriate for ions but can provide useful insight at high energies and can be solved analytically.

Other approximations can be applied to any of the three methods mentioned. Unitarity has a specific mathematical meaning and should be applied to the matrices derived from the ρ_{if}. Physically unitarity corresponds to conservation of flux in that the probabilities of all allowed scattering must sum to the initial flux. Naively it would seem that unitarity should always be satisfied but, if the elements of ρ_{if} are all small, it *may* not be essential to obtaining a good approximation.

Exchange has not been specifically mentioned. In principle, if all the wave functions in the expansion (4) and the potential (6) are fully antisymmetrized, exchange is explicitly included throughout the problem but with a great addition in the real work. Thus, numerous different approximations are in fact made in the way and extent to which exchange is included.

At this point it is instructive to glance ahead to Fig. 5 for excitation of Be^+. All of the various approximations discussed have been applied, and convergence of the approximations toward the experimental cross section reflects the quality of the approximations as inferred above.

2.4 Oscillator Strengths, Gaunt Factors, and Collision Strengths

If the general theoretical approach outlined above is followed but only for bound states, that is with trial wave functions Ψ which have no continuum electron function and which consequently obey $\Psi \xrightarrow[r\to\infty]{} 0$, then a related but much simpler bound-state problem is solved. Solutions of expression (5) result in oscillator strength rather than a reactance matrix. Seaton[11] and Van Regemorter[12] developed a specific expression for connection between excitation cross sections and oscillator strengths:

$$\sigma_{if} = \frac{8\pi}{\sqrt{3}} \frac{f_{if}\,\bar{g}}{E\,\Delta E_{if}}\,\pi a_0^2 \tag{10}$$

where E ($\equiv k_i^2$ earlier) is the incident electron kinetic energy (in Rydbergs), ΔE_{if} is the transition energy (in Rydbergs), f_{if} is the oscillator strength, and $\bar{g}$ is called the effective Gaunt factor which serves to adjust Eq. (10) in a systematic manner based on knowledge of ratios of cross sections to oscillator strengths. When large numbers of excitation cross sections are required, for example in plasma modeling,[13] Eq. (10) is generally used.

From early comparisons $\bar{g}$ was taken to be 0.2 near threshold with an explicit functional variation at higher energies. From an experimental viewpoint, Taylor et al.[14] found that for Ca^+, Ba^+, Hg^+, and N_2^+ measured cross sections, this value of $\bar{g}$ agreed with experiment to within a factor of two as originally advertised. For the simple $Be^+(2s-2p)$ case the $\bar{g}$ formula with $\bar{g} = 0.2$ at threshold fits the experimental data[15] as well as any theoretical result. However, for ions of charge greater than +1, taking $\bar{g} = 0.2$ seriously underestimates cross sections obtained from better theoretical approaches or the only experimental cases available.[16] Bely[17] and Blaha et al.[18] noted earlier that for multiply charged ions this value of $\bar{g}$ was probably too small. The most recent examination of the $\bar{g}$ approximation by Younger and Wiese[19] found that for multicharged ions and $\Delta n = 0$ transitions, $\bar{g}$ near 1.0 was best. For $\Delta n \neq 0$ and for optically forbidden transitions they found $\bar{g}$ varied from 0.05 to 0.7 for conditions appropriate to plasma modeling applications. These $\bar{g}$ predictions are not sufficiently reliable in many situations and rely on availability of oscillator strengths.

It is common in all theoretical predictions of excitation to present results as collision strengths, Ω_{if}. This quantity is the particle collision equivalent of the oscillator strength. The connection between collision strength and cross section is always

$$\Omega_{if} = w_i E \sigma_{if} \tag{11}$$

where E is again the incident electron kinetic energy *in Rydbergs*, w_i is the statistical weight of the lower level [$w_i = (2S_i + 1) \cdot (2L_i + 1)$ or $(2J_i + 1)$ if fine structure is taken into account, where S_i, L_i, and J_i are the usual total angular momentum quantum numbers of the state i], and σ_{if} is the cross section in units of πa_0^2.

3. EXPERIMENTAL TECHNIQUES

3.1 Plasma-Based Rate Measurements

Although this paper will deal only superficially with excitation rate measurements, it is worth noting that a few detailed studies have been performed. The 1972 review of Kunze[20] remains the standard reference for describing measurement techniques, and Gabriel and Jordan[21] have given a review of the interpretation for spectral intensities from plasmas. The intensity, I_{fj}, of radiation of a given transition from a well-characterized plasma, per unit volume, per steradian can be written

$$I_{fj}(t) = \frac{h\nu}{4\pi} \frac{A_{fj}}{\sum_r A_{fr}} N_e(t) N_i(t) \alpha_{if} \tag{12}$$

where N_e is the electron density, N_i is the ion density in the initial state, α_{if} is the rate coefficient for excitation from state i to f, A_{fj} is the transition probability for the observed radiation from state f to a specific state j, and A_{fr} is the transition probability from f to any state r. There are a number of assumptions implicit in Eq. (12):

1.   State f is populated only by transitions from a single initial state, i.

2.   For α_{if} to be the electron excitation rate, no other collision processes (electron capture, photo-excitation, cascade, etc.) can contribute to population of level f.

3. No radiation field, collision process, or electric or magnetic
    field can modify the transition probabilities A_{fr}.

4. Even though terms are explicitly time dependent, transport of
    the ions of N_i must be sufficiently slow so that those which
    are excited and radiate within a specific volume arrive from a
    region of constant plasma conditions — N_e and T_e.

Some of these assumptions are part of what is generally called
coronal equilibrium conditions. Condition 1 may be the most difficult to satisfy. Kunze discusses a case where an additional
metastable level also contributes to excitation population of state
f and points out that a unique solution for a given α_{if} is then not
possible, though comparison with theory for some mixture is still
possible. Thus, the first difficulty associated with excitation
rate measurements is assessing the extent to which Eq. (12) is
valid.

   If expression (12) is valid for the plasma then a rate coefficient, α_{if}, can be determined by measuring the electron density,
N_e; ion density, N_i; and absolute light intensity, I_{fj}, presuming
the A_{fr} are known. Of course the electron temperature, T_e, must
also be determined in order to have α_{if} as a function of electron
temperature. In practice the density N_i is inferred from the gas-
filling pressure, the time evolution of light from the plasma, and
a model (with some of the same assumptions about plasma conditions).
That is, the N_i is determined from ionization rate measurements.
N_e and T_e are generally determined by Thomson scattering of laser
light. Note that I_{fj} is per unit volume requiring determination of
the volume of plasma observed, which is nontrivial. If all other
conditions and measurements are satisfied the absolute intensity
calibration for light detection will still remain a difficult
problem with no reliable or readily transferable standard, which
has been a difficulty for all types of excitation experiments. The
experimental determination of excitation rate coefficients faces
both difficult assumptions and measurement, with resulting high
uncertainty, all to obtain a few data values in which much of the
physically meaningful information (structure in cross section vs
energy) has been averaged and lost. Nevertheless, a few excitation
rate coefficients have been measured (see Refs. 1 and 20 for references to original work and some comparisons). Those rate measurements provided the first tests of excitation theory for multicharged
ions.

   It is currently more common to view plasma physics research as
a consumer rather than a supplier of excitation cross sections.
Attempts are ongoing to incorporate the best available atomic data
to provide more accurate interpretation and modeling of plasmas
(Refs. 13 and 22 being typical examples).

It is conceivable to use trapped ions to study electron impact excitation as has been done for ionization. References 23 and 24 give reference to original ionization work. However, many of the difficulties which apply to plasma-rate measurements would apply to trapped-ion measurements, and no measurements have been reported.

3.2   Electron-Ion Beams Experiments

3.2.1   Overview of the technique.   Electron-ion beams experiments were extensively developed in the 1960s and 1970s and have been reviewed.[25,26]   In addition, some details are reviewed by other authors within the present volume so the comments given here are restricted to an overview and a few details associated with excitation experiments.

A schematic view of a representative electron-ion crossed-beams excitation experiment is shown in Fig. 2.[27]   The concept of the experiment is to collide electrons and ions and observe the photons of a given transition which are emitted by the decay of a particular state excited by the collision. The optical detection system consists of converging lenses, interference filter, and photomultiplier (the scintillator serves as a readily available check on relative variations in photomultiplier sensitivity). In other experiments a few different elements have been used, but typical total sensitivities of the optical system need to be about 10^{-4} counts/collision-produced photon in order for a typical experiment to be tractable.

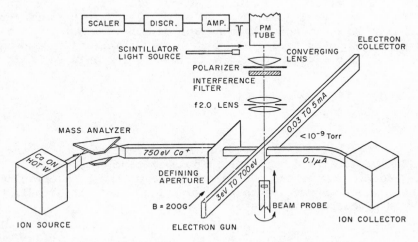

Fig. 2.   Schematic of a representative electron-ion crossed-beams excitation experiment (from Ref. 27).

The relationship which allows cross-section determination from experimental parameters is:

$$\sigma = \frac{S}{\eta} \frac{qe^2}{I_i I_e} VF \qquad (13)$$

where $\frac{S}{\eta}$ is the rate of production of photons by the colliding beams (signal rate, S, divided by detection efficiency, η), and the remaining terms express interacting particle density. Specifically, I_e/e is an electron rate number from the electron current, I_e, and similarly I_i/qe for the ions; V is the relative speed of the beams in the x-y plane,

$$V = \frac{v_i v_e \sin\theta}{\left(v_i^2 + v_e^2 - 2v_i v_e \cos\theta\right)^{1/2}}$$

which reduces to just v_i for crossed beams, $\theta = 90°$, with electrons much faster than ions; and the form factor, F, is the vertical (z axis) beams overlap,

$$F = \frac{\int i_e dz \int i_i dz}{\int i_e i_i \, \eta_o(z) \, dz}$$

where i_e and i_i are electron and ion beam vertical distributions. In Eq. (13) note that η is the probability of detecting a collision-produced photon which is dependent on where in space the photon was emitted. This vertical dependence is specifically included as $\eta_o(z)$ (set equal to 1 at a given height, z_0) in the form factor, and the x-y dependence has been neglected here which implies narrow crossed beams and excited-state decay times which are fast compared to motion of the excited ions. Additionally then, in Eq. (13)

$$\eta = \left(\frac{1 - P\cos^2\phi}{1 - \frac{1}{3}P}\right) D(z_0, \lambda)$$

where P is the polarization of the emitted radiation, ϕ is the direction of detection relative to polarization axis, and $D(z_0, \lambda)$ is the sensitivity of the photon detection system to isotropically emitted photons of wavelength λ originating at a central height (z_0) within the beams collision volume.

Thus, the concept of the experiment described in Fig. 2 and expression (13) is quite simple, but a number of detailed measurements have to be performed. The strength of the crossed-beams approach is that all of the parameters in Eq. (13) can be measured

and can be independently varied in systematic tests (with sufficient patience). In practice, the absolute detection efficiency,
$D(z_0,\lambda)$, is the most difficult parameter because it requires elaborate measurement procedures to relate it to absolute standards that
are difficult to reproduce in the laboratory with satisfactory
reliability.

3.2.2 <u>Details of some particular cases</u>. A primary issue in
any experiment is whether or not sufficient signal, in this case
photon count rate, S, will be obtained. Table 1 gives some of the
parameters of two experiments, one quite tractable[28] (Ba^+) and the
other quite difficult[16] (N^{+4}). In both cases what is given is
approximate "best" conditions achieved for an absolute measurement
near threshold. At first glance the parameters seem quite similar,
but the differences in cross-section magnitude, wavelength, and
charge state of the ion modify technical difficulties significantly.

The standard statistical uncertainty in the primary signal, S,
is given by the total accumulated counts according to

$$\Delta = \frac{\sqrt{(S + B)t}}{St}$$

where Δ is the uncertainty fraction, S and B are signal and background count rates, and t is the data collection time. Of course,
in actual measurements, the signal and background must be separately
accounted which generally has been accomplished by modulation of

Table 1. Parameters from Two Electron Impact Excitation Experiments.

Parameter	Ba^+ (Ref. 28)	N^{+4} (Ref. 16)
Electron energy	4 eV	15.5 eV
Electron current	10 μA	90 μA
Ion energy	750 eV	40 keV
Ion current	0.1 μA	1.0 μA
Pressure (both beams on)	1×10^{-9} torr	1.2×10^{-9} torr
Wavelength	455 nm	124 nm
Band pass	10 nm	30 nm
$D(z_0,\lambda)$	7.4×10^{-4}	4×10^{-4}
Signal (S) (both beams)	10 Hz	2 Hz
Background (B) (both beams)	3 Hz	90 Hz
Emission cross section	17.4×10^{-16} cm^2	2.7×10^{-16} cm^2

both beams and gating of counters which result in a real-time duty factor of about one-fourth. Thus, actual experimental time is typically 4t. For an arbitrary Δ of 0.03, 4t for one measurement is then 580 sec for the Ba^+ case and over 10^5 sec for the N^{+4} case. "Best" conditions are not maintained over extended times, and it is rarely satisfactory to measure one datum. Thus, for the N^{+4} case the accumulation of two absolute data points, one just below threshold (found to be zero) and the other at 15.5 eV, required about 2 x 10^5 sec of data collection time for a Δ = 0.06.

The N^{+4} case is useful for suggesting a few more of the difficulties associated with these experiments. Detailed systematic tests are not only possible with crossed beams, they can also be critical. Typically, cross sections are determined with deliberate variations of I_i, I_e, V, background pressure, and beams-chopping frequency because problems referred to generally as cross-beam modulation are revealed by such diagnostics. One of the beams may modify either the background or the other beam in such a way that spurious signal is produced which can either add to or subtract from the true signal. In this regard the most stringent single diagnostic is measurement of the cross section below threshold. Such problems are discussed in Ref. 25 and have been encountered specifically.[16,29-31]

During the N^{+4} study a problem was revealed by deliberate variation of I_e. The measured, below-threshold cross section was zero as it should be, and near threshold σ was found to be independent of I_e. However, at 52 eV a systematic variation of σ with I_e was observed (see Fig. 3). At first the problem was "guessed" to be due to redirection or focusing of the ion beam by the space charge of the electron beam. A component of ion-produced background due to ions striking surfaces had been previously identified from studies of ion noise vs background pressure, and in fact this source of background had been the cause of apparatus modifications. Nevertheless, some surface noise persisted, and it was suspected that the surface noise could be modulated by electron-beam space-charge redirecting ions. However, such a spurious signal should have been independent of electron energy even to below excitation threshold since the current density of the electron beam (which would control space-charge effects) changes only slightly between 10- and 50-eV energies. In fact the electron current density was deliberately changed by modifying the electron gun during the course of these studies. It was finally judged that the spurious signal was due to creation by the electron beam of some particular background target (probably an ion) which had a threshold for creation at energies above the N^{+4} excitation threshold. Light, within the band pass of the photon system, could then be produced by this target interacting with the ion beam. At any rate the problem persisted in spite of apparatus modification, and

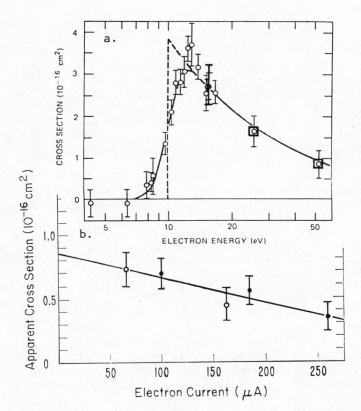

Fig. 3.   (a) Cross section for electron impact excitation of
          N^{+4}(2s-2p) from Ref. 16.   The solid point at 15.5 eV is
          the absolute measurement with outer error bar showing
          total absolute uncertainty at 90% confidence level —
          other points have 90% confidence level relative bars.
          The cross sections at 26 and 52 eV have been corrected
          for variation with electron current.   The dashed curve
          is coupled-state theory by van Wyngaarden and Henry (see
          Ref. 1).   (b) Apparent cross section at 52 eV as a func-
          tion of electron current.   For the solid points the
          electron current density was increased about 50% by
          decrease of electron beam height.   Error bars for (b)
          are 1 std. dev. statistics.

the value of the cross section at zero electron current (as extrapolated from Fig. 3) was finally accepted as the correct cross section at 52 eV.

3.2.3. <u>Directions of beams studies</u>. The work of Harrison and colleagues (see Ref. 25), Dolder and Peart[26] and, for excitation, particularly Taylor and Dunn,[27] advanced the crossed-beams technique significantly and set high standards for detail and accuracy. The spirit of these endeavors has been to measure cross sections with the highest achievable reliability in order to provide test data against which our understanding of physical processes could be meaningfully tested and developed. Others, particularly the Russian group at Uzhgorod,[32] have contributed significant data on excitation using crossed beams but have not provided sufficient detail to allow as high a level of confidence.

Experiments performed to date are nearly all similar to the description presented thus far. However, future progress may be achieved by a few departures from these conditions. Inclined beams ($\theta \simeq 8$ to $30°$) have been used extensively in related types of collisions by Dolder and Peart.[26] Kohl and colleagues[33] have adapted inclined beams to excitation observations and have attempted to study dielectronic recombination but have not yet reported cross sections. Inclined beams provide a greater beams overlap without compromising the ability to measure all of the parameters accurately.

Merged beams[34] are attractive from the viewpoint of beams overlap, which is a primary concern. However, application of merged beams to excitation experiments will require additional technical advances. The observation of photons appears to be impractical in merged beams because of the extensive volume in which they would be produced. However, energy loss of the electrons during the collision, which has not been applied as yet in ion-excitation studies, might be coupled with the merged-beams approach. Significant problems of angular dispersion of the electrons as a function of collision energy would have to be overcome to obtain detailed absolute cross sections from a merged-beams, energy-loss approach. However, the elimination of present problems associated with optical calibration $[D(z_0,\lambda)]$ is enticing, and the critical signal level would be increased.

3.3 Study of Excitation via Ionization

Crossed-beams studies of ionization cross sections have revealed features identified as excitation followed by autoionization. Subtracting the direct ionization cross section leaves only the excitation component. At present about as many excitation cross sections for ions can be estimated by this technique as have been measured directly.

There are a number of aspects of these excitation-autoionization
measurements which are important in analyzing the results as excitation measurements. The states excited are autoionizing levels which
are a particular class of highly excited states. The observed
excitation-autoionization is part of a total ionization cross section, and the possibility of an interference between direct ionization and the excitation component must be admitted. Additional
processes such as recombination-double-autoionization[35] may also
contribute to total ionization cross sections so that all of the
increase above direct ionization may not be attributable to excitation of a single well-defined level.

In spite of these issues it is possible that meaningful tests
of excitation theory can derive from excitation-autoionization measurements. A number of cases will be discussed here and some implications about excitation theory will be inferred based on the
assumption that excitation-autoionization simply adds to ionization
cross sections. Crossed-beams ionization measurements avoid problems of low-photon detection efficiency and its uncertainty and high
electron-produced background present in direct excitation experiments.

4. RESULTS AND DISCUSSION

4.1 Simple Systems

The simplest ion is He^+, and some of the earliest excitation
experiments were for this case. The 1s-2s excitation was studied
experimentally by Dance et al.[31] and by Dolder and Peart[36] who measured only relative cross sections due to problems of calibration at
30.4 nm, but obtained good agreement as to the shape of the cross
section. However, the comparison with theory[1,3,37-41] remains a
significant point of discrepancy in the field. Figure 4 shows the
results of the more recent experiment,[36] normalized to close-coupling
theory at high energy, compared with a number of theoretical results.
Presumably, close-coupling results should be best, but the agreement
among them is poor and all the theories are higher than experiment
near threshold.

The situation is better for the 1s-2p excitation in He^+ in
that the theories are in reasonable agreement,[1] and the experiment
of Dashchenko et al.[42] normalized to theory near 200 eV confirms
the predicted shape.

At this point it is interesting to speculate that the discrepancy in the 1s-2s case can be referred back to the Hamiltonian (2)
or the potential (6). The term $\frac{2}{r_{ij}}$ in Eq. (6) is expanded in a
multipole representation to do the calculations of cross section.

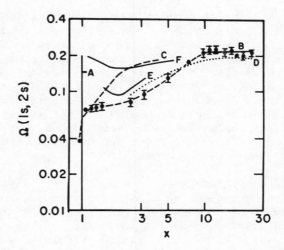

Fig. 4.   Collision strength for electron impact excitation of
          He$^+$(1s-2s) as a function of threshold energy — from Ref. 1.
          Solid points connected by dashed line are experiment
          (Ref. 36) normalized to theory B at high energy. Theore-
          tical results are: A - Three-state close-coupling (3cc)
          of Ref. 37; B - 3cc of Ref. 3; C - distorted wave of
          Ref. 38; D - distorted-wave second-order potential of
          Ref. 39; E - 5cc of Ref. 40; and F - 3cc of Ref. 41.

For 1s-2p the dipole term dominates while for 1s-2s only the monopole term contributes (except via exchange). At any rate the
speculation is that the theoretical representation overestimates
the monopole term. Another case will be presented later with
similar implications.

     The only He-like case studied experimentally is Li$^+$.   Rogers
et al.[43] measured the absolute emission cross section for 548.5 nm
which is excited by spin change (exchange), dominantly through the
$1^1S \rightarrow 2^3P$ transition. The cross section exhibits structure near
threshold attributed to interference involving autoionizing resonances of neutral Li (schematically illustrated in Fig. 1). The
available close-coupling calculations[1,43] do produce the observed
structure. However, none of the various coupled-state and distorted-
wave predictions represent the data better than about ±50% over the
experimental energy range 60-160 eV.

Other He-like ions have received significant theoretical attention[1,8,44] in part because of the importance of resonances in the intercombination (spin change) and optically forbidden transitions and partly because these ions persist over broad temperature ranges in high-temperature plasmas. Pradhan et al.[8] predict dramatic resonance structures in these transitions so that even energy-averaged rate coefficients may be increased by factors of six or so for intercombination (spin forbidden) transitions.

The Li-like ions are perhaps the best studied experimentally and provide instructive comparison between theory and experiment. The 2s-2p resonance lines of Be^+ (Ref. 15), C^{+3} (Ref. 45), and N^{+4} (Ref. 16) have been measured. The Be^+ results are shown in Fig. 5. None of the theories represent the experiment within experimental uncertainty (typically ±10% at high confidence, intended to be comparable to 98% confidence level statistics). However, with increasing power and complexity, the calculations[46,47] approach the experiment (neglecting the g result). This case nicely demonstrates the effects of electron correlation or coupling and shows that the rigid potential screening of the Coulomb-Born approach tends to overestimate the near threshold cross section.

By contrast Fig. 6 shows results near threshold for C^{+3}. The experiment is in excellent agreement with two-state close-coupling theory[48] if the experimental energy spread is folded with the theory. However, the Coulomb-Born[48] with exchange calculation is just as acceptable within the experimental uncertainty. For N^{+4} (Fig. 3) the situation is nearly identical to C^{+3}. The electronic structure is constant for these cases, but the relative spacing of energy levels is significantly different (see Fig. 7). This energy scaling alone probably does *not* account for the dramatic decrease in effects of coupling along the sequence. The electron correlation part of the potential [integral part of Eq. (6)] is generally weaker compared to the nuclear part as ionic charge increases along the isoelectronic sequence, and this is probably the dominant effect. In order to assess the extent to which distorted-wave or Coulomb-Born approximations are valid for highly charged ions, it remains important to measure cross sections along isoelectronic sequences other than Li-like where the scaling of energy levels will not correspond to Fig. 7.

The experimental study of ionization of Li-like ions[49,50] has produced information on excitation of the form $1s^22s \rightarrow 1s2s2\ell$ where ℓ = s or p. Figure 8 shows the N^{+4} case in detail, and Table 2 gives experimental cross sections deduced from simply

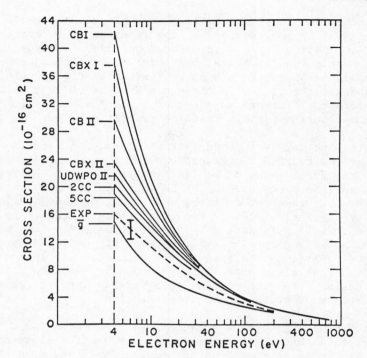

Fig. 5.   Electron impact excitation of Be$^+$(2s–2p) — from Ref. 15.
          Experiment is dashed curve with typical high confidence
          uncertainty indicated.   Theories from Refs. 46 and 47 are
          non-unitarized Coulomb–Born (CBI with CBXI including
          exchange); unitarized Coulomb–Born (CBII and CBXII);
          unitarized distorted_wave (UDWPOII); and close-coupling
          (2cc and 5cc).   The g approximation is according to Ref. 11.

subtracting the excitation-autoionization bumps from an extrapolated
smooth direct ionization curve near the excitation threshold and
comparing the experimental and theoretical results.[51,52]   The Be$^+$,
C^{+3}, and N^{+4} results compare reasonably with close-coupling theory,
but the O^{+5} case presents a significant discrepancy.   Uncertainties
of these experimental excitation cross sections are difficult to
assess because there are both assumptions and subtractions in
the determination.   However, the magnitude of inner-shell excitation,
the interaction of excitation and ionization, and the experiment for
O^{+5} are open to question.

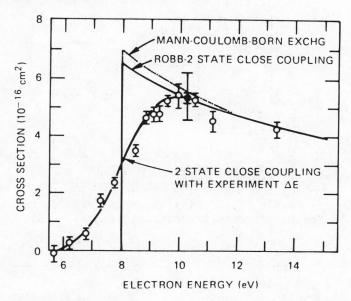

Fig. 6.   Electron impact excitation of $C^{+3}(2s-2p)$ near threshold
          from Ref. 45. Theoretical results are from Ref. 48. The
          solid curve through the data is the two-state close-coupling
          convoluted with the experimental energy spread. Error bars
          are 90% confidence level statistics except for the outer
          bar at 10.2 eV which is total absolute uncertainty, includ-
          ing absolute optical calibration, at 90% confidence level.

4.2   Complex Systems

     Some of the best experimental measurements have been for heavy
ions for which the wavelength and cross-section magnitude simplify
some of the technical difficulties. The excitation of the 4s-4p
resonance lines of Ca^+ by Taylor and Dunn[27] must rank as one of the
most detailed experiments to be accomplished in atomic collisions
studies. Crossed-beams experiments for Mg^+, Ca^+, Sr^+ (Ref. 32),
Ar^+, Kr^+ (Ref. 55), Hg^+ (Refs. 56 and 57), and Zn^+ (Ref. 58) have
been reported. The 6s-6p resonance transitions of Ba^+ have been
studied by the greatest number of experimenters.[28,32,59-61]

     For the present discussion Ba^+ will be taken as a representa-
tive case. Figure 9 shows the energy level distribution and the
$6^2S_{1/2} \rightarrow 6^2P_{3/2}$ excitation cross section derived from the 455.4 nm
emission cross section after corrections for cascade, metastable

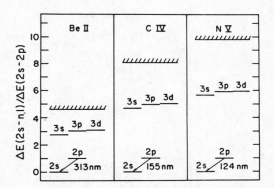

Fig. 7.   Energy levels of the Li-like ions scaled to the energy of
          the 2s-2p resonance transition, from Ref. 16.

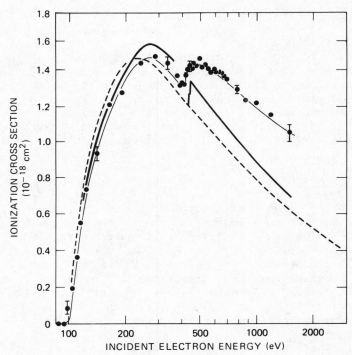

Fig. 8.   Electron impact ionization of N^{+4} showing contribution due
          to $1s^2 2s \rightarrow 1s2sn\ell$ inner-shell excitation-autoionization
          beginning at 415 eV — from Ref. 49.   The excitation calcu-
          lation of 6cc (Ref. 51) for $1s^2 2s \rightarrow 1s2s2\ell$ only is repre-
          sented by the solid curve beginning at 415 eV and has been
          added to the dashed curve which is scaled-Coulomb-Born
          direct ionization calculation of Ref. 54.   Solid curve
          below 350 eV is Coulomb-Born direct ionization of Ref. 53.

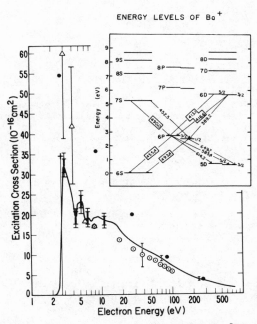

Fig. 9. Electron impact excitation of Ba^+ for $6^2S_{1/2} \to 6^2P_{3/2}$ and
energy levels of Ba^+ with the transitions studied experimentally indicated by boxes — from Ref. 28. Solid curve
is experiment of Ref. 28 with relative (inner) and absolute
(outer) error bars at 90% confidence level. Open circles
(Ref. 59) and open triangles (Ref. 60) are earlier measurements. Distorted-wave theory (Ref. 62) is indicated by
solid points.

Table 2.

Cross sections in 10^{-19} cm^2

Ion	Energy (eV)	Experiment[a]	Scaled Coulomb-Born[b]	Six-state close-coupling[c]	Ratio exp/6cc theory
Be^+	130	17	23	12.2	1.4
C^{+3}	340	3.2	3.7	2.15	1.5
N^{+4}	460	1.8	2.0	1.27	1.4
O^{+5}	612	2.8 (1.4)	1.1	0.74	3.8 (1.9)

[a] Cross sections determined by estimating the increase in total ionization cross section at energy 1.1 times excitation threshold. The Be^+
case is from Ref. 50 while C^{+3}, N^{+4}, and O^{+5} are from Ref. 49. For O^{+5}
the uncertainties are large and the value in parenthesis is the smallest
excitation cross section derived from the ionization data allowing for
the 90% confidence level error bars.

[b] From Ref. 52 which cautions that the technique is not appropriate
to low ionic-charge cases such as Be^+.

[c] From Ref. 51 with Be^+ case extrapolated according to $(Z - 1.4)^3$ $\sigma =$
constant for a given energy in threshold units.

ions in the beam, and branching ratio of the decay.[28] The effects
of resonances, schematically like those of Fig. 1-transition 3, are
clear in the experimental cross section and are even more obvious
in the polarization of the radiation. The coupled-channel technique
is probably appropriate for this case but has not been attempted
because of the complexity. The result that distorted-wave theory[62]
overestimates excitation for complex, singly charged ions appears
to be general for all the cases cited.

The magnitude of excitation-autoionization in observed *ionization* cross sections for the alkali-like ions increases with
increasing numbers of electrons[26,63] as shown in Fig. 10. The
excitation-autoionization is substantial for Ba^+, and the excitations
of the form $5p^66s \rightarrow 5p^55d6s$ can be estimated from the data. For
Ba^+ the first attempt at calculating the excitation cross sections[64]
did not account for the structure within the configurations and
neglected some of the important dynamic effects (exchange and
nondipole transitions). The result was artificially good agreement
of theory and the experiment for the excitation component. Hansen[65]
demonstrated that the original analysis was in error due to neglect
of structure so that the transition for which the cross section was
calculated[64] would not occur at the energy of excitation-

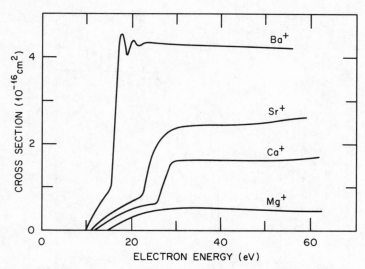

Fig. 10. Electron impact ionization of alkali-like ions showing
          significant enhancement due to excitation-autoionization
          at abrupt rises near 15, 23, and 27 eV for Ba$^+$, Sr$^+$, and
          Ca$^+$, respectively — from Ref. 63.

autoionization onset in the experiment. Ongoing structure calculations and distorted-wave prediction of cross sections[66] indicate
that excitation of a number of terms contributes and that exchange
and other complexities must be included to reasonably represent the
total excitation cross section for $5p^66s \rightarrow 5p^65d6s$ which can then
be more meaningfully compared with experiment.

Recent work on ionization of alkali-like ions Ti^{+3}, Zr^{+3},
and Hf^{+3} has observed that near threshold the ionization cross
sections are almost entirely due to excitation-autoionization.[67]
Figure 11 shows that for Ti^{+3} the excitation-autoionization is
about ten times the direct cross sections. Distorted-wave calculations of the excitations of $3p^63d \rightarrow 3p^53d^2$ have been divided by
2.5 to provide comparison with experiment and are also convoluted
with a 2-eV energy spread representative of the experiment. In
spite of the significant discrepancy in magnitude, the energy
location and relative contributions of different levels are in such
good agreement as to provide promise that theory can be developed
to predict such excitation cross sections for complex ions.

Experiments on Zn$^+$ for both direct excitation of bound states[58]
and excitation-autoionization[68] provide an additional test case
where accurate experimental data are available. The experimental

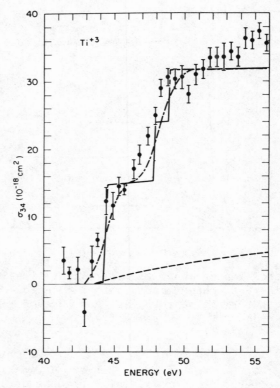

Fig. 11.   Electron impact ionization of Ti^{+3} from Ref. 67. Dashed
curve is the Lotz formula prediction taken to be the
direct ionization. The solid curve is a distorted-wave
dipole-only calculation *divided by 2.5* and added to the
Lotz prediction. Dot-dashed curve is the solid curve
convoluted with 2-eV energy spread for comparison to the
experiment.

results for Ga^+ ionization[68] also show distinct excitation-
autoionization contributions near threshold. Theoretical considerations for the Ga^+ case[69] have concluded that excitation of
ground-state Ga^+ $(3d^{10}4s^2)$ to the $(3d^94s^24p)^1P_1$ autoionizing level
is the most important transition.

This Ga^+ excitation-autoionization case was chosen by Pindzola
et al.[69] to examine the effects of quantum mechanical interference
between direct ionization and ionization through the $^1P_1$ autoioniz-

ing level. These calculations employ the general reaction theory
of Feshbach within the framework of the distorted-wave approximation. Little difference was found in the total ionization cross
section between simply adding the direct and indirect processes and
allowing for interference. However, a marked effect due to the
interference is predicted for the cross-section differential in the
ejected electron energy. Figure 12 illustrates the differential
cross sections for an incident electron energy of 22.0 eV, which is
3.5 eV above the ionization threshold and 1.2 eV above the
$(3d^94s4p)^1P_1$ excitation threshold. The dashed curve of Fig. 12
represents the angular distribution of electrons due to direct
ionization only. The solid curve illustrates the results of the
interference between direct and indirect ionization processes and
shows a dramatic resonance structure traditional for Feshbach-type
resonances. This calculation illustrates that the most sensitive
tests of interference are to be obtained from studies of differential cross sections rather than total cross sections. To date
there are no experimental studies of differential cross sections
for ions, but cases such as Ga^+ should be technically feasible.

Work on excitation-autoionization of Na-like ions[70-73] has
provided data on excitation of the configurations $2p^63s \rightarrow 2p^53s3\ell$.
This system is somewhat simpler than the more complex alkali-like
cases, but the excitations are $\Delta n = 1$ transitions rather than $\Delta n = 0$
so that the relative intensity is smaller. Figure 13 compares a
recent experiment[72] and distorted-wave theory[73] for Al^{+2}. It is
apparent that the $2p^63s \rightarrow 2p^53s3p$ transition, which is predicted to
be the largest component, is missing or quite small in the experiment. The same result is even more obvious in the Mg^+ case[70-73]
and occurs in both distorted-wave[73] and Coulomb-Born[71] approximations.
An interesting perspective is that the strength of the 2p-3p transition derives from the monopole term in the multipole expansion of
the potential (6) which is similar to the 1s-2s excitation of He^+
which remains in discrepancy.

One additional case of excitation-autoionization is cited
because it illustrates specific complications which could be dramatically important in particular cases. For the case of ionization of Fe^{+4} Griffin et al.[74] predict a strong dependence of
ionization cross section on the initial internal temperature of the
ion. For the ground configuration, the lowest term, $(3p^63d^4)^5D$,
has no strong dipole transitions for excitation to any of the 22
levels of configuration $(3p^53d^5)$ which are expected to be above the
ionization threshold for $Fe^{+4} \rightarrow Fe^{+5}$. However, for the slightly
higher energy terms of the $3p^63d^4$ configuration metastable states,
there are strong dipole transitions to some of these 22 levels of
$3p^53d^5$ which will autoionize. The consequence is that for internally "cold" or truly groundstate Fe^{+4} ions there is no excitation-
autoionization near threshold (within the dipole-only approximation),

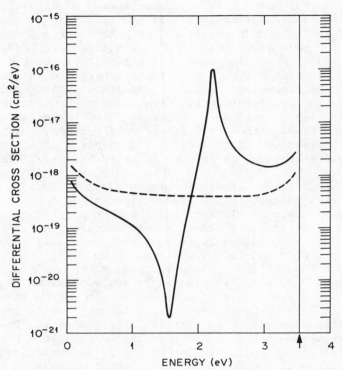

Fig. 12. The calculated cross section for electron impact ioniza-
          tion of Ga⁺ at 22.0 eV, differential in the ejected
          electron energy — from Ref. 69. The incident energy is
          3.5 eV above ionization threshold and 1.2 eV above the
          $^1P_1$ excitation threshold. Dashed curve is energy distri-
          bution for direct ionization only and solid curve allows
          interference between direct ionization and excitation-
          autoionization via the $^1P_1$ state.

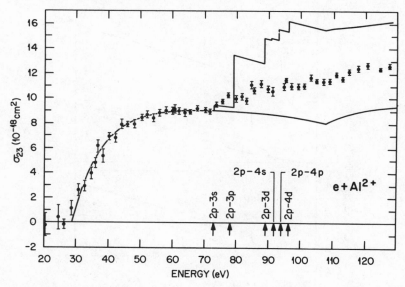

Fig. 13. Electron impact ionization of Al^{+2} showing distinct
contribution due to excitation-autoionization — from
Ref. 72. Solid curve is distorted-wave ionization theory
(normalized to the experiment at 70 eV) with distorted-
wave excitation of the form $2p^63s \rightarrow 2p^53sn\ell$ (Ref. 73)
added to the ionization theory.

while for internally "hotter" Fe^{+4} ions for which the metastable
terms are populated there should be significant excitation-
autoionization. Figure 14 illustrates the calculated ionization
cross sections for this case with initial internal ion temperatures
of 0.1 eV (ground state only), 1.0 eV (some metastable population),
and 5.0 eV (near full statistical population of ground-configuration
terms). Probably in most plasma or astrophysical environments the
5.0 eV temperature is most appropriate, but not all. Significant
differences in ionization rates can thus arise through the details
of atomic structure and excitation processes.

5. CONCLUSIONS AND DIRECTIONS

     Most of the experimental data on electron impact excitation of
bound states of ions have been for resonance lines of ions with one
electron outside a closed shell. Nevertheless, these experiments
are mostly for complex or many-electron cases, while initial detailed
theoretical work was for simple, few-electron systems, so that comparisons producing physical insight have been slow to develop.
Theory has attained a sufficient level of maturity and can undertake

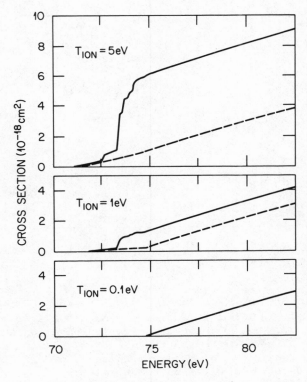

Fig. 14. Electron impact ionization of Fe^{+4} for different initial
ion temperatures corresponding to increasing population
of metastable terms of the ground configuration — from
Ref. 74. The excitation-autoionization from the initial
metastable states accounts for the dramatic increase in
ionization cross section with temperature. Dashed curves
show direct ionization only.

more complex systems, while recent experiments, such as those on
Li- and He-like ions, have provided more direct experimental data
for detailed comparison with theory.

However, excitation experiments are to some extent stalled at
present against technical difficulties of low signal and difficult
calibration. Experimental progress depends on development of new
approaches including new means of observing bound states as well as
observations like excitation-autoionization resonances. The possibility of interference between direct ionization and excitation-
autoionization complicates the comparisons and at the same time
presents a new physical problem. Theoretical techniques to study
interferences in the ionization channels are in development.[75]

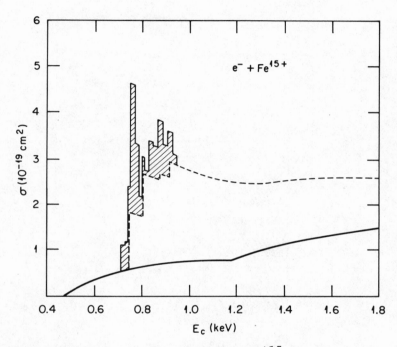

Fig. 15. Electron impact ionization of Fe^{+15} from Ref. 35. Solid
curve is Lotz formula for direct ionization with dashed
curve added to represent excitation-autoionization contribution and then the hatched area added to represent
dielectronic-recombination-double-autoionization contribution.

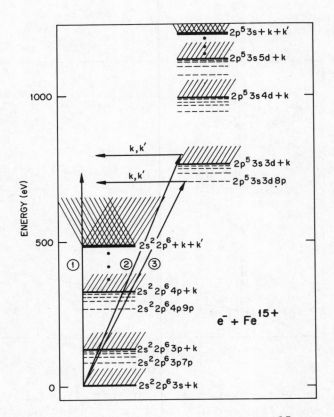

Fig. 16. Some of the energy levels of $(e^- + Fe^{+15})$ with the
initially free electron designated by k. For an ionization collision two free electrons, k and k´, result.
Transitions 1, 2, and 3 represent direct ionization,
excitation-autoionization, and dielectronic-recombination-
double-autoionization respectively.

The interactive nature of the processes of excitation, recombination, and ionization have become more obvious in theory, experiment, and applications of the basic data. These interactions are most apparent for multi-charged ions where the transition probabilities of these different processes become comparable. The current theoretical study of LaGattuta and Hahn[35] illustrates this issue. Figure 15 shows their present prediction of the electron impact ionization of Fe^{+15} which assumes simple addition of direct ionization, excitation-autoionization, and dielectronic-recombination-double-autoionization. The last process had not previously been proposed but arises through the resonances below each inner-shell excitation which are doubly excited states of the recombined ion Fe^{+14}. Figure 16 illustrates some of the atomic levels involved in the ionization of Fe^{+15}. This figure is qualitatively similar to Fig. 1 except that Fig. 16 attempts to illustrate the continua of both the incident and the ionized electrons. Excitation of Fe^{+15} by the incident electron is represented by solid lines with a single attached continua. Excitation of an inner (2p) electron results in excited states of Fe^{+15} which can decay by autoionization giving a net ionization event illustrated by transition 2 in the figure. The dashed levels are the recombined levels of Fe^{+14}, which in the present case are true resonance-dielectronic recombination levels for scattering since the incident electron (k) must stick in the orbital indicated as it excites an electron of the initial ion. A dominant mode of relaxation of these highly excited resonances is predicted to be double autoionization (transition 3) so that dielectronic recombination now makes an appearance in the ionization cross section.

It seems clear that a significant amount of interesting physics is not yet well understood in collisions of electrons and ions. However, it is a difficult challenge to experimental techniques to provide any data, especially for excitation, which can test understanding.

ACKNOWLEDGMENTS

Considerable recognition is due to P. O. Taylor, G. H. Dunn, R. A. Phaneuf, and D. C. Gregory without whom the author would have accomplished nothing in this field. D. C. Griffin and M. S. Pindzola were particularly helpful in discussions of theory during preparation of this manuscript.

At the completion of this manuscript it was discovered that a similar but broader review by G. H. Dunn[76] had been prepared for the conference on "Physics of Ionized Gases, 1980" held in Yugoslavia. Publication of those proceedings is forthcoming, and Dunn's paper should be interesting to anyone following the field.

REFERENCES

1. R.J.W. Henry, "Excitation of Positive Ions by Electron Impact,"
   Phys. Rep. 68, 1 (1981).

2. W. D. Robb in *Atomic and Molecular Processes in Controlled
   Thermonuclear Fusion*, M.R.C. McDowell and A. M. Ferendeci,
   eds., NATO Advanced Summer Institutes Series B, Vol. 53 (Plenum
   Press, New York and London, 1980), pp. 245-266.

3. M. J. Seaton, Adv. At. Mol. Phys. 11, 83 (1975).

4. O. Bely and H. Van Regemorter, Ann. Rev. Astron. Astrophys. 8,
   329 (1970).

5. A. Burgess, D. G. Hummer, and J. A. Tully, Philos. Trans. Soc.
   London 266A, 255 (1970).

6. A. L. Merts, J. B. Mann, W. D. Robb, and N. H. Magee, Jr.,
   *Electron Excitation Collision Strengths for Positive Atomic
   Ions: A Collection of Theoretical Data*, Report LA8267-MS
   (1980).

7. M. S. Pindzola and D. H. Crandall, *A Compilation of Theoretical
   Cross Sections for Electron Impact Excitation of Fe Atomic
   Ions*, ORNL/TM-7957 (1981).

8. A. K. Pradhan, D. W. Norcross, and D. G. Hummer, Phys. Rev. A
   23, 619 (1981).

9. L. I. Schiff, Chapter 5 in *Quantum Mechanics*, third edition
   (McGraw-Hill Book Co., 1968), p. 100ff.

10. P. G. Burke and W. D. Robb, Adv. At. Mol. Phys. 11, 143 (1975).

11. M. J. Seaton in *Atomic and Molecular Processes,* D. R. Bates, ed.
    (Academic Press, New York, 1962), p. 374.

12. H. Van Regemorter, Astrophys. J. 136, 906 (1962).

13. A. L. Merts, R. D. Cowan, and N. H. Magee, Jr., *The Calculated
    Power Output from a Thin Iron-Seeded Plasma*, Report LA6220-MS
    (1976).

14. P. O. Taylor, R. A. Phaneuf, D. H. Crandall, and G. H. Dunn,
    Abstracts of Contributed Papers of IX ICPEAC, J. S. Risley and
    R. Geballe, eds. (University of Washington Press, Seattle,
    1975), p. 391.

15. P. O. Taylor, R. A. Phaneuf, and G. H. Dunn, Phys. Rev. A 22,
    435 (1980).

16. D. Gregory, G. H. Dunn, R. A. Phaneuf, and D. H. Crandall, Phys. Rev. A 20, 410 (1979).

17. O. Bely, Proc. Phys. Soc. 88, 587 (1966).

18. M. Blaha, Astrophys. J. 157, 473 (1969). See also J. Davis, P. C. Kepple, and M. Blaha, J. Quant. Spectrosc. Radiat. Transfer 15, 1145 (1973).

19. S. M. Younger and W. Wiese, J. Quant. Spectrosc. Radiat. Transfer 22, 161 (1979).

20. H.-J. Kunze, Space Sci. Rev. 13 565 (1972).

21. A. H. Gabriel and Carole Jordan in *Case Studies in Atomic Collision Physics II*, E. W. McDaniel and M.R.C. McDowell, eds. (North-Holland Publishing, Amsterdam, 1972), p. 209.

22. H. P. Summers and R.W.P. McWhirter, J. Phys. B 12, 2387 (1979).

23. G. H. Dunn, IEEE Trans. Nucl. Sci. NS 23, No. 2, 929 (1976).

24. D. H. Crandall, Phys. Scr. 23, 153 (1981).

25. M.F.A. Harrison in *Methods of Experimental Physics, Vol. 7a, Atomic and Electron Physics - Atomic Interactions*, B. Bederson and W. L. Fite, eds. (Academic Press, New York and London, 1968). p. 95.

26. K. T. Dolder and B. Peart, Rep. Prog. Phys. 39, 693 (1976).

27. P. O. Taylor and G. H. Dunn, Phys. Rev. A 8, 2304 (1973); also Thesis of P. O. Taylor, Univ. of Colorado, 1972 (unpublished but available through University Microfilms, Ann Arbor, Michigan, and London, England, #73-18,602).

28. D. H. Crandall, P. O. Taylor, and G. H. Dunn, Phys. Rev. A 10, 141 (1974).

29. D. H. Crandall, W. E. Kaupilla, R. A. Phaneuf, P. O. Taylor, and G. H. Dunn, Phys. Rev. A 9, 2545 (1974).

30. D. H. Crandall, R. A. Phaneuf, and P. O. Taylor, Phys. Rev. A 18, 1911 (1978).

31. D. F. Dance, M.F.A. Harrison, and A.C.H. Smith, Proc. R. Soc. London A290, 74 (1966).

32. I. P. Zapesochnyi, V. A. Kel'man, A. I. Imre, A. I. Dashchenko, and F. F. Danch, Sov. Phys. JETP 46, 989 (1976) [original Russian Zh. Eksp. Teor. Fiz. 69, 1948 (1975)].

33. J. L. Kohl and G. Lafayatis, Center for Astrophysics at Harvard Univ., Cambridge, Mass. (private communication, 1981).

34. D. Auerbach, R. Cacak, R. Caudano, T. D. Gaily, C. J. Keyser, J. Wm. McGowan, J.B.A. Mitchell, and S.F.J. Wilk, J. Phys. B 10, 3797 (1977).

35. K. J. LaGattuta and Y. Hahn, Phys. Rev. A 24, 2273 (1981).

36. K. T. Dolder and B. Peart, J. Phys. B 6, 2415 (1973).

37. P. G. Burke and A. J. Taylor, J. Phys. B 2, 44 (1969).

38. M. A. Hayes and M. J. Seaton, J. Phys. B 10, L573 (1977).

39. B. H. Bransden and C. J. Noble, J. Phys. B 9, 1507 (1976).

40. R.J.W. Henry and J. J. Matese, Phys. Rev. A 14, 1368 (1976).

41. P. G. Burke, D. McVicar, and K. Smith, Proc. Phys. Soc. London 83, 397 (1964).

42. A. I. Dashchenko, I. P. Zapesochnyi, A. I. Imre, V. S. Bukstich, F. F. Danch, and V. A. Kel'man, Sov. Phys. JETP 40, 249 (1975) [original Russian Zh. Eksp. Teor. Fiz. 67, 503 (1974)].

43. W. T. Rogers, J. Olsen, and G. H. Dunn, Phys. Rev. A 18, 1353 (1978).

44. W. L. van Wyngaarden, K. Bahdra, and R.J.W. Henry, Phys. Rev. A 20, 1409 (1979).

45. P. O. Taylor, D. Gregory, G. H. Dunn, R. A. Phaneuf, and D. H. Crandall, Phys. Rev. Lett. 39, 1256 (1977).

46. M. A. Hayes, D. W. Norcross, J. B. Mann, and W. D. Robb, J. Phys. B 10, L429 (1977).

47. R.J.W. Henry and W. L. van Wyngaarden, Phys. Rev. A 17, 798 (1978).

48. The two-state close-coupling calculation is by W. D. Robb and the Coulomb-Born with exchange is by J. B. Mann. Both are reported in *Electron Impact Excitation of Carbon and Oxygen Ions* by N. H. Magee, Jr., J. B. Mann, A. L. Merts, and W. D. Robb, Report LA6691-MS (1977).

49. D. H. Crandall, R. A. Phaneuf, B. E. Hasselquist, and D. C. Gregory, J. Phys. B 12, L249 (1979).

50. R. A. Falk and G. H. Dunn, "Electron Impact Ionization of Be$^+$," submitted to Physical Review A.

51. R.J.W. Henry, J. Phys. B $\underline{12}$, L309 (1979).

52. D. H. Sampson and L. B. Golden, J. Phys. B $\underline{12}$, L785 (1979).

53. D. L. Moores, J. Phys. B $\underline{11}$, L403 (1978).

54. L. B. Golden and D. H. Sampson, J. Phys. B $\underline{10}$, 2229 (1977).

55. I. P. Zapesochnyi, A. I. Imre, A. I. Dashchenko, V. S. Vukstich, F. F. Danch, and V. A. Kel'man, Sov. Phys. JETP $\underline{36}$, 1056 (1973) [original Russian Zh. Eksp. Theor. Fiz. $\underline{63}$, 2000 (1972)].

56. D. H. Crandall, R. A. Phaneuf, and G. H. Dunn, Phys. Rev. A $\underline{11}$, 1223 (1975).

57. R. A. Phaneuf, P. O. Taylor, and G. H. Dunn, Phys. Rev. A $\underline{14}$, 2021 (1976).

58. W. T. Rogers, G. H. Dunn, J. Olsen, M. Reading, and G. Stefani, "Absolute Emission Cross Section for Electron Impact Excitation of Zn$^+$(4p^{2}P) and (5s^{2}S) Terms," accepted for Physical Review A.

59. F. M. Bacon and J. W. Hooper, Phys. Rev. $\underline{178}$, 182 (1969).

60. M. O. Pace and J. W. Hooper, Phys. Rev. A $\underline{7}$, 2033 (1973).

61. E. Hinnov, T. K. Chu, H. Hendel, and L. C. Johnson, Phys. Rev. $\underline{185}$, 207 (1969).

62. A. Burgess and V. B. Shorey, J. Phys. B $\underline{7}$, 2403 (1974).

63. B. Peart and K. Dolder, J. Phys. B $\underline{9}$, 56 (1975).

64. O. Bely, S. B. Schwartz, and J. L. Val, J. Phys. B $\underline{4}$, 1482 (1971).

65. J. E. Hansen, J. Phys. B $\underline{7}$, 1902 (1974) and J. Phys. B $\underline{8}$, 2759 (1975).

66. D. C. Griffin, M. S. Pindzola, and C. Bottcher (private communication, 1981).

67. R. A. Falk, G. H. Dunn, D. C. Griffin, C. Bottcher, D. C. Gregory, D. H. Crandall, and M. S. Pindzola, Phys. Rev. Lett. $\underline{47}$, 494 (1981).

68. W. T. Rogers, G. Stefani, R. Camilloni, G. H. Dunn, A. Z.
     Msezane, and R.J.W. Henry, "Electron-Impact Ionization of Zn^+
     and Ga^+," accepted for Physical Review A.

69. M. S. Pindzola, D. C. Griffin, and C. Bottcher, "Electron
     Impact Excitation-Autoionization of Ga II," accepted for
     Physical Review A, 1981.

70. S. O. Martin, B. Peart, and K. T. Dolder, J. Phys. B $\underline{1}$, 537
     (1968).

71. D. L. Moores and H. Nussbaumer, J. Phys. B $\underline{3}$, 161 (1970).

72. D. H. Crandall, R. A. Phaneuf, R. A. Falk, D. S. Belić, and
     G. H. Dunn, "Electron Impact Ionization of Na-like Ions - Mg^+,
     Al^{+2}, Si^{+3}," accepted for Physical Review A, 1981.

73. D. C. Griffin, C. Bottcher, and M. S. Pindzola, "The Contribu-
     tion of Excitation-Autoionization to the Electron Impact
     Ionization of Mg^+, Al^{+2}, Si^{+3}," accepted for Physical Review A,
     1981.

74. D. C. Griffin, D. Bottcher, and M. S. Pindzola, "Theoretical
     Calculations of the Contributions of Excitation-Autoionization
     to Electron Impact Ionizations in Ions of the Transition
     Series of Elements," accepted for Physical Review A, 1981.

75. H. Jakubowicz and D. L. Moores, "Electron Impact Ionization of
     Li-like and Be-like Ions," J. Phys. B $\underline{14}$, 3733 (1981); also
     Thesis of H. Jakubowicz, unpublished, University College,
     London (1980).

76. G. H. Dunn, "Electron-Ion Collisions," in *The Physics of
     Ionized Cases SPIG, 1980*, M. Matić, ed. (Boris Kidrić
     Institute of Nuclear Sciences, Beograd, 1981), pp. 49-95.

ELECTRON-IMPACT IONIZATION OF IONS

Erhard Salzborn

Institut für Kernphysik
Justus-Liebig-Universität Giessen
D-6300 Giessen, W.-Germany

1. INTRODUCTION

Electron-impact ionization of atoms and ions is one of the fundamental collision processes in all types of plasmas – in astrophysical objects as well as in laboratory discharges. Many of the plasma properties, for example the radiation emitted, depend on the state of ionization of ions in these plasmas. In order to interpret spectroscopic observations and to find theoretical models for the formation and evolution of the plasma, cross sections for electron-impact processes are needed.

The impetus for accurate data has been strongly intensified in recent years by fusion research aiming towards controlled thermonuclear plasmas as energy sources. In fusion plasmas impurity atoms released from wall surfaces are rapidly ionized by electron impact to very high charge states. The radiation emitted by these ions causes an enormous loss of plasma energy, and impurity concentrations beyond some tolerable levels can even prevent achieving energy break-even. Plasma modelling codes now demand the cross sections to be determined to accuracies better than a factor of two. For more detailed studies of plasma structures tolerances of ±25% are being proposed.

Whereas the electron-impact ionization of atoms has been investigated both experimentally (Lenard, 1894, 1904; Bloch, 1912) and theoretically (Thomson, 1912) since the turn of the last century, the first high-quality crossed-beams experiment measuring electron impact ionization of an ion (He^+) was performed only as recent-

ly as 1961 by Dolder et al.. The experimental study of electron-
ion ionization is difficult, especially for multiply charged ions.
Three major experimental techniques have been developed to study
these processes: the plasma spectroscopy method, arrangements employing trapped ions, and crossed-beams experiments. As a result,
there is now a fairly good data basis for the ionization of singly
charged ions. Accurate data for more than doubly charged ions have
been available since 1978, and they are still limited to ions in
relatively low charge states.

The theoretical treatment of electron-impact ionization turned
out to be fraught with difficulties, too. Additional to the many-
body nature of the problem there are a variety of possible ionization mechanisms which have to be considered. The most important are

$$A^{n+} + e \rightarrow A^{(n+1)+} + 2e \tag{1}$$

$$A^{n+} + e \rightarrow A^{(n+m)+} + (m+1)e \tag{2}$$

$$A^{n+} + e \rightarrow A*^{n+} + e$$
$$\quad \quad \hookrightarrow A^{(n+1)+} + e \tag{3}$$

$$A^{n+} + e \rightarrow A*^{(n+1)+} + 2e$$
$$\quad \quad \hookrightarrow A^{(n+m)+} + (m-1)e \tag{4}$$

Equations (1) and (2) represent direct knock-on ionization
in which the energy transferred from the incident electron to respectively one or several outer shell electrons exceeds the corresponding binding energies. Eq. (3) denotes the mechanism of excitation-autoionization in which the excitation of an inner-shell electron to a discrete state above the next ionization continuum is
followed by autoionization. Finally, eq. (4) refers to another two-
step process, inner-shell ionization-Auger emission, in which first
an inner-shell electron is removed followed by Auger emission of
additional electrons.

Commonly, channel (1) is assumed to be the most probable process in an electron-ion collision. However, experimental results
in recent years have shown that there may be significant, in some
cases even dominant, contributions from channel (3) processes. The
situation is similar for multiple ionization described by Eq. (2)
and contributions thereto originating from processes of type (4).

It is the purpose of this paper to give a survey of the characteristic features of electron-ion ionization processes with an
emphasis on recent experimental progress. The data, where possible,
are compared with theoretical predictions.

2. THEORETICAL APPROACHES

Because of its fundamental importance numerous efforts have been devoted towards a theoretical understanding of electron-impact ionization. However, a full solution of the longe-range three-body problem with two continuum electrons in the final state has not been possible and the results of various developed approximations based on classical, semiclassical, or quantum theory, are not yet as accurate as could be wished.

As early as 1912 Thomson had used classical theory to describe electron-impact ionization of neutral atoms. Considering the energy transfer via Coulomb interaction from the incident electron to one of the atomic electrons at rest he deduced that the total ionization cross section σ at collision energy E is given by

$$\sigma(E) = 4\pi a_o^2 \cdot E_H \cdot \frac{\xi}{E_I \cdot E} \left(1 - \frac{E_I}{E}\right) \tag{5}$$

with $a_o = 0.529 \; 10^{-10}$ m (Bohr radius) and $E_H = 13.6$ eV. ξ denotes the number of equivalent electrons of a shell with binding energy E_I.

Thomson's theory implies a scaling of cross sections of an atom, $\sigma_1(x)$, and an isoelectronic ion, $\sigma_2(x)$, according to

$$\frac{\sigma_1(x)}{\sigma_2(x)} = \left(\frac{E_2}{E_1}\right)^2 \tag{6}$$

where $x = E/E_I$ are reduced energies, i.e., the incident electron energies are expressed in terms of the respective ionization energies E_1 and E_2. The attractive Coulomb field of the ion however increases the kinetic energy of the incident electron and leads to a reduction in its impact parameter. Thus the cross section of the ion is enhanced for electron energies in the threshold region, whereas fast electrons are scarcely influenced.

The most serious inadequacy of the Thomson theory however is the incorrect high-energy behaviour of Eq. (5). Bethe's (1930) quantum approximation showed that at high energies the cross section behaves like

$$\sigma \underset{x \to \infty}{\sim} A \cdot \frac{\ln x}{x} + \frac{B}{x} \tag{7}$$

where $x = E/E_I$ is the reduced energy.

    Nevertheless, because of its simplicity, Thomson's approach
has formed the basis for many attempts aiming at improved classical
binary-encounter calculations: The theory has been modified (Thomson
and Garcia, 1969) for ionization of ions to incorporate the Coulomb
effect on the incident electron. The restriction that the target
electron is at rest prior to the collision was removed and various
distributions of initial velocities were taken into account
(Gryzinski, 1959, 1965a,b; Kingston, 1964, 1968; McDowell, 1966;
Abrines and Percival, 1966; Abrines et al., 1966; Friens and Bonson,
1968; Bell et al., 1970). And, a further modification of the model,
known as the 'exchange-classical' approximation, included certain
quantum features like electron exchange and interference (Burgess,
1963, 1964; Kumar and Roy, 1978).

    Another classical approach is based on the impact parameter
treatment of the ionization collision process (Alder et al., 1956;
Seaton, 1962; Burgess, 1964). This method reproduces the correct
high energy behaviour of the cross section given by Eq. (7) since
the binary nature of the collision is removed and many-body effects
are accounted for.

    Finally, in the 'exchange-classical impact-parameter' (ECIP)
method (Burgess, 1963, 1964) both previous approaches have been
combined to incorporate the complementary advantages of the two
theories. The 'exchange-classical' approximation is used to describe the close collisions ($r<R_0$) and the impact-parameter approximation is used to treat the distant collisions ($r>R_0$). Thus the
correct behaviour of the ionization cross section is obtained at
both low and high energies. The main difficulty of the method is
that it involves choosing the correct cutoff impact parameter R_0.

    All of the above classical and semiclassical approximations
only treat the direct knock-on ionization process. The effects of
inner-shell excitation or ionization followed by autoionizing transitions have to be accounted for by independent calculations. In an
ad hoc manner the appropriate cross sections are simply added to
yield the total ionization cross section (Salop, 1976).

    The first quantum calculation of ionization was performed by
Bethe (1930) using the Born approximation. Since that time numerous
quantum approaches have been carried out, most of them concerned
with the ionization of neutral atoms (Rudge, 1968; Peterkop, 1977).
The methods employed can be readily extended to the case of ionization of positive ions, but the theoretical procedures involved
in all these calculations are very complex and lengthy. The difficulties mainly arise from the long range Coulomb potential requiring a full solution of the three-body problem in the asymptotic
region. Various approximations have been developed to treat this
problem. Exchange and interference between the two continuum electrons have to be taken into account. For ionization of complex ions

additional problems arise from the need to find accurate wavefunctions for the target states before and after the collision.

The most commonly used and generally accepted quantum approach
to calculate electron-ion ionization is the Coulomb-Born approximation in which both the ionizing and ejected electron are described
by Coulomb wavefunctions. Several versions of the Coulomb-Born
approach have been developed differing mainly in their approximations to treat (or not) exchange effects. Coulomb-Born type calculations are available now for a number of ion species, preferentially of hydrogenic structure (Burke and Taylor, 1965; Rudge and
Schwartz, 1966; Moores and Nussbaumer, 1970; Moores, 1972, 1978;
Stingl, 1972; Golden and Sampson, 1977, 1980; Sampson and Golden,
1978; Golden et al., 1978; Moores et al., 1980; Jakubowicz and
Moores, 1980; Younger, 1980, 1981a,b). A scaling prescription for
ionization cross sections of hydrogenic ions with configurations
from 1s up through 4f was given, too (Golden and Sampson, 1977,
1980; Sampson and Golden, 1978; Golden et al., 1978; Moores et al.,
1980).

The Coulomb-Born approximation has been modified to account
for screening effects close to the target ion by using distorted
waves instead of Coulomb waves to describe the ionizing electron
(Younger, 1980, 1981a,b). On the other hand, if the charge of the
target ion is neglected the plane wave Born approximation applies.
This gives, involving only relatively easy computational efforts,
a good representation of the high energy behaviour of the cross
section. At low energies however the cross section is underestimated, particularly for highly charged ions.

The necessity to take indirect processes like excitation-autoionization into account further complicates the theoretical treatment. Considerable complexity results from the presence of configuration interaction and intermediate coupling in the autoionizing
state. For highly charged ions it is important to account for non-
unit branching ratios for autoionization versus radiative decay.
Usually direct and indirect ionization mechanisms are assumed to
take place independently. In this case the cross section for excitation (multiplied by the branching ratio for autoionization) is
merely added to the cross section for direct ionization to yield
the total cross section.

Only recently, a method of calculating ionization cross sections was developed (Jakubowicz and Moores, 1980; Jakubowicz, 1980;
Moores, 1981) in which close-coupling wavefunctions are used to
describe the target states within the Coulomb-Born approximation.
By including suitable target configurations in the close-coupling
expansion, it is possible to allow for effects such as simultaneous
excitation and ionization, inner-shell ionization and ionization
via an autoionizing level directly within the calculation of the io-

nization amplitude. Thus, for the first time, the importance of interference between direct and indirect processes can be addressed.

In view of the very complex theoretical approaches and motiva-
ted by the need to provide reasonably accurate estimates of unknown
cross sections in a very simple fashion, several semi-empirical formulas have been proposed (Post, 1961; Drawin, 1961, 1963; Seaton,
1964; Percival, 1966; Lotz, 1967, 1968, 1970). The most successful
one, which has been used extensively in the calculation of plasma
properties, was given by Lotz (1967):

$$\sigma(E) = \sum_{\nu=1}^{N} a_\nu \xi_\nu \frac{\ln(E/E_\nu)}{E \cdot E_\nu} \left\{ 1 - b_\nu \cdot \exp\left(- c_\nu (E/E_\nu - 1) \right) \right\} \tag{8}$$

where E_ν is the binding energy of an electron in the νth of N subshells ($\nu=1$ means the outermost shell) containing ξ_ν equivalent
electrons. The coefficients a_ν, b_ν and c_ν are tabulated individual
constants, which have been fitted to best reproduce experimental
data. The error in the formula is estimated to less than $\pm 40\%$. For
ions in charge states $n \geqq 4$, the values $a_\nu = 4.5 \cdot 10^{-14}$ cm^2 ^{30} (eV)^2
and $b_\nu = c_\nu = 0$ give results which agree within 20% with theoreti-
cal calculations of Rudge and Schwartz (1966) for hydrogen-like,
sodium-like, and magnesium-like ions with high Z-numbers in the
Coulomb-Born-exchange approximation.

All empirical formulas only include direct ionization processes. In view of recent experimental results showing that indirect processes may give significant contributions to the total ionization cross section any attempt to represent all ionization cross
sections by a universal analytic function must fail.

3. EXPERIMENTAL TECHNIQUES

In order to determine cross sections for electron-ion ionization it is necessary to measure the reaction rate in an experimental arrangement where the densities of the electrons and ions and
the relative impact velocities are known. Three major techniques
are used.

3.1 Plasma spectroscopy method

This method is based on intensity measurements of the time-
resolved spectral emission of a well-diagnosed transient plasma
(Gabriel and Jordan, 1972; Kunze, 1972). Ionization rate coefficients
$\langle \sigma \cdot v \rangle$, i.e., the product of cross sections and initial electron
velocity averaged over the electron-velocity distribution function,
can be deduced from a set of coupled rate equations if the plasma

density and temperature as well as the time evolution of the plasma
volume in the direction of observation are known. Using a computer
program the ionization rate coefficients are varied until the calculated time histories of properly chosen lines emitted by the desired ions agree with the observed ones. The obtained effective
rate coefficients include ionization from excited states as well as
from inner shells; however, under suitable conditions these contributions are negligible.

Although no absolute line intensities are required the results
of this indirect and model dependent method are generally considered
less accurate than ionization rate coefficients obtained by integrating cross sections from crossed-beams experiments. Whereas for hydrogen-like ions very good agreement is found between plasma-derived
rates and those derived from crossed-beams measurements, for helium-
like ions the plasma-derived rates are uniformly below the crossed-
beams data by about 15%, and for lithium-like ions the discrepancy
increases to approximately 40%. On the other hand, by using the plasma spectroscopy technique ionization rates for highly charged ions
(e.g., Fe^{9+}, Datla et al., 1975) could be measured which up to the
present time are inaccessible to crossed-beams experiments.

Since detailed information on the energy dependence of ionization cross sections cannot be inferred from plasma spectroscopy
measurements, this method will not be further discussed here.

3.2 Trapped-ion method

This technique is based on the ionization of ions held in a
trap and bombarded by electrons. Cross sections for successive ionization to higher charge states can be deduced from the charge state
analysis of extracted ions by means of a mass spectrometer or a
time of flight method.

Several experimental approaches (see e.g., Dunn, 1976) have
been developed for the trapping of ions. Most fruitful was the concept which makes use of the space charge of an electron beam constrained by an axial magnetic field (Baker and Hasted, 1966). Since
the trapping time can be increased considerably by applying proper
potential wells to the ends of the beam (Redhead, 1967; Redhead
and Feser, 1968), the electron energy dependence of currents of
multiply charged Cs^{n+} and Ba^{n+} ions (up to n = 10) could be measured
(Redhead and Gopalaraman, 1971).

In recent years Donets and Ovsyannikov (1977) have further
developed this concept to a high degree of perfection. By variation
of the trapping time the charge state distribution of ions extracted from the electron beam ion source (EBIS) can be controlled. In
Fig. 1 the time evolution of the charge state distribution of argon
ions is shown.

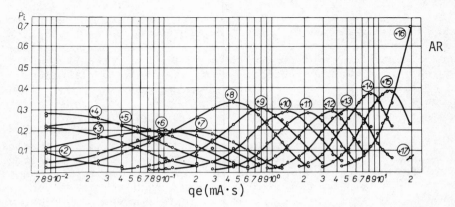

Fig. 1. Normalized charge state distribution of argon ions extracted from the EBIS source as a function of increasing trapping time (which is multiplied on the x-axis by the electron beam current)(Donets and Ovsyannikov, 1977).

        Measured charge state distributions can be reproduced by cal-
culations in terms of a simple balance model in which the ionization cross sections are treated as fit parameters. In general only
cross sections for single ionization of ground state ions (Eq. (1))
are taken into account. Other processes, e.g., multiple ionization,
charge exchange, recombination and ion losses from the trap, are
neglected. In spite of the simplicity of the model, the obtained
ionization cross sections are in good agreement with crossed-beams
results where comparison is possible. By this method cross sections
have only been determined for electron impact energies above approximately 2 keV. Nevertheless, this is the only experiment which has
provided cross section functions for ions in charge states n > 5.
Fig. 2 shows cross sections for Ar^{n+} ($n \geq 8$) ions deduced from EBIS
measurements by Donets and Ovsyannikov (1977) as a function of electron impact energy. The data are compared with theoretical calculations of Salop (1976) based on the classical binary-encounter
approximation.

3.3 Crossed-beams method

        The technique of intersecting electron-ion beams has provided
the most accurate and detailed information on impact ionization

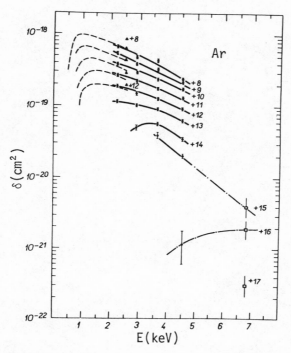

Fig. 2.   Cross sections for Ar^{n+} $(n \geq 8)$ ions deduced from EBIS
measurements by Donets and Ovsyannikov (1977). Dashed
curves are calculations of Salop (1976) based on classical
binary-encounter theory.

mechanisms described by Eqs. (1) to (4). The general features and/or
results of crossed-beams experiments have been discussed and reviewed by several authors (Harrison, 1966, 1968; Dolder, 1969,
1977; Dunn, 1969, 1979, 1980; Dolder and Peart, 1976; McGowan, 1979;
Crandall, 1981a,b).

     The principle of this method is shown schematically in Fig. 3.
A collimated beam of ions of given species, charge state and energy
is bombarded by a monoenergetic electron beam at an angle θ. Parent
and product ions are separated downstream by the field of a second
magnetic (or electric) analyzer and detected by collectors.

     This technique seems to work straightforwardly. However, there
are many experimental difficulties and sources of error (Harrison,

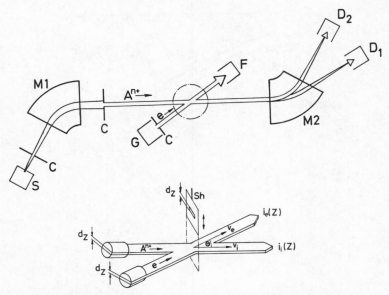

Fig. 3.   Schematic arrangement of a crossed-beams experiment: S,
          ion source; C, collimator; M_1, beam selector magnet; G,
          electron gun; F, electron collector; M_2, beam analyzer
          magnet; D_1, parent ion detector; D_2, product ion detector.
          In the lower part of the figure the collision region is
          shown in more detail; Sh, slotted shutter used for form
          factor determination.

1968; Dolder and Peart, 1976). Very recently experimental progress
has been achieved towards the solution of two major problems involved in these experiments. The first one is related to the measurement of beam density profiles in the interaction region, which must
be known in order to deduce absolute cross sections from the observed reaction rate. The second problem arises from the small signal to background ratio. It could be considerably enhanced by the
use of a high intensity electron gun. Both advances shall be discussed in more detail.

In a crossed-beams configuration the ionization cross section
σ can be determined from the relation (Harrison, 1966)

$$\sigma = \frac{R\,ne^2}{I_e\,I_i} \cdot \frac{v_e v_i \sin \Theta}{(v_e^2 + v_i^2 - 2\,v_e v_i \cos \Theta)^{1/2}} \cdot F \qquad (9)$$

where R denotes the reaction rate; I_e, I_i, v_e, v_i, are the total
currents and velocities, respectively, of electrons and n-fold
charged ions, and Θ is the angle of intersection. The form factor F
takes account of the spatial overlap of the non-uniform current densities within the beams, and is given by

$$F = \frac{\int i_e(z)\,dz \cdot \int i_i(z)\,dz}{\int i_e(z) \cdot i_i(z)\,dz} \qquad (10)$$

where $i_e(z)$ and $i_i(z)$ are the currents flowing in elements of the
beam of height dz. The factor F can be determined by sliding a
slotted shutter along the z-axis through both beams, so that the
beam density distributions can be measured simultaneously. Usually
both beams intersect at $\Theta = 90°$, and since mostly $v_i \ll v_e$, the
cross section σ can be expressed as

$$\sigma = \frac{ne^2 \cdot v_i}{I_e \cdot I_i} \cdot R \cdot F \qquad (11)$$

Defrance et al. (1981) very recently used a new experimental
approach to circumvent the cumbersome procedure of measuring
the form factor F. They swept the electron beam,while keeping it
parallel to its initial axis,back and forth through the ion beam.
The distortion-free 'see-saw' motion occured at a constant speed u
and   the displacements were large enough that, at the extreme positions, the beams no longer overlapped. The   instantaneous reaction
rate R was stored as a function of time in a multichannel analyzer
used in the multiscaling mode. The product $R \cdot F$ in Eq. (11) then
follows from the relation

$$R \cdot F = u \cdot K \qquad (12)$$

where K denotes the total number of reactions during time T which
is given by

$$K = \int_{-T/2}^{+T/2} R(t)\,dt \qquad (13)$$

This technique has been employed to remeasure the cross section for ionization of He^+ in the energy range 55 - 74 eV. The results are displayed in Fig. 4. The comparison with previous measurements (Dolder et al., 1961; Peart et al., 1969a) shows a clear improvement in accuracy although the error bars have different meanings for different authors. It can be expected that the sweep method

is intrinsically more accurate than the standard one which depends
on a direct determination of F.

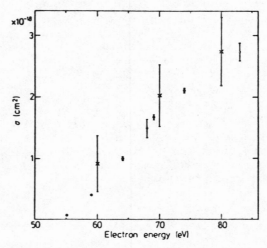

Fig. 4.   Measured cross sections for ionization of He[+]
          ● Defrance et al. (1981), 63%; x Dolder et al. (1961), 95%;
          o Peart et al. (1969a), 90%. The percentages given are the
          confidence limits of the random error. From Defrance et
          al. (1981).

        Crossed-beams experiments usually suffer from small signal to
background ratios. Stripping collisions of parent ions with residual gas atoms give rise to a background reaction rate which, even
at gas pressures below 10^{-8} Torr, may exceed the true signal rate
by orders of magnitude. Usually beam pulsing techniques (Harrison,
1968; Dolder and Peart, 1976) are applied to separate signal from
background events. Another useful approach is to provide a large
electron density and a long interaction path for the ions in order
to maximize the ratio of signal to background. The potential in the
interaction region however should be constant to allow for a well
defined electron impact energy and to avoid defocusing of the ions.

        In Fig. 5 a schematic view of a high current intensity electron
gun is shown which was developed by Becker et al. (1977) to meet
these requirements. The design is based on modified versions of
computer codes (Kirstein and Hornsby, 1963; Hermannsfeldt, 1973)
which solve Poisson's equation for arbitrary electrode systems with
an emitting surface and the related equations of motion.

        The electrons are emitted from a cylindrical cathode of 6 cm
length along the ion beam direction. Pierce angle electrode design

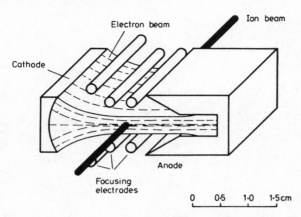

Fig. 5.   Perspective view of a high current intensity electron gun
          (Becker et al., 1977) which was used in crossed-beams ex-
          periments by Müller et al. (1980a,b,c).

at all boundaries of the cathode,together with six rods mounted in
parallel to the cathode, and connected to proper electric potentials,
provide an electron beam of 9.5 μA$\cdot$V$^{-3/2}$ perveance. Thus a total
electron current of 300 mA could be achieved at a voltage of 1000 V
between cathode and anode.

This electron gun was used in crossed-beams experiments with
multiply charged Ar^{n+} (n=1,...,5) ions by Müller et al. (1980a,b).
Without making use of the beam pulsing technique cross sections as
low as 10^{-20} cm^2 could be measured due to a high signal to background ratio.

The design of this electron gun offers an additional advantage.
Since the distributions of electron current densities j_e are almost
constant (at least for low electron energies) within the cross section of the traversing ion beam, the form factor F is simply given
by

$$F = \frac{I_e}{1 \cdot j_e} \qquad (14)$$

where 1 is the length of the ion path across the electron beam. The
electron current density j_e can either be measured by using a move-
able scanner or determined from calculations. Both methods agreed
to better than 5% for all electron energies.

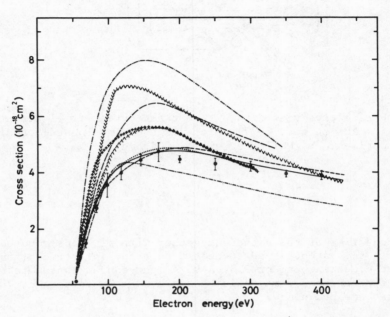

Fig. 6.   Cross sections for ionization of He[+] as a function of elec-
          tron impact energy. The experimental data (Peart et al.,
          1969a) are compared with calculations in the following
          approximations:
          — — —   Born (Massey, 1956)
          —·—·—   Coulomb-Born (Burgess and Rudge, 1963)
          •••••   Coulomb-Born-exchange (Rudge and Schwartz, 1965)
          +++++   Born-Oppenheimer (Burke and Taylor, 1965)
          ⌄⌄⌄⌄⌄   Born-exchange (Rudge and Schwartz, 1966a)
          ————   distorted wave-exchange (Younger, 1980a)
          ^^^^^   close-coupling (Burke and Taylor, 1965)
          —·—··   semiclassical (Kunc, 1980)
          — — —   Lotz formula (Lotz, 1967)

4.   EXPERIMENTAL RESULTS AND DISCUSSION

4.1   Single ionization

4.1.1 Singly charged ions

Accurate measurements of cross sections for single ionization
of singly charged ions by electron impact have been performed in
many experiments, mostly employing the crossed-beams technique.
Fig. 6 shows measured cross sections (Peart et al., 1969a) for the
simplest ion, He^+, as a function of electron impact energy. The data
are compared with the results of calculations in various quantum
approximations, with a semiclassical theory, and with the Lotz formula. As can be seen from Fig. 6, most theories overestimate the
cross section, especially for impact energies below the cross section maximum. The Coulomb-Born-exchange (Rudge and Schwartz, 1965)
and distorted wave-exchange (Younger, 1980a) calculations, as well
as the Lotz (1967) semiempirical formula, give the best approach to
the data; the agreement is almost within the experimental uncertainty.

The ionization cross sections for He^+ ions and neutral hydrogen atoms may be compared in order to test Thomson's classical scaling law (Eq. (6)). Fig. 7 shows both cross sections as a function
of reduced impact energies. The cross section for H^0 (Fite and
Brackmann, 1958) has been multiplied by the scaling factor $(E_H/E_{He^+})^2$.
In the threshold energy range the cross section for He^+ is greater
than for H^0 as a result of the focusing of the incident electron
by the attractive Coulomb field of the ion, but at higher energies
both curves coincide exactly.

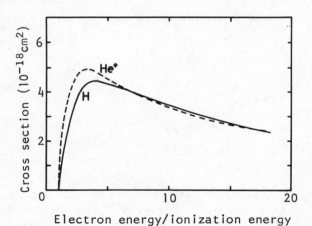

Electron energy/ionization energy

Fig. 7.   Measured cross sections for He^+ (Peart et al., 1969a) compared with classically scaled measurements for H^0 (Fite
and Brackmann, 1958) as a function of reduced impact energy.

In Fig. 8 cross sections for a number of ions with configuration $1s^2 2s^2 2p^q$ are compared with Coulomb-Born, without exchange, calculations of Moores (1972). The theory generally overestimates the cross sections in the peak region, but the agreement is good in all cases at high energies.

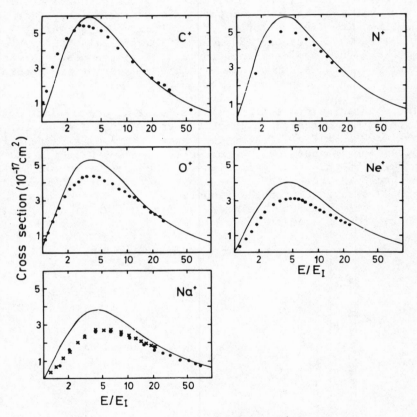

Fig. 8.   Cross sections for ionization of C^+ (Aitken et al., 1971), N^+ (Harrison et al., 1963b), O^+ (Aitken and Harrison, 1971), Ne^+ (Harrison et al., 1963a), and Na^+, circles (Peart and Dolder, 1968), crosses (Hooper et al., 1966) compared with Coulomb-Born (less exchange) calculations of Moores (1972).

A remarkably different energy dependence of the ionization cross section has been observed in many experiments with heavier ions. The mechanism of excitation -autoionization (Eq. (3)) may give large, sometimes dominant, contributions to the ionization cross sections. Excitation to autoionizing states can produce abrupt increases in the cross section function since the excitation cross section is non-zero at its threshold.

Fig. 9 shows measurements for the singly-charged, alkali-like ions Mg^+, Ca^+, Sr^+, and Ba^+ (Peart and Dolder, 1975). Anomalies in the cross sections due to inner-shell excitations are apparent for all but Mg^+ ions. The contributions from excitation-autoionization exceed the direct ionization cross sections by factors ranging from 1.5 for Ca^+ to about 4 for Ba^+. Most of the autoionization comes from dipole-allowed, $\Delta n=0$ transitions $(np) \rightarrow (nd)$ where n=5,4, and 3 for Ba^+, Sr^+, and Ca^+, respectively (Hansen, 1975). The different result for Mg^+ is very plausibly explained by the non-existence of 2d-states. A predicted (Bely, 1968) enhancement by a factor of 2 for Mg^+, mainly due to the $\Delta n=1$ excitation $(2p^63s) \rightarrow (2p^53s3d)$, is not seen in the experimental data.

Contributions from excitation-autoionization are most impor-tant when a many-electron inner-shell underlies a sparsely popula-ted valence shell. Very recently, considerable structure and en-hancement (by factors of up to ~2.5) due to excitation-autoioniza-tion has been found by Rogers et al. (1981) in the ionization cross

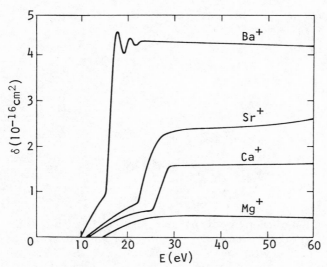

Fig. 9.   Measured cross sections for the ionization of Mg^+, Ca^+. Sr^+ and Ba^+ ions (Peart and Dolder, 1975).

sections of Zn$^+$ and Ga$^+$ ions in the energy region between one and
two times threshold. In Fig. 10 the measured cross section for Ga$^+$
is plotted, and its derivative dσ/dE which shows excitation-autoionization thresholds as peaks. Immediately above the ground state
ionization threshold at 20.51 eV a large number of statistically
significant structures can be identified, which are attributed to
the excitation of autoionizing 3d^{9}4s^2 nl levels.

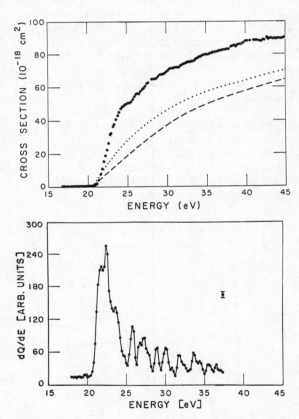

Fig. 10. Measured cross sections for the ionization of Ga$^+$ (Rogers
          et al., 1981) compared with semi-empirical Lotz formula
          (dotted curve) and scaled plane wave Born calculations of
          McGuire (1977) (dashed curve). The lower figure shows the
          derivative with respect to energy of the ionization cross
          section indicating excitation-autoionization contributions
          as peaks.

4.1.2 Multiply charged ions

Crossed-beams measurements of ionization cross sections for multiply charged ions are rather scarce. Although the basic experimental techniques employed in these measurements are the same as for singly charged ions, suitable ion sources for the production of low energy beams of multiply charged ions became only available in the past several years. In 1969 the first accurate data for a doubly charged ion (Mg^{2+}, Peart et al., 1969b) and in 1978 for ions in charge states $n = 3$ and 4 (C^{3+}, N^{4+}, Crandall et al., 1978) have been published. The highest charge state up to now investigated in crossed-beams experiments is $n = 5$ (O^{5+}, Crandall et al., 1979; Ar^{5+}, Müller et al., 1980a).

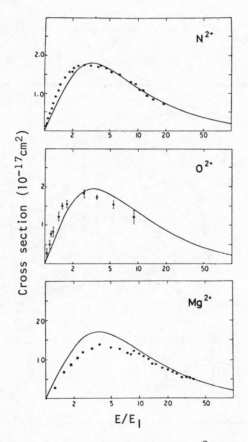

Fig. 11.   Cross sections for ionization of N^{2+} (Aitken et al., 1971), O^{2+} (Aitken and Harrison, 1971), and Mg^{2+} (Peart et al., 1969b) compared with Coulomb–Born (less exchange) calculations of Moores (1972).

Some results of early measurements for N^{2+} (Aitken et al., 1971), O^{2+} (Aitken and Harrison, 1971), and Mg^{2+} (Peart et al., 1969b) are shown in Fig. 11. The data are compared with calculations of Moores (1972) using the Coulomb-Born approximation, without exchange. The theory includes contributions from ejection of both 2s and 2p electrons, but neglects contributions from 2s excitation followed by autoionization. The agreement between calculated and measured cross sections is fairly close.

The results of a systematic study of cross sections for multiply charged Ar^{n+} (n=1,...,5) ions by Müller et al. (1980a) are shown in Fig. 12. The solid lines are cross sections calculated from the simple empirical formula

$$\sigma_{n,n+1}(E) = A \cdot \frac{\ln (E/E_n)}{E \cdot E_n} \tag{15}$$

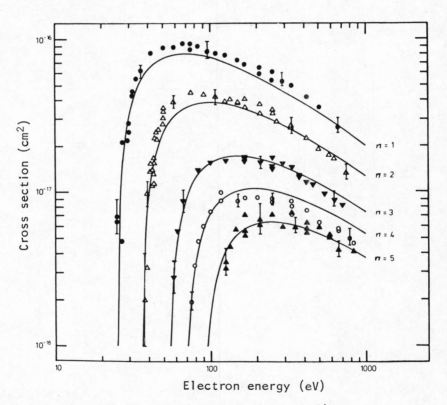

Fig. 12.   Cross sections for ionization of Ar^{n+} (n=1,...,5) ions (Müller et al., 1980a). The solid curves are calculated from formula (15) with $A = 1.4 \cdot 10^{-13}$ cm^2 $(eV)^2$ for all n.

where E_n are the respective ionization energies of argon ions in charge states n and A is one common fit parameter. Taking $A = 1.4 \cdot 10^{-13}$ cm^2 (eV)^2 the deviations of measured and calculated cross sections are within a limit of ±20% for all n. Contributions to the cross sections of Ar^{4+} near 250 eV due to the excitation of a 2p electron followed by autoionization have been predicted by Salop (1976) in classical binary-encounter calculations. The experimental resolution however does not allow to identify possible structures unambiguously.

Significant contributions of excitation-autoionization have been observed by Crandall et al. (1979) in the ionization cross sections of Li-like C^{3+}, N^{4+} and O^{5+} ions. In Fig. 13 the measured data are compared to Coulomb-Born-exchange calculations of Sampson and Golden (1979) which include contributions from autoionizing (1s^{2}2s)→(1s2s̄nl) excitations. It can be seen that the relative contribution from excitation-autoionization processes strongly increases with Z, however, it is predicted that this effect should moderate for higher Z along the Li-isoelectronic sequence due to the increasing probability for radiative decay.

Falk et al. (1981) have shown very recently in crossed-beams experiments with alkali-like Ti^{3+}, Zr^{3+}, and Hf^{3+} ions that excitation-autoionization can completely dominate the direct ionization as predicted by the Lotz formula (Eq. (8)). The experimental results are shown in Fig. 14. Beginning just above the thresholds the enhancement of the cross sections due to the indirect processes reaches factors of about 10 for Ti^{3+} and Hf^{3+} and 20 for Zr^{3+}. The increase is attributed to dipole-allowed, $\Delta n=0$ transitions of the type (np^{6}nd)→(np^{5}nd^2) with n = 3,4, and 5 for Ti^{3+}, Zr^{3+}, and Hf^{3+}, respectively. Calculated positions of excitation resonances using the distorted wave approximation agree quite well with the experimentally observed structures; the calculated excitation-autoionization cross sections however are too large by a factor of 2.5 in all cases.

Another interesting example with large contributions from indirect processes has been observed by Gregory et al. (1981) for Xe^{3+} (Fig. 15). For energies up to 90 eV the measured cross section rises more rapidly than the direct ionization cross section predicted from the Lotz formula by including ejection of 4d, 5s, and 5p electrons. A dominant contribution from the dipole-allowed, $\Delta n=0$ transition (4d^{10}5s^{2}5p^3)→(4d^{9}5s^{2}5p^{3}4f) is expected near the cross sections maximum at 95 eV. However, just beyond the maximum the cross section decreases stronger than the function (lnE)/E which characterizes a dipole-allowed excitation. Presumably other excitation processes are contributing to the cross section enhancement, too. This case illustrates that, for heavy, partially stripped ions, the ionization cross sections are not simply predicted.

Very recently, a novel mechanism for electron impact ioniza-

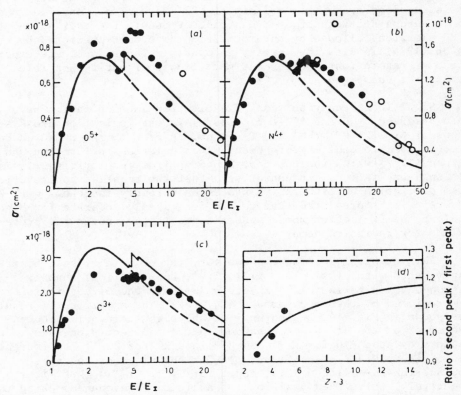

Fig. 13. Cross sections for the ionization of O^{5+}, N^{4+}, and C^{3+}
ions, respectively, as a function of electron impact ener-
gy in threshold units. The full circles are data of
Crandall et al. (1979), open circles are data of Donets
and Ovsyannikov (1977). The broken curves in parts (a),
(b) and (c) represent calculated direct ionization from
the 2s sublevel, the solid curves give the calculated
total cross section including contributions from excitation-autoionization. Part (d) shows calculated (solid
line) and experimental (full circles) ratio of the second
peak (direct plus excitation-autoionization) to the first
peak (direct ionization only) of the cross section as a
function of initial ionic charge. Here the broken curve
is the limiting value for the ratio as $Z \rightarrow \infty$ when radiative decay is neglected (Sampson and Golden, 1979).

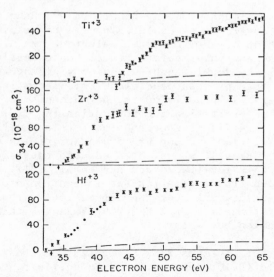

Fig. 14. Cross sections for electron-impact ionization of Ti^{3+}, Zr^{3+}, and Hf^{3+} ions (Falk et al., 1981) compared to the direct ionization cross section as predicted by the Lotz formula. Error bars show one standard deviation of the mean statistical uncertainty.

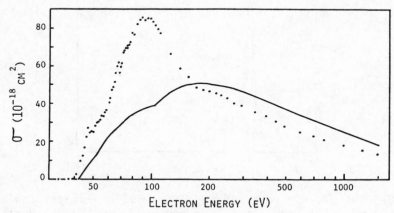

Fig. 15. Cross section for electron-impact ionization of Xe^{3+} ions (Gregory et al., 1981). The solid curve is the Lotz prediction (Eq. (8)) for direct ionization from the 5p, 5s and 4d subshells.

tion of ions has been considered by LaGattuta and Hahn (1981). Here, the incident electron is first captured by the target ion, with si-multaneous excitation of an inner-shell electron, resulting in a doubly excited resonant intermediate state. Subsequently, this state decays sequentially by double Auger emission yielding two free elec-trons. This three-step process of dielectronic-recombination-double autoionization operates in addition to the mechanism of direct ioni-zation (Eq. (1)) and excitation-autoionization (Eq. (3)). It is ex-pected to yield a significant contribution for ions with a large probability for radiationless capture and Auger yields close to unity, e.g., Fe^{15+}. Fig. 16 shows the predicted total ionization cross section for this ion (LaGattuta and Hahn, 1981). Besides do-minant contributions from excitation-autoionization processes there is a significant enhancement over the direct ionization cross sec-tion in the energy range 720 eV $\lesssim$ E $\lesssim$ 960 eV due to this additional capture-autoionization mechanism.

Several other predictions for the ionization cross section of Fe^{15+} are available (Bely, 1968; Kim and Cheng, 1978; Cowan and Mann, 1979) which differ up to an order in magnitude. Here, experi-mental data would be most desirable not only to test the theoretical

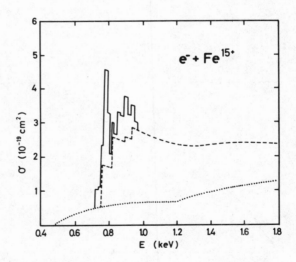

Fig. 16.   Calculated cross sections for electron-impact ionization of Fe^{15+} ions (LaGattuta and Hahn, 1981). The dotted curve gives the Lotz prediction (Eq. (8)) for direct ionization, the dashed curve adds the contributions from excitation-autoionization processes, and the solid curve represents contributions from capture-autoionization processes.

approaches for really high charge states and to search for the predicted new ionization mechanism but also to meet the data needs of
the fusion community.

4.2 Multiple ionization

Multiple ionization by electron-impact (Eqs. (2) and (4)) has
not been extensively studied. Only very few absolute measurements for
ions have so far been performed and almost all of the theoretical
work has been confined to the double ionization of atomic helium.

Two possible mechanisms have been considered by Mittleman
(1966): In the first, the incident electron initially ionizes one
of the bound ones and then either of the two outgoing electrons
further ionizes within the same atom. This double direct process is
likely to be significant only for large atoms. It can be neglected
for helium. Especially for high enough energies of the incident electron the probability of a second interaction will be small for the
two continuum electrons since most probably they will have high
energies, too. In the second mechanism, the initial ionization process takes place in a time which is much shorter than the orbital
periods of the other electrons being merely spectators. Consequently, they remain in states which are not eigenstates of the ion and
in the process of relaxation there is a finite probability of a
second transition to the continuum.

The second 'sudden approximation' predicts that at sufficiently high energies the ratio of cross sections for single and double
ionization σ_+/σ_{++} is constant and independent of the impact energy.
For helium this ratio was calculated by Mittleman (1966) to be 198;
however, it was pointed out by Bryon and Joachain (1966, 1967) that
calculations of multiple ionization very critically depend on the
choice of target wavefunctions, in particular on the terms involving
electron correlations.

4.2.1 Singly charged ions

The simplest example of multiple ionization of a positive ion
is the case

$$\mathrm{Li}^+ + e \rightarrow \mathrm{Li}^{3+} + 3e \tag{16}$$

for which the cross section has been measured by Peart and Dolder
(1969). The results are shown in Fig. 17. As is seen from the inset
of the figure, the ratio σ_+/σ_{++} comes fairly close to the value of
350 quoted for Li^+ (Peart and Dolder, 1969), but it does not seem
to be constant at high energies as predicted by the 'sudden approximation' theory.

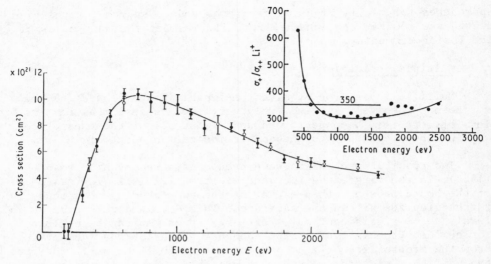

Fig. 17. Cross section σ_{++} for the double ionization of Li^+ ions
(Peart and Dolder, 1969). The inset shows the ratio of
cross sections for single (σ_+) and double (σ_{++}) ionization plotted versus electron energy.

Fig. 18 shows the results of another measurement of double ionization (Peart et al., 1971) involving the simplest negative ion, H^-,

$$H^- + e \rightarrow H^+ + 3e \qquad (17)$$

In this case the ratio σ_+/σ_{++} is nearly constant for energies above
100 eV. Since the first ionization potential of H^- is only 0.74 eV
an almost negligible interaction is sufficient to ionize the first
electron without perturbing the second one. Thus the assumption of
the 'sudden approximation' is more closely fulfilled than in the
Li^+ case. However, a recent remeasurement of the cross section σ_{++}
for H^- by Claeys and Defrance (1981) yielded data which are much
lower than those of Peart et al. (1971). This discrepancy, which
increases from a factor 5 up to 15 with increasing energy, removes
the independence on energy of the ratio σ_+/σ_{++} for $E \gtrsim 100$ eV.

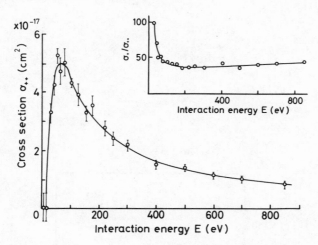

Fig. 18.   Cross section σ_{++} for double detachment of electrons from
           H^- (Peart et al., 1971). The inset shows the ratio of
           cross sections for single and double detachment plotted
           versus interaction energy.

Recently, absolute cross sections for the double, triple, and
quadruple ionization of Ar^+ ions by electron impact have been measured by Müller and Frodl (1980b). The results are shown in Fig. 19.
Whereas the cross section $\sigma_{1,3}$ for double ionization is a smooth
function of the electron energy within the experimental accuracy,
the cross section $\sigma_{1,4}$ for triple ionization abruptly increases at
about 250 eV electron energy which is the ionization threshold of
the $L_{2,3}$ shell of Ar^{1+}. The observed enhancement of $\sigma_{1,4}$ of up to
one order in magnitude is attributed to contributions from the mechanism of inner-shell ionization-Auger emission (Eq. (4)), in
which first an $L_{2,3}$-shell electron of Ar^+ is ionized followed by
Auger-emission of two additional electrons. The deexcitation after
L-shell ionization takes place via several different rearrangement
pathes.

Very recently, absolute cross sections $\sigma_{1,3}$, $\sigma_{1,4}$, and $\sigma_{1,5}$
for multiple ionization have also been measured for Rb^+ ions by
Hughes and Feeney (1981). The results are shown in Fig. 20. In
these data structure due to inner-shell ionization-Auger emission
can not be conclusively regarded as present.

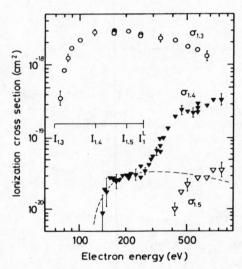

Fig. 19.   Cross sections for double ($\sigma_{1,3}$), triple ($\sigma_{1,4}$), and
           quadruple ($\sigma_{1,5}$) ionization of Ar^+ ions (Müller and Frodl,
           1980b). The ionization thresholds $I_{1,f}$ of the f outermost
           electrons and I_1^L of the $L_{2,3}$ shell of Ar^+ are indicated.
           The broken line gives an estimate of the cross section
           for direct three-electron ionization. It is obtained by
           fitting the data below the L threshold with the function
           $\sigma = A/(E_i \cdot E) \cdot \ln(E/E_i)$ where A is a fit parameter and E_i
           is the sum of the ionization potentials of the three
           outermost electrons.

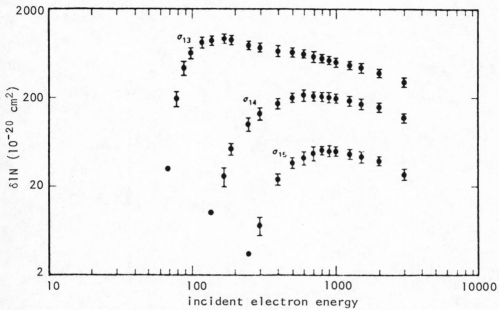

Fig. 20.   Cross sections for double ($\sigma_{1,3}$), triple ($\sigma_{1,4}$), and
           quadruple ($\sigma_{1,5}$) ionization of Rb^+ ions (Hughes and Feeney,
           1981).

4.2.2 Multiply charged ions

There is only a single crossed-beams experiment by Müller and
Frodl (1980b),which investigates the multiple ionization of a multi-
ply charged ion. The measured cross sections $\sigma_{2,4}$, $\sigma_{3,5}$, and $\sigma_{2,5}$
for double and triple ionization of Ar^{2+} and Ar^{3+} ions, respectively, are shown in Fig. 21. The experimental data give strong evidence for dominant contributions to the direct multiple ionization
mechanism by inner-shell ionization-Auger emission processes. Abrupt
increases are observed in the cross sections at the respective
threshold energies for the ionization of the $L_{2,3}$ shell. Compared
with the direct ionization of two or three electrons the contribution of the two-step mechanism becomes more and more important for
increasing charge state of the target ion and increasing number of
released electrons.

The inner-shell contributions $\sigma_{2,4}(L)$ and $\sigma_{3,5}(L)$ can be determined by subtracting the partial cross sections for the direct
removal of the two outermost electrons from the measured total cross
sections. Estimates of the partial cross sections are provided by
fitting the data below the L threshold with the function $\sigma = A/(E_i \cdot E)$
$\cdot \ln(E/E_i)$ where A is a fit parameter and E_i the sum of the ionization potentials of the two outermost electrons. The obtained partial

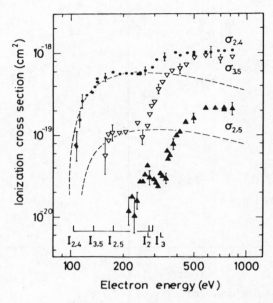

Fig. 21. Cross sections $\sigma_{2,4}$, $\sigma_{3,5}$, and $\sigma_{2,5}$ for double and triple
          ionization of Ar^{2+} and Ar^{3+} ions, respectively (Müller
          and Frodl, 1980b). The ionization thresholds $I_{2,4}$, $I_{3,5}$,
          and $I_{2,5}$ for the outermost electrons and I_2^L, and I_3^L of
          the $L_{2,3}$ shell of Ar^{2+} and Ar^{3+}, respectively, are indi-
          cated. The dashed curves represent estimates of the cross
          sections for direct double ionization (see text).

cross sections for direct double ionization are shown by the dashed
curves in Fig. 21. The extracted cross sections $\sigma_{2,4}(L)$ and $\sigma_{3,5}(L)$
for the contributions resulting from L-shell ionization are compared
in Fig. 22 with the cross section $\sigma_{0,2}(L_{2,3})$ for the single Auger
process $L_{2,3}$-MM including shakeup transitions during the Auger process, i.e. $L_{2,3}$-MMM*, initiated by electron impact, which has been
determined from Auger-spectroscopy measurements (Christofzik, 1970)
with neutral Ar atoms. Fig. 22 shows that the three cross sections
for the twofold ionization of Ar^{n+} (n=0,2,3) via L-shell vacancy
production do not differ very much from each other. This seems to
be intelligible if only the binding energies of the $L_{2,3}$ electrons

in Ar^{n+} are considered which are within 270 ± 20 eV for charge states
n ranging from 0 to 3. However, for a more detailed understanding
the various deexcitation pathes following the L shell vacancy production have to be taken into account (Müller and Frodl, 1980b). Assuming that the branching ratios for these processes do not depend
on the charge state of the parent Ar, it is possible to calculate
$\sigma_{n,n+2}(L)$ for Ar^{n+} (n=0,2,3) if the pertinent cross sections σ_{2s}
and σ_{2p} for pure ionization of the 2s and 2p-subshells, respectively, are known. The latter can be obtained from the classical binary-
encounter approximation (Gryzinski, 1965a). Fig. 33 shows that the
results of these calculations are in surprisingly good agreement
with the data.

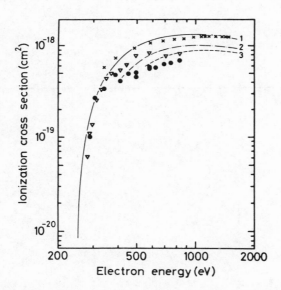

Fig. 22.   Cross sections $\sigma_{n,n+2}(L)$ for double ionization of Ar^{n+}
           (n=0,2,3) via L-shell vacancy production by electron im-
           pact (Müller and Frodl, 1980b):
           $\sigma_{0,2}(L_{2,3})$, crosses (Christofzik, 1970); $\sigma_{2,4}(L)$ closed
           circles; $\sigma_{3,5}(L)$ triangles. The curves 1, 2, and 3 are
           calculations of the cross sections $\sigma_{0,2}(L)$, $\sigma_{2,4}(L)$, and
           $\sigma_{3,5}(L)$, respectively, based on classical binary-encounter
           theory (Gryzinski, 1965a).

4.3 Molecular ions

To the authors knowledge only a single crossed-beams experiment
has been performed (Müller et al., 1980c) investigating the electron-impact ionization of a molecular ion. The measurement involved
singly charged ions of the carbon dioxide molecule which is an important constituent of the global and stellar atmospheres. The cross
section measured for the process

$$CO_2^+ + e \rightarrow CO_2^{2+} + 2e \tag{18}$$

is shown in Fig. 23. The shape of the cross section function does
not give evidence for additional ionization channels besides Eq. (18)
or for molecular effects like dissociative ionization. By choosing
different energies for the CO_2^+ ions it was checked experimentally
that contributions to the measured ionization cross section from
the population of a metastable state of the CO_2^+ ion, which dissociates with a half life of 2.3 ± 0.2 µs (Newton and Sciamanna, 1964)
according to

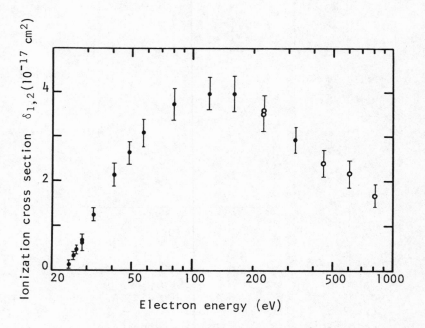

Fig. 23.   Cross section for the ionization of CO_2^+ ions (Müller et
al., 1980c). •, 5 keV ion energy; o, 10 keV ion energy.

$$CO_2^+ + e \rightarrow (CO_2^{2+})^* + 2e \rightarrow CO^+ + O^+ + 2e \qquad (19)$$

are smaller than the experimental error. Hence, the measured ionization cross section can be regarded as a partial cross section for the production of chemically stable CO_2^{2+} ions.

In ionization measurements involving molecular ions additional experimental problems may arise from the need to separate the ionization products from a background formed by dissociation fragments. Furthermore, the problem of initial excitation of the parent ion is even more severe for molecular ions. The production of adequate ($\gtrsim 10^{-9}$ A) molecular ion beams with known initial vibrational excitation is the most serious obstacle to further accurate crossed-beams measurements.

5. CONCLUSIONS

Despite the experimental activities initiated two decades ago by the first precise crossed-beams measurement, accurate data for the electron impact ionization of ions are only available for a limited number of species. The data basis is even more sparse for ionization of multiply charged ions, for multiple ionization by impact of a single electron, and for ionization of molecular ions. Of course, the number of possible cases is too large to investigate only a significant fraction of them. However, there should be more experiments to test the theoretical approaches (both sophisticated as well as semi-empirical ones); to increase our understanding of the various ionization mechanisms, their systematics and interplay; and to meet the data need of users.

The techniques for these measurements are now reasonably well developed. In combination with suitable ion sources which produce intense beams of highly charged ions, e.g., the ECR source (Geller et al., 1980),interesting new results for the electron impact ionization of ions are to be expected in the near future.

Acknowledgements

The author would like to thank his coworker Dr. A. Müller for helpful discussions. Considerable recognition is due to Drs. D. Crandall, G. Dunn, and their coworkers of the ORNL-JILA collaboration not only for fruitful conversations but also for allowing me to quote their results prior to publication.

References

Abrines, R., and Percival, I.C., 1966,
     Proc. Phys. Soc., 88: 861
Abrines, R., Percival, I.C., and Valentin, N.A., 1966,
     Proc. Phys. Soc., 89: 515
Aitken, K.L., and Harrison, M.F.A., 1971,
     J. Phys. B: Atom. Molec. Phys., 4: 1176
Aitken, K.L., Harrison, M.F.A., and Rundel, R.D., 1971,
     J. Phys. B: Atom. Molec. Phys., 4: 1189
Alder, K., Huus, T., Mottelson, B., and Winther, A., 1956,
     Rev. Mod. Phys., 28: 432
Baker, F.A., and Hasted, J.B., 1966,
     Phil. Trans. R. Soc., A261: 33
Becker, R., Frodl, R., and Klein, H., 1977,
     Workshop on EBIS and Related Topics, Darmstadt 1977, GSI-
     Report P-33-77, p. 75
Bell, K.L., Freeston, M.W., and Kingston, A.E., 1970,
     J. Phys. B: Atom. Molec. Phys., 3: 959
Bely, O., 1968,
     J. Phys. B: Atom. Molec. Phys., 1: 23
Bethe, H., 1930,
     Ann. d. Phys., 5: 325
Bloch, S., 1912,
     Ann. d. Phys., 38: 559
Bryon, F.W., and Joachain, C.J., 1966,
     Phys. Rev. Lett., 16: 1139
Bryon, F.W., and Joachain, C.J., 1967,
     Phys. Rev., 164: 1
Burgess, A., 1963,
     Proc. of the 3rd Intern. Conference on the Physics of Elec-
     tronic and Atomic Collisions, London 1963, ed. M.R.C.McDowell,
     North-Holland Publ. Comp. (1964), p. 237
Burgess, A., 1964,
     Proc. Symp. on Atomic Collision Processes in Plasmas, Culham
     1964, A.E.R.E. Report No. 4818 (1964), p. 63
Burgess, A., and Rudge, M.R.H., 1963,
     Proc. Roy. Soc., A273: 372
Burke, P.G., and Taylor, A.J., 1965,
     Proc. Roy. Soc., A287: 105
Christofzik, H.-J., 1976,
     Diplomarbeit, Universität Münster, unpublished
Claeys, W., and Defrance, P., 1981,
     European Conference on Atomic Physics, Heidelberg 1981, Book
     of Abstracts, Vol. 5A, Part II, p. 753, eds. J. Kowalski,
     G. zu Putlitz, and H.G. Weber, European Physical Society
Cowan, R.D., and Mann, J.B., 1979,
     Astrophys. J., 232: 940
Crandall, D.H., Phaneuf, R.A., and Taylor, P.O., 1978,
     Phys. Rev., A18: 1911

Crandall, D.H., Phaneuf, R.A., Hasselquist, B.E., and Gregory, D.C.,
        1979, J. Phys. B: Atom. Molec. Phys., 12: L249
Crandall, D.H., 1981a,
        Physica Scripta, 23: 153
Crandall, D.H., 1981b,
        'Electronic and Atomic Collisions',Invited papers and progress
        reports, ICPEAC XII, Gatlinburg 1981, ed. S. Datz, to be
        published
Datla, R.U., Blaha, M., and Kunze, H.J., 1975,
        Phys. Rev., A12: 1076
Defrance, P., Brouillard, F., Claeys, W., and Van Wassenhove, G.,
        1981, J. Phys. B: Atom. Molec. Phys., 14: 103
Dolder, K.T., Harrison, M.F.A., and Thonemann, P.C., 1961,
        Proc. Roy. Soc., A264: 367
Dolder, K.T., 1969,
        'Case Studies in Atomic Collision Physics', Vol. I, eds.
        E.W. McDaniel and M.R.C. McDowell, North-Holland Publ. Comp.
        (1969), p. 249
Dolder, K.T., and Peart, B., 1976,
        Rep. Progr. Phys., 39: 693
Dolder, K.T., 1977,
        'Electronic and Atomic Collisions', Invited papers and pro-
        gress reports, ICPEAC X, Paris 1977, ed. G. Watel, North-
        Holland Publ. Comp. (1978), p. 281
Donets, E.D., and Ovsyannikov, V.P., 1977,
        Reprint P7-10780, Joint Institute for Nuclear Research,
        Dubna/USSR
Drawin, H.W., 1961,
        Z. Physik, 164: 513
Drawin, H.W., 1963,
        Z. Physik, 172: 429
Dunn, G.H., 1969,
        'Atomic Physics', eds. V.W. Hughes, V.W. Cohen, and F.M.J.
        Pichanick, Plenum Press (1969), p. 417
Dunn, G.H., 1976,
        IEEE Trans. Nucl. Sci., NS-23: 929
Dunn, G.H., 1979,
        Proc. of the Nagoya Seminar on Atomic Processes in Fusion
        Plasmas, Reprint IPPJ-AM-13, Institute of Plasma Physics,
        Nagoya University, Japan 1979, p. 57
Dunn, G.H., 1980,
        'The Physics of Ionized Gases', Invited lectures and progress
        reports of SPIG'80, Dubrovnik 1980, ed. M. Màtič, Boris
        Kidrič Institute of Nuclear Sciences, Beograd (1980), p. 49
Falk, R.A., Dunn, G.H., Griffin, D.C., Bottcher, C., Gregory, D.C.,
        Crandall, D.H., and Pindzola, M.S., 1981, Phys. Rev. Lett.,
        47: 494
Fite, W.L., and Brackmann, R.T., 1958,
        Phys. Rev., 113: 1141
Friens, L., and Bonson, T.F.M., 1968,
        J. Phys. B: Atom. Molec. Phys., 1: 1123

Gabriel, A.H., and Jordan, C., 1972,
    'Case Studies in Atomic Collision Physics', Vol. 2, p. 211,
    eds. E.W. McDaniel and M.R.C. McDowell, North-Holland Publ.
    Comp. (1972)
Geller, R., Jacquot, B., and Pauthenet, R., 1980,
    Revue Phys. Appl., 15: 995
Golden, L.B., and Sampson, D.H., 1977,
    J. Phys. B: Atom. Molec. Phys., 10: 2229
Golden, L.B., Sampson, D.H., and Omidvar, K.J., 1978,
    J. Phys. B: Atom. Molec. Phys., 11: 3235
Golden, L.B., and Sampson, D.H., 1980,
    J. Phys. B: Atom. Molec. Phys., 13: 2645
Gregory, D.C., Dittner, P.F., and Crandall, D.H., 1981,
    Proc. of the XII Intern. Conference on the Physics of Elec-
    tronic and Atomic Collisions, Gatlinburg 1981, Abstracts of
    contributed papers, p. 465, ed. S. Datz
Gryzinski, M., 1959,
    Phys. Rev., 115: 374
Gryzinski, M., 1965a,
    Phys. Rev., A138: 305, 322, 336
Gryzinski, M., 1965b,
    Phys. Rev. Lett., 14: 1059
Hansen, J.E., 1975,
    J. Phys. B: Atom. Molec. Phys., 8: 2759
Harrison, M.F.A., 1966,
    Brit. J. App. Phys., 17: 371
Harrison, M.F.A., Dolder, K.T., and Thonemann, P.C., 1963a,
    Proc. Roy. Soc., A274: 546
Harrison, M.F.A., Dolder, K.T., and Thonemann, P.C., 1963b,
    Proc. Phys. Soc., 82: 368
Harrison, M.F.A., 1968,
    'Methods of Experimental Physics', Vol. 7A, eds. B. Bederson
    and W.L. Fite, Academic Press (1968), p. 95
Herrmannsfeldt, W.B., 1973,
    SLAC 166 Electron trajectory program
Hooper, W., Lineberger, W.C., and Bacon, F.M., 1966,
    Phys. Rev., 141: 165
Hughes, D.W., and Feeney, R.K., 1981,
    Phys. Rev., A23: 2241
Jakubowicz, H., 1980,
    Thesis, University College, London
Jakubowicz, H., and Moores, D.L., 1980,
    Comm. At. Mol. Phys., 9: 55
Kim, Y.-K., and Cheng, K.-T., 1978,
    Phys. Rev., A18: 36
Kingston, A.E., 1964,
    Phys. Rev., A135: 1537
Kingston, A.E., 1968,
    J. Phys. B: Atom. Molec. Phys., 1: 559
Kirstein, P.T., and Hornsby, J.S., 1963,
    CERN-Report-63-16

Kumar, A., and Roy, B.N., 1978,
        Phys. Lett., 66A: 362
Kunc, J.A., 1980,
        J. Phys. B: Atom. Molec. Phys., 13: 587
Kunze, H.J., 1972,
        Space Sci. Rev., 13: 565
LaGattuta, K.J., and Hahn, Y., 1981,
        Phys. Rev., A24: 2273
Lenard, P., 1894,
        Wied. Ann., 51: 225
Lenard, P., 1904,
        Ann. d. Phys., 15: 485
Lotz, W., 1967,
        Z. Physik, 206: 205
Lotz, W., 1968,
        Z. Physik, 216: 241
Lotz, W., 1970,
        Z. Physik, 232: 101
Massey, H.S.W., 1956,
        'Encyclopaedia of Physics', 36: 354
McDowell, M.R.C., 1966,
        Proc. Phys. Soc., 89: 23
McGowan, J.W., 1979,
        'Electronic and Atomic Collisions', Invited papers and progress
        reports, ICPEAC XI, Kyoto 1979, eds. N. Oda and K. Takayanagi,
        North-Holland Publ. Comp. (1980), p. 237
McGuire, E.J., 1977,
        Phys. Rev., A16: 62, 73
Mittleman, M.H., 1966,
        Phys. Rev. Lett., 16: 498
Moores, D.L., and Nussbaumer, H., 1970,
        J. Phys. B: Atom. Molec. Phys., 3: 161
Moores, D.L., 1972,
        J. Phys. B: Atom. Molec. Phys., 5: 286
Moores, D.L., 1978,
        J. Phys. B: Atom. Molec. Phys., 11: L403
Moores, D.L., Golden, L.B., and Sampson, D.H., 1980,
        J. Phys. B: Atom. Molec. Phys., 13: 385
Moores, D.L., 1981,
        'Electronic and Atomic Collisions', Invited papers and pro-
        gress reports, ICPEAC XII, Gatlinburg 1981, ed. S. Datz,
        to be published
Müller, A., Salzborn, E., Frodl, R., Becker, R., Klein, H., and
        Winter, H., 1980a,
        J. Phys. B: Atom. Molec. Phys., 13: 1877
Müller, A., and Frodl, R., 1980b,
        Phys. Rev. Lett., 44: 29
Müller, A., Salzborn, E., Frodl, R., Becker, R., and Klein, H.,
        1980c, J. Phys. B: Atom. Molec. Phys., 13: L221

Newton, A.S., and Sciamanna, A.F., 1964,
    J. Chem. Phys., 40: 718
Peart, B., and Dolder, K.T., 1968,
    J. Phys. B: Atom. Molec. Phys., 1: 240
Peart, B., and Dolder, K.T., 1969,
    J. Phys. B: Atom. Molec. Phys., 2: 1169
Peart, B., Walton, D.S., and Dolder, K.T., 1969a,
    J. Phys. B: Atom. Molec. Phys., 2: 1347
Peart, B., Martin, S.O., and Dolder, K.T., 1969b,
    J. Phys. B: Atom. Molec. Phys., 2: 1176
Peart, B., Walton, D.S., and Dolder, K.T., 1971,
    J. Phys. B: Atom. Molec. Phys., 4: 88
Peart, B., and Dolder, K.T., 1975,
    J. Phys. B: Atom. Molec. Phys., 8: 56
Percival, I.C., 1966,
    Nucl. Fusion, 6: 182
Peterkop, R.K., 1977,
    'Theory of Ionization of Atoms by Electron Impact', Colorado
    Associated University Press (1977), Colorado/USA
Post, R.F., 1961,
    Plasma Physics, 3: 273
Redhead, P.A., 1967,
    Can. J. Phys., 45: 1791
Redhead, P.A., and Feser, S., 1968,
    Can. J. Phys., 46: 865
Redhead, P.A., and Gopalaraman, C.P., 1971,
    Can. J. Phys., 49: 585
Rogers, W.T., Stefani, G., Camilloni, R., Dunn, G.H., Msezane, A.Z.,
    and Henry, R.J.W., 1981, Phys. Rev., to be published
Rudge, M.R.H., 1968,
    Rev. Mod. Phys., 40: 564
Rudge, M.R.H., and Schwartz, S.B., 1965,
    Proc. Phys. Soc., 86: 773
Rudge, M.R.H., and Schwartz, S.B., 1966a,
    Proc. Phys. Soc. (London), 88: 563
Rudge, M.R.H., and Schwartz, S.B., 1966b,
    Proc. Phys. Soc. (London), 88: 579
Salop, A., 1976,
    Phys. Rev., A14: 2095
Sampson, D.H., and Golden, L.B., 1978,
    J. Phys. B: Atom. Molec. Phys., 11: 541
Sampson, D.H., and Golden, L.B., 1979,
    J. Phys. B: Atom. Molec. Phys., 12: L785
Seaton, M.J., 1962,
    Proc. Phys. Soc., 79: 1105
Seaton, M.J., 1964,
    Planet. Space Sci., 12: 55
Stingl, E.J., 1972,
    J. Phys. B: Atom. Molec. Phys., 5: 1160

Thomson, J.J., 1912,
        Phil. Mag., 23: 449
Thomas, B.K., and Garcia, D.J., 1969,
        Phys. Rev., 179: 94
Younger, S.M., 1980a,
        Phys. Rev., A22: 111
Younger, S.M., 1980b,
        Phys. Rev., A22: 1425
Younger, S.M., 1981a,
        Phys. Rev., A23: 1138
Younger, S.M., 1981b,
        Phys. Rev., A24: 1272, 1278

EXPERIMENTAL STUDIES OF ELECTRON-ION RECOMBINATION

J. B. A. Mitchell and J. Wm. McGowan

Department of Physics
The University of Western Ontario
London, Ontario, N6A 3K7   Canada

HISTORICAL INTRODUCTION

Electron-ion recombination is one of the most fundamental concepts in atomic and molecular physics and plays a major role in the chemistries of the upper atmosphere, interstellar space, combustion and thermonuclear fusion. Indeed, it has been from studies of these fields that much of our knowledge concerning the mechanisms of recombination has come. These studies have been backed up by a considerable theoretical and experimental effort in an attempt to gain a quantitative understanding of recombination as well as to elucidate the fine details underlying the electron capture and stabilization process for different ion species. In turn great advances have been made in these areas as a result of our improved understanding of the physics of recombination. This trend will undoubtedly continue.

The early history of the study of electron-ion recombination has been elegantly covered by D.R. Bates (1974) and so only a brief outline is presented here.

Radiative recombination, namely the process:

$$X^+ + e \longrightarrow X + h\nu$$

in which the excess energy of the electron-ion pair is dispersed via photon emission was first examined theoretically by Kramers in 1923. Because of the long time scale (typically a few nanoseconds) for photon emission as compared to the collision time (10^{-15} secs) this process has a very small rate coefficient and so far has not been studied in the laboratory. In the 1940's, however, measurements of the diurnal variation of electron densities in the low energy plasma of the ionosphere indicated that much more rapid electron-ion recombination processes were dominating and this prompted theoreticians to propose alternative electron-ion recombination mechanisms.

The first such alternative was proposed by Sayers (1943), who suggested that an inverse autoionization process;

$$e + X^+ ---> X^{**} ---> X^* + h\nu$$

proceeding through the formation of an intermediate doubly excited state, would have a higher rate coefficient than direct radiative recombination.

This process, later renamed Dielectronic Recombination, was discussed in detail by Massey and Bates (1943) who concluded that in fact it would be unimportant in the ionosphere. The rate coefficient would still be several orders of magnitude smaller than that required to explain the rapid disappearance of electrons mentioned earlier.

Interest in dielectronic recombination was eventually renewed when calculations by Burgess (1964) showed that the rate coefficients for highly stripped ions were sufficient to explain the discrepancies between the measured values of the temperature of the solar corona and values calculated from the ionization balance.

The study of dielectronic recombination has of course received even greater stimulus over the last few years since it was appreciated that it plays a major role in the emission of radiation from highly stripped impurity ions in thermonuclear plasmas leading to severe energy loss problems.

Returning to the ionosphere problem it was eventually proposed by Bates and Massey (1947), that molecular ions present in the ionosphere would be lost much more rapidly than atomic species because of an additional stabilization mechanism, namely dissociative recombination.

$$e + XY^+ ---> X + Y$$

If the electron could be captured into a repulsive neutral state which subsequently dissociated, then the excess energy could be carried away as kinetic energy or excitation of the dissociation products. Unlike the case for radiative and dielectronic recombination, experimental justification for this proposal was not long in coming. In 1949, Biondi and Brown published results indicating a recombination rate coefficient in a helium afterglow plasma of $\sim 10^{-8} cm^3 sec^{-1}$. Bates (1950), used these results to elaborate on his physical model for dissociative recombination involving the capture of the electron into a doubly excited repulsive state crossing the ground state of the recombining ion.

Following this, Biondi carried out an extensive series of experiments which proved conclusively that dissociative recombination was an accurate representation of the recombination process for molecular ions. See Bardsley and Biondi (1970).

Ironically the first measurement involving helium later proved to be erroneous, the decay of the plasma actually being dominated by ambipolar diffusion to the walls of the apparatus. In fact He_2^+ appears to be one of the few molecular ions which does not exhibit a large recombination coefficient in its normally occurring form, due to a lack of suitable curve crossings.

DISSOCIATIVE RECOMBINATION MEASUREMENT TECHNIQUES

Laboratory studies of dissociative recombination have been performed by many groups using a wide variety of techniques and yet definitive measurements of some cross sections have not yet been reported. This is due in part to the lack of definition of the initial and final states of the recombining system. New techniques are being developed to examine the microscopic details of the recombination process and as these mature we can expect that our understanding will advance dramatically over the next few years.

The approaches to the measurement of electron—ion recombination can be divided into two separate classes of experiment, one in which the rate coefficient for the recombination process is determined by observing the decay of a plasma species (usually electrons); the other, where cross sections are obtained by studying the formation of neutrals in intersecting beam experiments.

The rate coefficient for an electron—ion recombination process is defined according to the relationship.

$$\frac{dn}{dt} = n_i n_e \alpha \qquad ---(1)$$

where n_i and n_e are ion and electron number densities and α is related to the cross section σ as follows:

$$= \int \sigma \, v \, f(v) \, dv \qquad ---(2)$$

where v is the electron velocity and f(v) is the velocity distribution.

PLASMA DECAY MEASUREMENTS

Afterglows

The afterglow technique was used to provide the first measurements of dissociative recombination more than thirty years ago and is still actively pursued. In the experiments, fig. (1) a glow discharge plasma is formed in a reaction vessel and when the exciting mechanism is turned off the decay of the plasma electrons is observed as a function of time by studying the reflection of low energy microwaves. The change in the electron density is given by:

$$\frac{\partial n_e}{\partial t} = \Sigma P_i - \Sigma L_i - \nabla \cdot \Gamma_e \quad - - - (3)$$

where P_i = formation rate
  L_i = Loss rate due to atomic processes
  $\nabla \cdot \Gamma_e$ = Loss rate due to diffusion.

Care has to taken to make sure that all contributions to each of the three terms on the RHS of equation (3) are taken account of. For example, metastable collisions can give rise to associative ionization thus increasing P_i, if the pressure, is too high then 3 body effects contribute to L_i.

One must also consider the importance of electron heating processes such as superelastic collision of electrons with excited atoms and molecules.

Ambipolar diffusion losses are usually minimized by using an appropriate buffer gas. Under ideal conditions:

$$\Sigma P_i = 0$$
$$\nabla \cdot \Gamma_e = 0$$
$$L_i = \alpha\, n_i n_e$$
$$n_i = n_e$$

Then

$$\frac{dn_e}{dt} = -\alpha n_e^2 \quad - - - (4)$$

and so

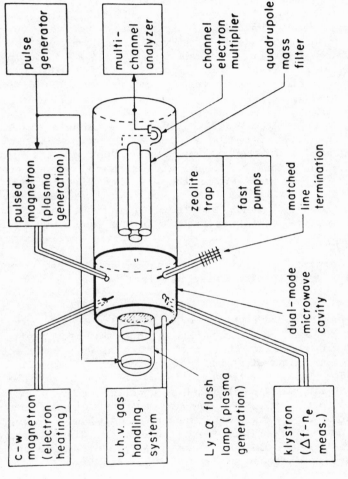

Fig. 1 Schematic Diagram of Biondi's microwave afterglow apparatus

$$\frac{1}{n_e} (r,t) = \frac{1}{n_e} (r,o) + \alpha t \quad - - - (5)$$

Hence the rate coefficient α can be determined from the slope of the $1/n_e$ vs. t plot. In practice, however, ambipolar diffusion giving rise to non uniform electron distributions cannot be ignored requiring computer generated solutions from equation (3). Under normal conditions, $T_e = T_i = T_{gas}$, that is, the system has time to thermalize. It is possible however, to raise T_e, the electron temperature by microwave heating and so study the temperature dependance of the dissociative recombination rate coefficient. In this case T_i remains equal to T_{gas}. The results of measurements using this technique will be discussed later.

The major problems associated with this technique are concerned with the identification of the ion under study and its excitation state. Clustering processes occur in the plasmas under certain conditions and these give rise to species such as $H_3O^+ \cdot (H_2O)_n$, $(N_4)^+$, etc. which compete for electrons in the plasma with the primary ion under study in the recombination process. The use of a mass spectrometer for the monitoring of plasma conditions is mandatory for the proper measurement of individual recombination rates of the various ion species.

A more complete description of a microwave afterglow apparatus employing mass spectrometric sampling may be found in Mehr and Biondi (1969).

Determination of the state of excitation of the ion under study, is a major problem in recombination measurements. The problem lies in the fact that molecular ions can be electronically, vibrationally and rotationally excited. These states are often long lived compared to the lifetime of the ion in the system. In the past most experimenters have inferred the presence or absence of excitation by studying the rate coefficient for the recombination as the experimental conditions are varied. For example inert buffer gases are used in the plasma afterglow technique and the pressure of these gases can determine the rate of vibrational relaxation of the ion under study.

Recently Zipf (1980a,b) however has introduced the use of laser induced photofluorescence for the determination of initial ion excitation in afterglow measurements. In this technique, a 1 MW tuneable dye laser is used to excite a particular transition from a given initial vibrational state (e.g v = 0) and the fluorescence arising from the relaxation of the upper state is measured. Since the absolute laser flux entering the cavity is

known then the population of the state under study can be determined directly.

Both Zipf (1970, 1980a,b) and Shiu et al (1977a,b,78) have
further extended the afterglow technique to the examination of
the excitation states of the dissociation products by measuring
the light emitted when these states relax. Measurements such as
these are vital to our eventual understanding of the mechanisms
for dissociative recombination. Unfortunately, however, plasma
techniques are capable of measuring rate coefficients only, because the observations are averaged over the electron velocity
distribution of the plasma. They are therefore not as sensitive
as the intersecting beam methods capable of measuring cross sections.

SHOCK TUBES

In this technique, a shockwave caused by overpressure burs-
ting of a thin diaphragm is made to pass through a region containing a plasma formed by microwave discharge. The temperature
of the plasma is rapidly raised due to passage of the shock wave
and within $\sim 500\,\mu s$, a temperature equilibrium is reached so that
$T_e \simeq T_{gas} \simeq T_{ion}$.

The resulting value for T_{gas} is determined by the shock
velocity which may be varied by altering the thickness of the
diaphragm and by changing the driver gas.

The number density of plasma ions is measured using multiple Langmuir probes and the recombination coefficient is
determined by monitoring the decay of the ions with time. It
should be noted that mass spectrometric interpretation of the
recombining ion species is not performed due to the difficulties
involved and so care has to be taken to model the plasma conditions using known gas phase rate coefficients in order to ensure that the ion being investigated is dominant.

Further details concerning the shock tube technique may be
found in papers by Ogram et al. 1980, Fox and Hobson 1966 and
Cunningham and Hobson 1969. More will be said about this technique later in this chapter.

FLAME SAMPLING

This has had very limited application with most studies being concerned with H_3O^+, the dominant ion in flames. Two different approaches have been used to study the decay of ions
formed in the reaction zone namely, Langmuir probe measurements
of electron density and mass spectrometric sampling of the ion

under study. (See Hayhurst & Telford 1974 and references therein). In both cases, measurements are made as a function of
height (and therefore time) above the reaction zone where the
ions are formed. The recombination rate coefficient can be
determined from the slope of the measured ion density vs time
curve.

Although in principle a simple measurement, in practice
many corrections have to be applied to the data. Extensive analysis has been performed by Hayhurst on this technique (Burdett
and Hayhurst, 1979 and references therein) taking account of the
disturbance of the flame by the sampling system, the cooling of
the sample due to the supersonic expansion of gases into the
mass spectrometer region, the formation of cluster ions during
this expansion and the scattering of sampled ions leading to
losses in the input region of the mass spectrometer.

Despite the very impressive theoretical modelling the
results obtained by Hayhurst using this method are in complete
disagreement with those obtained by other means.

INTERSECTING BEAM EXPERIMENTS

These are experiments in which a beam of electrons is made
to collide with a beam of ions and the products of the recombination are subsequently measured. Three variations have been
used namely crossed beams, inclined beams and merged beams.

The theory behind the crossed beam and inclined beam tech-
niques has been dealt with in depth by Harrison (1966) and
Dolder (1969) and that for the merged beam was discussed by
Auerbach et al (1977).

Considering the case of a monenergetic ion beam with energy
E_i intersecting a monenergetic electron beam of energy E_e at
some angle θ, then the collision energy in the centre of mass
frame for this system is given by:

$$E_{cm} = \tfrac{1}{2}\mu(\bar{v_i} - \bar{v_e})^2 = \mu\,[E_i/m_i + E_e/m_e - 2(E_iE_e/m_im_e)^{\tfrac{1}{2}}\cos\theta]$$

$$- - - (6)$$

where m_i, m_e, are the ion and electron masses.

$\mu = m_em_i/(m_e + m_i) \simeq m_e$ = the reduced mass of the system.

v_r = relative velocity of the ion and the electron.

We can define a quantity called the reduced ion energy E_+ by:

$$E_+ = (m_e/m_i)E_i \qquad\qquad \text{- - - (7)}$$

so that equation (6) becomes simplified to:

$$E_{cm} = E_+ + E_e - 2(E_+E_e)^{\frac{1}{2}} \cos \theta \qquad\qquad \text{- - - (8)}$$

When θ is small, this reduces to

$$E_{cm} \simeq (E_+^{\frac{1}{2}} - E_e^{\frac{1}{2}})^2 + (E_+E_e)^{\frac{1}{2}}\theta^2 \qquad\qquad \text{- - - (9)}$$

Hence when $E_+ = E_e$, the minimum achievable centre of mass energy is limited by the value of θ.

In fig. (2) the crossed ($\theta=90°$), inclined (θ small) and merged ($\theta=0$) beam techniques are compared from a kinematic standpoint. It can be seen that in principle the merged beam configuration allows zero centre of mass collision energy to be achieved.

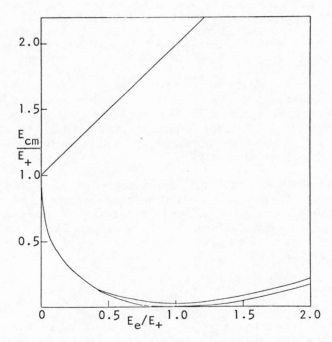

Fig. 2   Plot of E_{cm}/E_+ vs E_e/E_+ (see Text) for merged, inclined and crossed beams.

By differentiating equation (6) with respect to E_i, E_e and θ one can obtain the following expression for the energy resolution ΔE_{cm}. (Assuming Gaussian distributions for ΔE_i, ΔE_e, and $\Delta \theta$).

$$\Delta E_{cm} = \{\{[1 - (E_+/E_e)^{\frac{1}{2}}]\Delta E_e\}^2 + \{[1 - (E_e/E_+)^{\frac{1}{2}}]\Delta E_+\}^2 + [2(E_e E_+)^{\frac{1}{2}}\theta \Delta \theta]^2 \}^{\frac{1}{2}} \qquad --- (10)$$

When $E_+ = E_e$ the contributions due to ΔE_e and ΔE_i become negligible and the energy resolution is dominated by the angular term. This enables one to achieve very high resolution in the merged beam case provided θ can be made small. This is discussed in more detail by Mul et al (1981c).

Three groups have thus far used intersecting beams for the study of dissociative recombination. Dunn, at JILA, used a crossed beam experiment for the study of the formation of D(2p) and D(n=4) atoms in $e + D_2^+$ recombination in the energy range from 0.6 - 7eV. Dolder, at Newcastle upon Tyne extended the energy range down to 0.3eV by using the inclined beam technique for $e + D_2^+$, H_2^+ and H_3^+ collisions. The most extensive studies of dissociative recombination using the intersecting beam technique have been performed by McGowan's group at the University of Western Ontario, using a merged beam apparatus. To date more than 30 different species have been studied. McGowan et al (1979).

In Dunn's experiments, the recombination was measured by detecting the photons emitted during the decay of the excited dissociation products. This is technically difficult to do but detailed information concerning the decay channels of the process can be obtained in this way.

Dolder and McGowan have both concentrated on measuring the number of neutrals formed as a result of the electron ion recombination. Because space does not permit detailed description of each of the apparatuses, only a brief outline of the MEIBE I (Merged Electron Ion Beam Experiment) at the University of Western Ontario will be given here.

MEIBE I

Ions are produced in an rf ion source mounted in the terminal of a 400 Kev Van de Graaff accelerator. After focusing and mass analysis, the ion beam enters the ultra high vacuum experimental chamber where after being offset electrostatically to remove neutrals it passes through the interaction region. After collision with the electrons, the ion beam is analysed electro-

statically to remove neutrals formed as a result of both electron-ion and background collisions and the primary ions are collected in a Faraday cup. The neutrals are allowed to strike a
surface barrier detector and are subsequently counted.

A major reduction in background noise levels can be achieved by exploiting the energy resolving properties of the surface
barrier detector. Neutral atoms arising from dissociative collisions of the primary beam with the background gas giving ion-
atom pairs, arrive with only a fraction of the beam energy and
so can be distinguished from other processes such as dissociative recombination and dissociative charge exchange where the
resulting neutrals carry the total beam energy.

Separation of these latter two processes is then accomplished by modulating the electron beam and counting the neutrals in and and out of phase with the modulation using gated
scalers.

The electron beam is formed in a Pierce type electron gun
and is subsequently merged with the ion beam using a trochoidal
analyser. This device operates by having a magnetic field axial
to the electron beam and an electric field perpendicular to it.
The electrons undergo a precise spiralling motion in the analyzer and when they emerge, if the correct conditions are present,the input and output vectors of the beam will be identical
but the axis of the beam will be shifted to a new axis offset

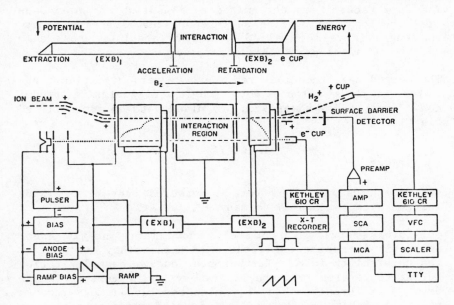

Fig. 3 Schematic of McGowan's MEIBE I apparatus.

from the original. This new axis is made to coincide with the axis of the ion beam so that merging occurs.

After the interaction with the ion beam, the electrons are then "demerged" using a second trochoidal analyzer before being collected in a Faraday cup. A schematic diagram of the apparatus is given in fig. (3).

The recombination cross section may be obtained using the following expression:

$$\sigma(E_{cm}) = \frac{C_n e^2 F}{I_e I_i L} \left| \frac{v_i \cdot v_e}{v_i - v_e} \right| \simeq \frac{C_n e^2 F}{I_e I_i L} \left| \frac{v_i v_e}{[(v_i - v_e)^2 + v_i v_e \theta^2]^{\frac{1}{2}}} \right|$$

$$- - - (11)$$

where C_n = neutral count rate
I_e, I_i = Electron and Ion beam currents
v_e, v_i = Electron and Ion beam velocities
θ = intersection angle
L = intersection length

$$F = \left(\iint_{S_e} i_e(x,y) \, dxdy \iint_{S_i} i_i(x,y) \, dxdy \right) \left(\iint_{S} i_i(x,y) dxdy \right)^{-1}$$

$$- - - (12)$$

The form factor F is determined by measuring the beam density distributions and overlap at three places along the intersection length (Keyser et al 1979) and so absolute cross sections may be obtained using this apparatus.

The merged beam technique represents a very significant advance over previous methods for studying dissociative recombination. Some of its advantages and disadvantages are outlined in Table (I).

TABLE I

ADVANTAGES

1. Absolute cross sections can be measured over a very wide energy range, a few meV to many eV.

2. Collision energy is accurately known since nearly monoenergetic beams are used, the only uncertainty coming from the uncertainty in θ. Furthermore, energy deamplification during conversion from the laboratory to the centre of mass frame of reference makes very high resolution measurements possible (Few meV). Lower limit is determined primarily by intersection angle.

3. Signal to background ratio higher than for inclined beam experiments.

4. Use of nuclear counting techniques allow the neutral products to be studied. Measurement of final state excitation and branching ratios are possible by improving the energy resolution of the counting system and exploiting the energy amplification upon transforming from the centre of mass to the laboratory frame of reference.

DISADVANTAGES

1. So far limited to relatively light species due to limitation on minimum electron energy ($\sim 13eV$) and maximum ion energy ($\sim 450KeV$). Higher energy accelerator would lessen this restriction greatly.

2. Mass spectrometric identification of the recombining ion can cause problems when dealing with different species with the same molecular weight. Common to all measurements.

3. Excitation state of primary ion beam uncertain. Much work has to be done to characterize the beams better. Excitation state can be controlled to some degree in the ion sources by using buffer gases, rf trap sources etc.

HYBRID TECHNIQUES

Quadrupole Ion Trap

The quadrupole Ion Trap, (Walls and Dunn, 1974 a,b, Heppner et al. 1976) is something of a hybrid device. A cloud of ions is formed inside the trap fig. (4) and a beam of electrons with known energy and intensity is subsequently fired through the ion cloud. The decrease in the density of ions in the trap due to recombination with the electrons is measured yielding the recombination cross section.

The actual trapping of the ions is accomplished by using a powerful magnetic field (11.75KG) to confine them in the ($r-\theta$) plane together with a quadrupole electric field to give z-axis confinement. In this way a harmonic well is formed along the axis within which the ions oscilate with frequency:

$$w_z = \{ 4V_0(q/m)[1/(R_0^2 + 2Z_0^2)]\}^{\frac{1}{2}} \qquad \text{- - - (13)}$$

where V_0 = Voltage applied between the ring and the end caps. ($\sim 1V$)

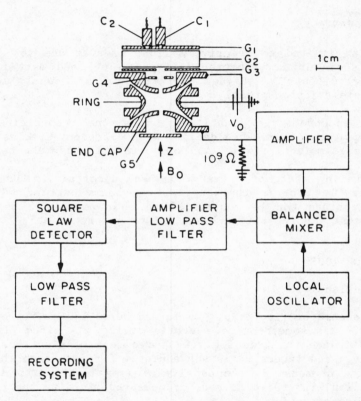

Fig. 4   Schematic of Dunn's Quadrupole ion trap apparatus

$R_0 \sim 0.625$
$Z_0 \sim 0.385$ Inside dimensions of the trap.

The number of ions in the trap is determined by measuring
the noise power in the image currents on one end of the trap.
The spectrum of this noise contains peaks corresponding to the
oscillating frequencies W_z of different ion species, ie., different (q/m). The area under these peaks is proportional to the
product of n_i and T_i, if one of these quantities is kept constant then the other can be studied.

After the ions are formed, usually by low energy electron
impact, coulomb collisions set up a Maxwellian energy distribution in $\sim 10^{-2}$ secs and evaporation of ions causes cooling so
that the kinetic temperature of the ions reaches room temperature with a few minutes.

Perhaps the main advantage of the ion trap is the fact that
the ions can in some cases be stored for a period sufficient to
reach their ground state. This is not true for homopolar molecules such as O_2^+ but is true for heteropolar molecules like NO^+
where the relaxation time is of the order of a few milliseconds. For inert species such as NH_4^+ the residence time in
the trap can be quite long (~ 24 hrs.).

For more reactive species however this time rapidly decreases so that for example, NO^+ can be stored for only 40
mins. Cross section measurements are made by first observing the
natural decay time τ_n of the trap and then measuring the decay
time with the electrons. The cross section is obtained from the
equation:

$$\sigma = (e/i_e)A^- [(1/t_e) \ln (N_1/N_2) - 1/\tau_n)] \qquad - - - (14)$$

where i_e = electron circuit
t_e = time electron current is on
N_1, N_2 = Number of ions before and after electrons are on.
A^- = Area of ion cloud as seen by electron beam.

This last quantity is difficult to measure with reasonable
accuracy and in practice measurements have been normalised to
previous results of other workers.

Regarding the energy resolution of this technique, this is
largely determined by the energy resolution of the incident
electron beam. Two different electron guns have been used, one
with low and one with high energy resolutions ($\Delta E \sim 0.110$ eV and
$\Delta E \sim 0.03$ eV respectively). The energy range of the latter varied from 0.045 to 0.200meV.

Hollow Electron Beam Trap

A different form of ion trap has been used by (Mathur et al 1978, 1979) for the measurement of dissociative recombination of H^+_n ions, (n=2,3,5,---). In this case a high current hollow electron beam is formed within which a potential well of ~ 1 V depth is set up due to space charge. A molecular beam is passed through the side of the high current electron beam and ions thus formed become trapped inside. A second, axial electron beam is used for the electron-ion interactions and sampling of the ions is performed using a quadrupole mass spectrometer. Only relative cross sections can be measured in this apparatus, normalization being required to produce absolute values.

DISSOCIATIVE RECOMBINATION RESULTS

Experimental measurements made prior to 1970 have been discussed in detail by Bardsley and Biondi (1970). Consequently, less will be said about them here. The biggest advances since then have been the advent of the intersecting beam ion trap and laser excitation techniques which have provided much more sensitive and wide ranging approaches to recombination studies than previously available. These techniques have opened up their own debates such as for example the question of effectiveness of indirect dissociative recombination as opposed to the direct mechanism and the eventual fate of the dissociating state in the stabilization process.

An up-to-date summary of experimentally measured dissociative recombination rate coefficients and cross sections is given in Table II.

DIATOMIC IONS

H_2^+

The dissociative recombination of the simplest molecular ion, H_2^+ is of great interest from a theoretical standpoint. Before the advent of intersecting beam techniques however it was not possible to obtain experimental information about H_2^+ because in a plasma it rapidly converts to H_3^+, H_5^+ and higher molecular weight clusters.

Peart and Dolder (1974a) made the first measurements of the total cross section for e + H_2^+ recombination and found cross sections an order of magnitude higher than those measured by Vogler and Dunn (1969, 1975) for the process:

TABLE II

Summary of measured rate coefficient and cross sections for dissociative recombination.

SPECIES	RATE COEFFICIENT $cm^3 \ sec^{-1}$	CROSS SECTION cm^2	REF.	COMMENTS
DIATOMICS				
H_2^+	$2.3 \times 10^{-7}[300/T_e]^{0.4}$	$8.0 \times 10^{-15} [0.1/E_e]^{0.92}$	(1)	
HD^+	$2.3 \times 10^{-7}[300/T_e]^{0.4}$	$8.0 \times 10^{-15} [0.1/E_e]^{0.92}$	(1)	
D_2^+	$1.56 \times 10^{-7}[300/T_e]^{0.3}$	$5.5 \times 10^{-15} [0.1/E_e]^{0.8}$	(1)	
CH^+	$3.0 \times 10^{-7}[300/T_e]^{0.4}$	$8.0 \times 10^{-15} [0.1/E_e]^{0.92}$	(2)	
NH^+	$8.6 \times 10^{-8}[300/T_e]^{0.5}$	$2.5 \times 10^{-15} [0.1/E_e]^{1.0}$	(1)	
OH^+	$7.5 \times 10^{-8}[300/T_e]^{0.5}$	$2.0 \times 10^{-15} [0.1/E_e]^{1.0}$	(3)	
C_2^+	$6.0 \times 10^{-7}[300/T_e]^{0.5}$	$1.4 \times 10^{-14} [0.1/E_e]^{1.0}$	(4)	
N_2^+	$3.5 \times 10^{-7}[300/T_e]^{0.5}$	$1.0 \times 10^{-14} [0.1/E_e]^{1.0}$	(4)	Merged Beam
	$1.8 \times 10^{-7}[300/T_e]^{0.39}$		(6)	Afterglow
NO^+	$2.3 \times 10^{-7}[300/T_e]^{0.5}$	$5.0 \times 10^{-15}[0.1/E_e]^{1.0}$	(4)	Merged Beam
	$4.3 \times 10^{-7}[300/T_e]^{0.37}$		(7)	Afterglow
O_2^+	$1.9 \times 10^{-7}[300/T_e]^{0.5}$	$4.5 \times 10^{-15} [0.1/E_e]^{1.0}$	(4)	
RARE GAS IONS				
He_2^+	$10^{-8} - 10^{-10}$		(8)	competing processes make identification of dissociative recombination difficult
Ne_2^+	$1.8 \times 10^{-7}[300/T_e]^{0.43}$		(9)	Shock tube results display $T_e^{-1.5}$ dependence above 1000K (9a)

(continued)

TABLE II

SPECIES	RATE COEFFICIENT $cm^3 \, sec^{-1}$	CROSS SECTION cm^2	REF.	COMMENTS
Ar_2^+	$9.1 \times 10^{-7}[300/T_e]^{0.61}$		(10)	Shock tube results display $T_e^{-1.3}$ dependence above 1000K (9a)
Kr_2^+	$1.6 \times 10^{-6}[300/T_e]^{0.55}$		(11)	
Xe_2^+	$2.66 \times 10^{-6}[300/T_e]^{0.6}$		(12)	$T_e^{-0.33}$ below 1300K $T_e^{-0.7}$ above 1300K
POLYATOMICS				
H_3^+	$2.3 \times 10^{-7}[300/T_e]^{0.5}$	$1.0 \times 10^{-14}[0.1/E_e]^{1.0}$	(13) (1)	
HD_2^+		$0.8 \times 10^{-14}[0.1/E_e]^{1.0}$	(1)	
D_3^+		$0.6 \times 10^{-14}[0.1/E_e]^{0.84}$	(1)	
CH_2^+	$5 \times 10^{-7}[300/T_e]^{0.5}$	$1.3 \times 10^{-14}[0.1/E_e]^{1.0}$	(14)	
CH_3^+	$7.0 \times 10^{-7}[300/T_e]^{0.5}$ Below 2000K	$1.6 \times 10^{-14}[0.1/E_e]^{1.0}$ Below 0.15 eV $6.0 \times 10^{-15}[0.1/E_e]^{1.6}$ Above 0.15 eV	(14)	Rate coefficient tends to $T_e^{-1.0}$ above 10^4K.
CH_4^+	$7.0 \times 10^{-7}[300/T_e]^{0.5}$ Below 1000K	$1.4 \times 10^{-14}[0.1/E_e]^{1.0}$ Below 0.07 eV $5.6 \times 10^{-15}[0.2/E_e]^{1.5}$ Above 0.07 eV	(14)	Rate coefficient tends to $T_e^{-1.2}$ above 10^4K.
CH_5^+	$7.0 \times 10^{-7}[300/T_e]^{0.5}$ Below 1000K	$1.4 \times 10^{-14}[0.1/E_e]^{1.0}$ Below 0.08 $5.6 \times 10^{-15}[0.2/E_e]^{1.5}$ Above 0.08	(14)	Rate coefficient tends to $T_e^{-1.2}$ above 10^4K.
NH_2^+		$3 \times 10^{-14}[0.04/E_e]^{1.0}$ Below 0.05 eV $9 \times 10^{-15}[0.1/E_e]^{1.35}$ Above 0.05 eV	(15)	
NH_3^+		$3.1 \times 10^{-14}[0.07/E_e]^{1.0}$ Below 0.09 eV $6.5 \times 10^{-15}[0.2/E_e]^{1.8}$ Above 0.09 eV	(15)	

Table II

SPECIES	RATE COEFFICIENT $cm^3\ Sec^{-1}$	CROSS SECTION cm^2	REF.	COMMENTS
NH_4^+	$1.3 \times 10^{-6}[410/T_e]^{0.5}$		(16)	
		$1.08 \times 10^{-14}[0.1/E_e]^{1.4}$ Below 0.5 eV	(17)	
		$5.0 \times 10^{-16}[0.5/E_e]^{3.43}$ Above 0.3 eV		
H_2O^+		$1.8 \times 10^{-14}[0.1/E_e]^{1.0}$ Below 0.15 eV	(3)	
		$6 \times 10^{-15}[0.2/E_e]^{2.0}$ Above 0.15 eV		
H_3O^+/D_3O^+	$6.3 \times 10^{-7}[300/T_e]^{0.5}$ Below 1000K	$1.8 \times 10^{-14}[0.08/E_e]^{1.0}$ Below 0.1 eV	(4)	Merged Beams Rate coefficient tends to T_e -1.1 dependence above 10^4K.
		$5 \times 10^{-15}[0.2/E_e]^{2}$ Above 0.1 eV		
	$7.5 \times 10^{-7}[800/T_e]^{0.5}$		(4a)	Shock tube. Slight increase in T_e dependence above 4000K.
C_2H^+	$5.4 \times 10^{-7}[300/T_e]^{0.5}$	$1.4 \times 10^{-14}[0.1/E_e]^{1.0}$	(2)	
$C_2H_2^+$	$5.4 \times 10^{-7}[300/T_e]^{0.5}$	$1.4 \times 10^{-14}[0.1/E_e]^{1.0}$ Below 0.1	(2)	
		$6 \times 10^{-14}[0.2/E_e]^{1.2}$ Above 0.1		
$C_2H_3^+$	$9 \times 10^{-7}[300/T_e]^{0.5}$	$2 \times 10^{-14}[0.1/E_e]^{1.0}$ Below 0.1 eV	(2)	
		$8.4 \times 10^{-15}[0.2/E_e]^{1.3}$ Above 0.1 eV		
HCO^+	2.0×10^{-7} At 300K		(19)	
N_2H^+/N_2D^+	$7.5 \times 10^{-7}[300/T_e]^{0.5}$	$1.2 \times 10^{-14}[0./1E_e]^{1.0}$ Below 0.1 eV	(20)	Rate coefficient tends to T_e -1.1 above 10^4K
		$4.4 \times 10^{-15}[0.2/E_e]^{1.3}$ Above 0.1 eV		
CO_2^+	3.8×10^{-7} At 300K		(15)	

(continued)

Table II (continued)

SPECIES	RATE COEFFICIENT cm^3 Sec^{-1}	CROSS SECTION cm^2	REF.	COMMENTS
CLUSTER IONS				
H_3^+ (H_2)	3.6×10^{-6} $T_e = 205K$		(13)	
	2.5×10^{-6} $T_e = 300K$		(22)	
$H_3O^+(H_2O)$	2.2×10^{-6} $T_e = 415K$		(23)	No observed energy dependence
$H_3O^+(H_2O)_2$	3.8×10^{-6} $T_e = 300K$		(23)	No observed energy dependence
$H_3O^+(H_2O)_3$	4.9×10^{-6} $T_e = 300K$		(23)	
$H_3O^+(H_2O)_4$	6.0×10^{-6} $T_e = 205K$		(23)	
$H_3O^+(H_2O)_5$	7.5×10^{-6} $T_e = 205K$		(23)	
$H_3O^+(H_2O)_6$	$< 10^{-5}$ $T_e = 205K$		(23)	
$NH_4^+(NH_3)$	$2.82 \times 10^{-6}[300/T_e]^{0.147}$		(16)	
$NH_4^+(NH_3)_2$	$2.68 \times 10^{-6}[300/T_e]^{0.05}$		(16)	
$NH_4^+(NH_3)_3$	3×10^{-6} $T_e = 200K$		(16)	
$NH_4^+(NH_3)_4$	3×10^{-6} $T_e = 200K$		(16)	
$N_2^+(N_2)$	$1.4 \times 10^{-6}[300/T_e]^{0.41}$		(24)	
$CO^+(CO)$	$1.3 \times 10^{-6}[300/T_e]^{0.34}$		(25)	
$CO^+(CO)_2$	$1.9 \times 10^{-6}[300/T_e]^{0.33}$		(25)	
$O_2^+(O_2)$	2.3×10^{-6} $T_e = 205K$		(26)	

(continued)

Table II (continued)

NOTES

1. Generally rate coefficients are indicated at 300K and cross sections at 0.1 eV. Exceptions occur where measurements have not been made at these values.

2. Temperature/Energy dependencies are indicated except in cases where only data point exists.

3. In some cases, cross sections exhibit abrupt changes as noted. Rate coefficients, however, exhibit more gradual change from a constant low temperature dependence (typically below 1000K) to a constant high temperature dependence (above 10,000K).

4. Intersecting beam techniques measure cross sections from which rate coefficients may be calculated. Plasma techniques measure rate coefficients only.

REFERENCES TO TABLE II

1. McGowan et al 1979.
2. Mitchell and McGowan 1978.
3. Mul et al 1981b.
4. Mul and McGowan 1980.
5. Mul and McGowan 1979a.
6. Mehr and Biondi 1969.
7. Huang et al 1975.
8. Maruyama et al 1981.
9. Phillbrick et al 1969.
9a. Cunningham and Hobson 1969.
10. Shiu and Biondi 1978.
11. Shiu and Biondi 1977b.
12. Shiu et al 1977a.
13. Leu et al 1973b.
14. Mul et al 1981a.
15. McGowan et al (unpublished).
16. Huang et al 1976.
17. Dubois et al 1978.
18. Ogram et al 1980.
19. Leu et al 1973C .
20. Mul and McGowan 1979b.
21. Weller and Biondi 1967.
22. Mathur et al 1979.
23. Leu et al 1973a.
24. Whitaker et al 1981b.
25. Whitaker et al 1981a.
26. Kasner & Biondi 1968.

$$e + H_2^+ \longrightarrow H\ (2p) + H.$$

Later measurements by Phaneuf et al (1975) for

$$e + H_2^+ \longrightarrow H\ (n=4) + H.$$

also yielded cross sections an order of magnitude lower.

Merged beam results (Auerbach et al 1977) indicated the presence of resonance structure in the total cross sections for $e + H_2^+$ recombination see fig (5), unfortunately recent remeasurement of this process (D'Angelo 1978) has failed to reproduce this structure. However modifications made to the apparatus in the meantime served to increase its sensitivity but decrease its energy resolution.

At the time that the earlier measurements were made, McGowan et al. (1976) proposed that the resonant structure was a manifestation of the indirect process for dissociative recombination. In this model, originally proposed by Bardsley (1968) and Chen and Mittelman (1967) the incoming electron can vibrationally excite the ion core and having lost its energy, be captured into a vibrationally excited Rydberg state of the neutral molecule. See fig (6). This state can then either autoionize or predissociate. Calculations by Giusti (1980), Derkits & Bardsley (1979) have shown that this process should indeed give rise to narrow resonances (Width $\sim$ few meV) in the dissociative recombination cross section, however, these may be too narrow to have been observed in the Auerbach experiment.

An additional feature of this approach is that when the energy of the incoming electron is sufficient to cause a vibrational excitation of the ion core, the cross section exhibits a step downwards and this leads to an overall change of the energy dependance above the threshold for this process (Typical vibrational spacings for molecular ions are $\sim$0.1eV). McGowan et al. (1979) have shown that the cross sections for a number of polyatomic ions exhibit such an energy dependance change above 0.1eV and possibly this can be attributed to this effect.

O'Malley (1981) however, has proposed an alternative approach to $e + H_2^+$ recombination which examines the fate of the diabatic repulsive state into which the electron is captured in the direct mechanism.

This state can undergo Landau Zener curve crossings with Rydberg states lying below the H_2^+ ion state and these states can subsequently dissociate. There is experimental evidence that such crossings do in fact occur (see Next Section), for the

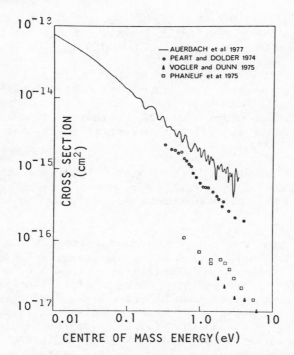

Fig. 5   Results for dissociative recombination of H_2^+.
(Auerbach's curve is displaced for sake of clarity.)

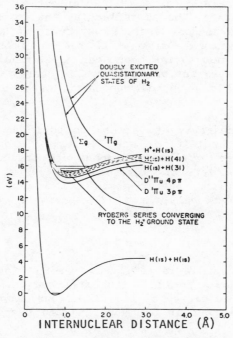

Fig. 6   Potential energy curves for H_2 and H_2^+.

ultimate decay channel for this state, namely $H^+ + H^-$ formation is found to represent a very small fraction of the total recombination cross section (Peart and Dolder, 1975).

These Rydberg states can be treated as resonances and analysis by O'Malley has shown that curve crossings to these resonant states will give rise to resonant structure in the dissociative recombination cross section. Moreover, the widths of this structure should be $\sim 30 - 50$ meV, similar to that seen by Auerbach et al. (1977).

Modification of the MEIBE I apparatus is planned that will greatly enhance its energy resolution and a careful reexamination of these resonant structures will be made in the near future.

One of the problems in comparing experimental results for H_2^+ with theory is the fact that being homopolar, H_2^+ has no dipole moment and so excited vibrational states are very long lived ($>10^6$ secs). A Frank-Condon analysis of the ionization of molecular hydrogen predicts that H_2^+ ions are formed with all 18 possible vibrational states capable of being excited. A careful experimental study by Von Busch and Dunn (1972) who analysed the photodissociation of H_2^+ indicated a slightly different distribution weighted more towards the lower excited states.

Calculations (Derkits et al,1979, Bottcher 1976) have shown that the cross sections for H_2^+ recombination varies greatly with vibrational state, indeed reflecting the Frank-Condon overlap between the particular state's wavefunction and the repulsive doubly excited state into which the electron is captured. This variation has not so far been seen experimentally, rather a smeared out energy variation with an E^{-1} energy dependance has been found. This is expected when all states are summed together.

Efforts are underway at Western to build an ion source of the type designed by Teloy in West Germany (Teloy and Gerlich, 1974). Ions formed by electron impact are trapped by a radiofrequency field for tens of microseconds in which time deactivating collisions with buffer gases can occur leading to H_2^+ ions with low vibrational states. For example, with Neon, the process

$$H_2^+ + N_e \longrightarrow N_eH^+ + H$$

is endothermic, occuring rapidly only for H_2^+ ions with $v > 1$. Hence, H_2^+ ions should dominate the extracted beam in the $v = 0$ or 1 states. Determination of the vibrational population of the H_2^+ beam is carried out in our apparatus by measuring the dissociative excitation process:

$$e + H_2^+ \quad ---> \quad H^+ + H + e$$

and modelling the resulting cross section using calculated curves (Peek, 1967, 1974) for the excitation from individual vibrational states (D'Angelo, 1978).

Isotope Effects

Peart and Dolder (1973, 1974a) in their studies of $e + D_2^+$
and $e + H_2^+$ showed that these cross sections had slightly different energy dependancies and absolute values. McGowan et al
have subsequently measured $e + H_2^+$, HD^+, D_2^+, H_3^+, HD_2^+, D_3^+,
H_3O^+, D_3O^+, N_2H^+ and N_2D^+ (McGowan et al 1979, Mul et al 1981b,
1979b) in an effort to examine isotopic effects on recombination
cross sections. This has particular relevance to the chemical
fractionation of isotopes in astrophysics Watson (1977). Their
studies have shown that the light ions H_2^+, HD^+ and D_2^+ and H_3^+,
HD_2^+ and D_3^+ exhibit small changes in the absolute cross sections and energy dependancies but no differences were found for
heavier species such as N_2H^+/N_2D^+ and H_3O^+/D_3O^+. This reflects
the relative influence that the added mass of the deuterated
species have on the vibrational energy states of the molecules.

Ion-Pair Formation

This interesting process, namely

$$e + H_2^+ \quad ---> \quad H^+ + H^-$$

has been studied by Peart and Dolder (1975) who showed that the
maximum cross section for this reaction varies from 5×10^{-18} cm^2
to 1.16×10^{-18} cm^2 over the measured range from 0.4 eV to 5 eV.
It is in fact the process with the largest cross section for the
formation of H^- and has attracted considerable attention from
the thermonuclear fusion community interested in producing high
current H^- ion sources for neutral beam injectors. As well as
practical applications, this measurement has great significance
for the theory of dissociative recombination.

The $^1\Sigma_g^+$ $(1\sigma u^2)$ state into which the electron is captured,
is repulsive but does not dissociate into neutral fragments.
Rather it eventually leads to $H^+ - H^-$ ion pair formation at
large internuclear separation. (O'Malley 1969). Dolder's measurements however, show that this process is by no means the dominant end point for dissociative recombination. Instead neutral
atom formation is much more important. We now know that many
curve crossings occur to some of the Rydberg states of H_2 lying
below the H_2^+ potential energy curve and dissociating to excited

atom pairs. See fig. (6). This has great significance for the
mechanism of dissociative recombination particularly with regard
to branching ratios for final states about which very little is
known. Furthermore it will be seen later that many polyatomic
molecules have large cross sections which are very similar possibly due to the multitude of Rydberg states available for eventual decay following the curve crossings from the repulsive capture state.

RARE GAS IONS

Reviews of the recombination of rare gas ions have been
published by Oskam (1969), Bardsley and Biondi (1970) and more
recently by Maruyama et al. (1981). These processes have
received considerable attention not only because of their great
importance to the physics of Excimer lasers but also because of
the controversies underlying their measurement.

He_2^+
_____

The helium dimer ion mentioned in the historical
introduction would appear to have the smallest dissociative
recombination rate coefficient of any molecular ion yet
studied. Indeed in a helium plasma, the recombination is
primarily collisional radiative. The reason for this is the
lack of a suitable curve crossing in the vicinity of the lower
vibrational levels of the ground electronic state (Mulliken,
1964). The most recent experimental measurements (See Review by
Maruyama et al. 1981) suggest a rate coefficient in the range
from $10^{-8} - 10^{-10} cm^3$ sec. $^{-1}$. All measurements so far have been
made in plasmas where it is difficult to keep account of the
various competing three body recombination processes. So far no
beam measurements have been made for He_2^+ but are planned when
the new MEIBE II apparatus being constructed at the University
of Western Ontario is operational.

Ne_2^+, Ar_2^+, Kr_2^+, Xe_2^+
_____

Central to any discussion of the dissociative recombination
of diatomic ions must be the question of vibrational state
dependance of the rate coefficient. A strong controversy arose
over this subject in the late 1960's when the afterglow
technique and the Shock Tube Method produced widely differing
variations of the recombination rate coefficient as a function
of Te for Ne_2^+, Ar_2^+ and Kr_2^+ ions. Figs (7a) and (7b).
Whereas the afterglow results indicated a constant $Te^{-1/2}$
temperature dependance over the entire measurement range (300 -
10,000K), results taken with the shock tube apparatus indicated

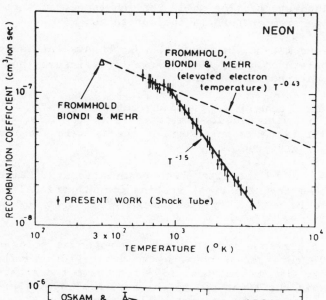

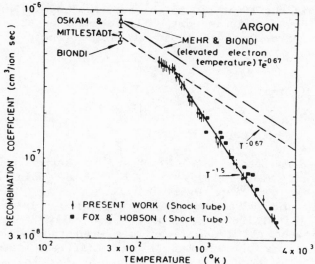

Figs 7a and 7b   Results for dissociative recombination for (a)
                 Ne$_2^+$ and (b) Ar$_2^+$ (taken from Cunningham and
                 Hobson, 1969).

an abrupt change to a $T_e^{-3/2}$ dependence above 1000K. The major difference in the two techniques is that in the case of the afterglow measurement, the electron temperature, T_e, is raised by microwave heating while the ion temperature remains at the ambient gas temperature in the apparatus.

In the shock tube, however, the shock wave initially raises the ion temperature T_i and the electron temperature then follows via collisional thermalization.

This means that as T_i (and therefore T_e) is raised, the probability of vibrational excitation of the ions increases. Hence the vibrational populations of the ions in the two experiments will be different.

In order to rationalize these observations, O'Malley et al (1972) proposed that the cross section for vibrationally excited states should be negligibly small compared to that for the $v = 0$ state.

A possible reason for this was suggested as being the lack of a suitable curve crossing with the upper vibrational levels. More recently, Cunningham et al (1981a,b) have reevaluated these observations in the light of present day knowledge. Before discussing this however, it is worthwhile to consider the other diatomics that have received great attention namely the atmospheric ions.

ATMOSPHERIC GAS IONS

O_2^+
___

This ion has been studied by afterglow (Kasner and Biondi, 1968, Mehr and Biondi, 1969). Shock tube (Cunningham and Hobson, 1972b), Ion Trap (Walls and Dunn, 1974), Merged Beam (Mul and McGowan, 1979a) and by the afterglow / Laser Induced Fluorescence (Zipf, 1980a,b) methods. Generally there is good agreement between all these techniques even though the vibrational populations encountered in the different methods are probably quite different. This indicates that there is very little vibrational state dependance of the total recombination rate coefficient for O_2^+. Zipf (1980a) however, has found that the percentage of $O(^1S)$ atoms formed due to the dissociative recombination of O_2^+ is critically dependant on the initial vibrational state. Contradictions between this experimental observation and theoretical treatments of the vibrational distribution of O_2^+ in the ionosphere coupled to observations of $O(^1S)$ have led Bates and Zipf (1980) to propose an alternative mechanism for the recombination of O_2^+ involving an initial capture into the $^3\Pi$ bound state which subsequently predissociates yielding $O(^1S)$ atoms.

N_2^+

Once again N_2^+ has been studied by a very wide variety of techniques, Mehr & Biondi (1969), Kasner (1967), Sayers (1956), Mul & McGowan (1979), Cunningham & Hobson (1972a) Zipf (1980b) and again there is fairly good agreement between the results although those of Mehr and Biondi lie some 30% below those of Mul & McGowan. There is however quite a controversy involving N_2^+. Results by Orsini et al (1977) based upon chemical modelling of the ionosphere indicated that the rate coefficient for N_2^+ in the v = 1 and 2 states were 22 and 35 times that for the v = 0 level respectively.

This is in total contradiction to theoretical calculations of the dissociative recombination rate coefficients by Michels (1972) who found that the v = 1 and 2 rate coefficients were respectively one-half and one-fiftieth of the v = 0 value.

Subsequent laboratory measurements by Zipf (1980b) have shown that, in fact, the v = 0, 1 and 2 rate coefficients for N_2^+ are all approximately equal. This is a clear example of how unreliable chemical modelling and theoretical calculation can be when dealing with such a sensitive problem as dissociative recombination.

NO^+

NO^+ is the third member of the important family of diatomic atmospheric ions and has received considerable attention both from the theoreticians (Bardsley, 1968, Michels, 1975, Lee, 1977) and from experimentalists, more recent studies being made by Walls and Dunn, 1974, Mul and McGowan, 1979a, Huang et al, 1975, and Hayhurst and Kittelson, 1978.

Although there is excellent agreement between the ion trap and merged beam results, those of Huang et al using the afterglow technique are higher by a factor of 1.7 at 500K, and at 2000K, the flame measurements of Hayhurst and Kittelson are larger by a factor of four. Furthermore, the afterglow results display a lesser temperature dependance, $T_e^{-0.37}$ as opposed to the ion trap and merged beam studies with a $T_e^{-0.5}$ dependence. These differences cannot be easily explained in terms of a vibrational dependance of the rate coefficient as the ion trap and merged beam experiments are expected to have quite different vibrational state distributions.

The apparent contradiction between the vibrational dependence of
the rare gas recombination rate coefficients and the independence of the atmospheric gas ions has been the subject of a recent theoretical analysis by Cunningham et al (1981 a.b.)

Comment On Diatomics

Since the early measurements which initiated the afterglow
vs shock tube controversy two new pieces of information have become available. Accurate potential energy curves for the rare
gas ions are now available (Michaels et al, 1978) and state by
state measurements by Zipf for O_2^+ and N_2^+ have proven the vibrational state independance of the total rate coefficients. As
mentioned earlier, however, Zipf did find that the amount of
$O(^1S)$ formed in the recombination of O_2^+ did depend critically
upon the initial vibrational state indicating that several dissociating states are involved.

Analysis by Cunningham et al has shown that when there is a
large number of such crossings, then in the absence of autoionization the total recombination rate coefficient will be independant of v. However, neglecting autoionization assumes that
the system spends very little time between the initial capture
point and the stabilization point, fig (8) ie where the potential energy lies below the ion state. Examination of the potential energy curves for the atmospheric ions shows that the dissociating states have very steep slopes in the region between
capture and stabilization and so the neglect of autoionization
is justified. The rare gas ions however, have much shallower
slopes and so the probability of autoionization before stabilization is reached becomes much greater particularly for higher
vibrational states. This model would appear to provide a satisfactory explanation for this puzzling controversy which has
overshadowed the field for nearly a decade.

CH^+

This ion deserves special mention because of its importance
to interstellar chemistry. The rate coefficient for the dissociative recombination of CH^+ has been a subject of controversy
since 1951 when Bates and Spitzer showed that the observations
of CH^+ in space could not be easily reconciled with a large destruction rate coefficient since the mode of formation of CH^+ in
cold low density diffuse clouds is via the very slow radiative
association of C^+ ions and H atoms.

Despite the development of more complex steady state chemistry schemes based on molecular Hydrogen, CH^+ continued to be
an enigma leading Solomon and Klemperer (1972) to propose that
the dissociative recombination of CH^+ should be abnormally slow.

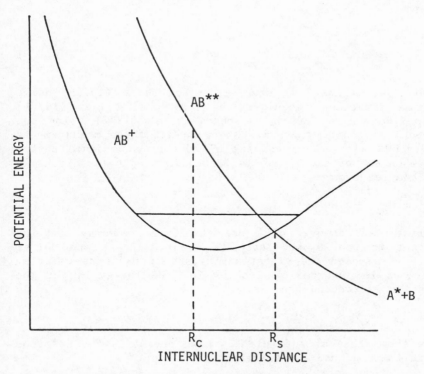

Fig. 8   Schematic representation of ion state and diabatic state
          showing capture point Rc and stablization point Rs.

Theoretical calculations by Bardsley and Junker (1973) and
Krauss and Julienne (1973) indicated the recombination coefficient at 100°K should be $\sim 10^{-7}$ cm^3 sec.$^{-1}$. Later studies however, by Giusti-Suzor and Lefebre-Brion, 1977,indicated that
distortion of the dissociating state in the vicinity of the CH$^+$
ion state would lead to an avoided crossing. They calculated a
recombination coefficient of 10^{-9} cm^3 sec^{-1} at 100°K. More
recent calculations by Guisti-Suzor (private communication) however, indicated a somewhat higher rate, 10^{-8} cm^3 sec.$^{-1}$.

Measurements by Mitchell and McGowan (1978) indicated that
the rate coefficient at 100°K was actually 5×10^{-7} cm^{3}Sec^{-1}, ie.
500 times greater than the maximum allowed for by steady state
chemistry. Since then new chemistry schemes based on transitory
shock wave heating have been developed (Elitzer and Watson 1978,
1980) and these are capable of reproducing the observed densities of CH$^+$ even with a large destruction rate coefficient.
Recent observations by Frisch and Jura (1980), and by Federman
(1980) give support to this hypothesis.

POLYATOMICS

H_3^+
___

This species has been studied by Leu et al (1973a) using the plasma afterglow technique, by Peart and Dolder (1974b) using inclined beams and by Auerbach et al (1977) using the merged beam approach. Agreement between all three measurements is very good* and a line showing E^{-1} energy dependance passes through each set of data. H_3^+ is formed in a hydrogen discharge through the reaction

$$H_2^+ + H_2 \ ---> \ H_3^+ + H$$

Unlike H_2^+ however H_3^+ has a dipole moment allowing transitions to the ground state to occur. Peart and Dolder (1974a) have studied the vibrational state of H_3^+ ions emerging from various ion sources by looking for the threshold for the dissociative excitation process

$$e + H_3^+ \ ---> \ H_2^+ + H$$

and have shown that at sufficiently high pressures the H_3^+ ions emerge in their ground $v = 0$ state.

H_3^+ recombination is important to the heating of inter- stellar clouds by X-rays from hot stars. These X-rays cause ionization of molecular hydrogen, the major component of dense clouds, followed by H_3^+ formation via reaction with H_2. H_3^+ is subsequently destroyed via dissociative recombination, namely:

$$H_3^+ + e \ ---> \ H + H + H \ \ - - - A$$
$$---> \ H_2 + H \ \ \ \ - - - B$$

Branching path A deposits $\sim$5ev of kinetic energy into the fragments whilst reaction B deposits $\sim$ 9ev. These fragments be- come thermalized in the cloud leading to an overall gain in temperature. Obviously if reaction A dominates over reaction B the temperature of the clouds would be considerably lower than if the reverse was true. There is in fact quite a controversy here because interstellar clouds appear to be anomalously warm. (Glassgold & Langer 1973). To date, no measurements of this branching ratio have been made.

---

* The results shown in Auerbach's paper are actually low by a factor of 2 due to an error in the form factor measurements.

Ion Pair Formation

The process:

$$e + H_3^+ \longrightarrow H_2 + H^-$$

has been quoted (Hiskes et al. 1979) as being an important mechanism for the production of H^- ions in ion sources. Measurements of Peart, Forest and Dolder (1979) however, have indicated
that in fact the maximum cross section is only 1.6×10^{-18} cm^2,
occurring at 8 eV and so ion pair production from H_3^+ is therefore unlikely to contribute significantly to H^- production.

H_3O^+

H_3O^+ is a fascinating ion being widely found in terrestrial
and extraterrestrial plasmas. It is the dominant ion in flames
being formed via the reactions:

$$CH + O \longrightarrow CHO^+ + e \qquad\qquad I$$

(the dominant ionization mechanism in flames), followed by:

$$CHO^+ + H_2O \longrightarrow H_3O^+ + CO \qquad\qquad II$$

and finally decaying via:

$$H_3O^+ + e \longrightarrow H_2O + H \qquad\qquad IIIa$$

$$\longrightarrow H_2 + OH \qquad\qquad IIIb$$

The branching ratio for process (III) is not known although
it is of great significance to interstellar chemistry. Despite
the fact that H_3O^+ is widely dispersed throughout dense interstellar clouds, H_2O is found only in small localized maser regions, suggesting possibly that IIIb is the dominant dissociation
channel.

H_3O^+ recombination has been measured by all the techniques
described in the experimental section of this chapter, the most
recent measurements being made by Ogram et al (1980) (shock
tube), Heppner et al 1976 (ion trap) Hayhurst et al (1974)
(flame), Leu et al (1973b) (afterglow) and Mul et al (1981) merged beams.

There is considerable disagreement between the various results regarding the temperature dependence of the rate coefficient. Most puzzling of all, are the flame results of Hayhurst
et al (1974) which actually exhibit a positive temperature

dependence. Admittedly the temperature range over which the measurements were made is small (2000-2500K) and the errors are large (25-30%). The magnitude of the rate coefficient lies within the range found in other measurements.

This result might be overlooked but for the very extensive theoretical modelling underlying the method by Hayhurst in seven different papers. (See Burdett and Hayhurst 1979, and references therein). Mass spectrometic sampling of flames is an important diagnostic technique for studying combustion chemistry and the very great discrepancy occuring in this instance might have serious repercussions throughout the field.

$CH_2^+ - CH_5^+$, $C_2H^+ - C_2H_4^+$

A number of other polyatomics have been studied and these are summarized in Table II.

Two main points of interest emerge.

1.   The energy dependence of these ions and all polyatomics thus far studied undergoes an abrupt break around 0.1eV possibly reflecting the opening up of the excitation channel of the ion and the subsequent opening up of new auto-ionization channels of the ion-electron complex, McGowan et al 1979.

2.   As the complexity of the ion increases the recombination cross section tends to a constant value, irrespective of species. This is a most unusual finding but can probably be explained in terms of the multitude of dissociation channels available within a complex molecule. This means that the cross section is entirely dominated by the coulomb capture probability of an electron and a singly charged ion.

These effects are illustrated in fig (9)

CLUSTER IONS

These intriguing compounds which are found under medium density plasma conditions are made up of one or more neutral molecules clustered around an ionic core, the whole being held together by polarization forces. Leu et al (1973a,b), Huang et al (1976) Whittaker et al (1981) have made a study of the dissociative recombination of several of these compounds namely H^+ $(H_2)_n$, $H_3O^+(H_2O)_n$, NH_4^+ $(NH_3)_n \cdot$, $CO^+(CO)_n$ and have found some very interesting results. According to Biondi (1976) the more

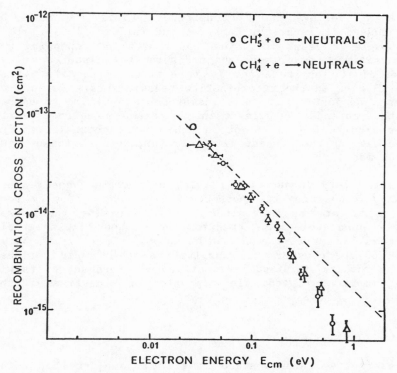

Fig. 9   Results for the dissociative recombination of CH_4^+ and
         CH_5^+ showing similarity of cross sections and the abrupt
         change in the energy dependence near 0.1eV.

complex compounds appear to have very large rate coefficients
$(10^{-5} cm^3 Sec^{-1})$ which might be expected as their complexity
allows for many more dissociation channels.   It is the temperature dependence or rather the independence of the rate coefficient with T_e which is most suprising.   Over the range for
which measurements have been made namely 200–600K, virtually no
variation of the rate coefficient is observed. $CO^+(CO)_n$ is the
exception here, exhibiting a $T_e^{-0.33}$ temperature dependence.
Biondi has proposed that a different recombination mechanism is
operating in the case of cluster ions, namely one in which
energy is lost to internal motions of the cluster with a subsequent dissociation occurring due to the reduced binding energy
of the neutralized species.   It is difficult to see however why
this should give rise to the observed temperature independence.

An alternative suggestion might be given, namely, that some
of the cluster ions are formed in the mass spectrometer region
due to the expansion occurring across the inlet aperture.   Such
an expansion leads to cooling of the gas thus encouraging clust-

tering which follows a T_e^{-1} dependence. This phenomenon has been discussed in detail by Hayhurst and Telford (1971). It is not a simple process, the amount of clustering depending upon the composition, temperature and pressure of the source gas and the size of the aperture nor is it entirely clear what effect this would have on the recombination measurements. Perhaps the best means of justifying these measurements would be to compare them with independant studies using for example the intersecting beam technique which would not be subject to uncertainties from this source. At the present time, however, no such measurements have been made.

Another very fundamental problem underlying these measurements is the question of whether or not such a loosely bound fragile structure as a cluster ion can survive the passage through a quadrupole mass spectrometer without disintegrating (Cunningham 1981; private communication). If, in fact, only a fraction of the clusters did survive then this would seriously alter the results of Biondi's experiments. Because of its wide reaching implications this is a problem that deserves further study.

FUTURE DIRECTIONS AND GOALS

Dissociative recombination is still a subject that offers great challenges to both the experimentalist and the theoretician. Furthermore its great importance in diverse fields such as astrophysics, combustion research, lasers and aeronomy make its understanding imperative. Most recent studies tend to indicate that direct dissociative recombination is the major mechanism operating, although the importance of the indirect process for complex species is still an open question. A satisfactory demonstration of the effects of the initial excitation state of the recombining ion has yet to be made. The most obvious candidate for this study is H_2^+, where only one low lying repulsive state dominates and the poor overlap of the v = 0, 1 and 2 states with this curve should give rise to very different energy dependancies (Derkits et al (1979). The measurement of branching ratios for the dissociation channels of polyatomic molecules is a field which is completely open, no measurements having been made to date. Measurement of the excitation state of the fragments is still in its infancy although some work has been performed by Shiu et al (1977a,b, 1978), Zipf (1980a,b) and by Dunn's Group at JILA.

DIELECTRONIC RECOMBINATION

The measurement of dielectronic recombination rate coef-

ficients and cross sections presents quite a different set of
problems from those involved in dissociative recombination.
Since dielectronic capture occurs into resonant states lying
several or even ten's of ev above the ground state of the recombining ion the problem of achieving very low energies in intersecting beam experiments is no longer relevant. On the other
hand, the low temperature afterglow experiments are not applicable and plasma type measurements require the use of high temperature fusion type devices. The major difference between atomic ion and molecular ion recombination however is of course in
the magnitude of the rates of the two processes. While typical
rates for dissociative recombination lie in the 10^{-5}-10^{-7}
$cm^3 sec^{-1}$ range, those for dielectronic recombination have been
calculated to be from 10^{-9}-$10^{-11} cm^3 sec^{-1}$. Typical cross sections
might lie in the range from 10^{-18}-$10^{-22} cm^2$. This presents serious signal recovery problems due to low signal count rates and
high signal to background noise ratios.

Before examining in detail some of the approaches that are
being made to this problem, it is worthwhile to examine the general process namely:
$$e + A^{z+} \longrightarrow A(z-1)^{+**} \longrightarrow A(z-1)^{*+} + h\nu$$

This is illustrated diagrammatically in fig. (10). The incoming electron excites a target electron to a higher level and

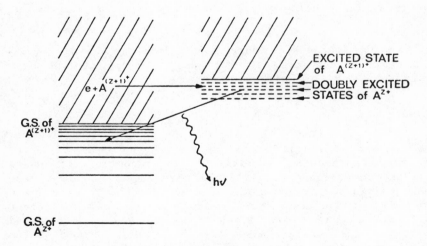

Fig. 10   Delectronic Recombination

is itself captured in the process most probably into a high
Rydberg state. The resulting electron-ion complex can be thought
of as consisting of an excited ion core with a satellite electron
partner which perturbs the core very little. This complex or
resonance lies above the ionization limit of the daughter ion but
below the energy of the corresponding excited state of the ion
core.

Two things can then happen. Either the captured partner
electron is detached from the complex i.e. autoionization occurs,
or the excited ion core can relax, emitting a photon which carries away the excess energy.

This brings the energy of the complex below the ionization
limit of the daughter ion and thus constitutes a stabilization of
the recombination. It should be noted, however, that the possibility of reionization of the resulting high rydberg daughter ion
by interaction with its environment is quite high and so this
stabilization mechanism is not as secure as that for dissociative
recombination.

Because the energy levels of the ion core are affected very
little by the presence of the partner electron the stabilizing
transition corresponds closely to a conventional resonance transition of the ion. In fact these transitions have been observed
as satellite lines adjacent to resonance lines in solar and terrestrial high temperature plasmas (see review by Dubau et al,
1980).

In summary then, dielectronic recombination is characterized
by the appearance of a daughter ion and the emission of a photon
with energy slightly below a resonance transition of the parent
ion.

Experimental techniques currently being employed to study
this process therefore are seeking to detect either the daughter
ion, or the photon or in some cases both.

PLASMA TECHNIQUES

Two separate groups have published results on rate
coefficient measurements of dielectronic recombination using
plasma techniques (Brooks et al 1978, Breton et al 1978). These
methods depend upon the comparison of measured and calculated
line emission intensities from impurity ions in a transient
plasma whose electron temperature and density have been measured
independantly. All ions are assumed to be in their ground states
and a set of coupled rate equations shown below is solved.

$$\frac{dN_z}{dt} = Ne\left[S_{z-1}N_{z-1} + \alpha_{z+1} - (S_z + \alpha_z)\,N_z)\right] \qquad - - - \quad (15)$$

where N_z is number density of ions with charge Z
N_e is number density of electrons
S_z is ionization rate coefficient
α_z is recombination rate coefficient
t is time.

The method of solution is as follows. The time dependence of the number density of a particular ion N_z is measured spectroscopically and the coefficients S_z and α_z are varied in equation (15) in order to match the measured value of dN_z/dt. Details of the techniques are contained in the papers cited above.

Brooks et al have published rate coefficients for the dielectronic recombination of Fe IX, Fe X and Fe XI ions with a stated accuracy of 30% and these are listed in table (III). It should be noticed that agreement with calculations by Jacobs et al (1977) is within a factor of two whilst that with those of Burgess (1965) is considerably worse particularly for Fe IX.

Breton et al do not actually specify their measured values for the ions Mo^{30+}, Mo^{31+}, Mo^{32+} which they studied but simply state that their values are between 0.5 and 1 times those of Burgess (1965). Plasma measurements of dielectronic recombination rate coefficients are obviously less than satisfactory from the point of view of making detailed studies of the process. The information obtained depends heavily on the modelling of the plasma and on the rate coefficients for competing processes. Furthermore, it is difficult to obtain information on temperature dependancies and the overall accuracy of the measurement is low.

Nevertheless, they do yield information on highly charged species difficult to produce as ion beams and for those interested in an order of magnitude estimate of the recombination rates in fusion devices they do provide in situ answers.

TABLE III

Measured and calculated Dielectronic Recombination Rate Coefficients.

	EXP	JACOBS (1977)	BURGESS (1965)
Fe IX	1.1×10^{-10}	2.7×10^{-10}	6.1×10^{-10}
Fe X	1.6×10^{-10}	2.3×10^{-10}	3.5×10^{-10}
Fe XI	2.2×10^{-10}	2.2×10^{-10}	3.3×10^{-10}

INTERSECTING BEAM METHODS

At least four groups are presently attempting to measure di-electronic recombination cross sections using intersecting beam techniques in North America, along with two groups in Japan and one in West Germany. As of this writing no results have been published from any of these groups and so detailed descriptions of the techniques are not available in the literature. Not enough is known of the techniques being used overseas to comment on them here but outlines of the four North American experiments are given in the following sections.

J.Wm.McGowan - THE UNIVERSITY OF WESTERN ONTARIO

Two attacks on the problem of dielectronic recombination are being made at UWO. The first, is to use the MEIBE I apparatus, described previously for examining the recombination of singly charged atomic ions. In order to do this the system has been up-graded by installing closed-cycle helium cryopumps which have reduced the pressure to 1×10^{-10} torr. As of this writing experiments are underway for the study of C^+ recombination and pre-parations are being made to examine Mg^+.

The second approach has been to construct an entirely new merged beam apparatus, MEIBE II which although using the same merging-demerging principle as MEIBE I has a radically improved pumping system and a capability for studying multiply charged ions. This apparatus is shown in Fig (11). It is designed to operate at pressures better than 1×10^{-12} torr and incorporates both liquid helium and closed cycle helium cryopumping. Separation of the small product signals from the primary ion beam is accomplished using a two stage electrostatic analyser. MEIBE II is presently undergoing its finishing touches and initial testing is about to begin.

J. Kohl - HARVARD-SMITHSONIAN CENTER FOR ASTROPHYSICS

This apparatus which has been under development for a number of years uses the inclined beam principle pioneered by Peart and Dolder. In its present form, the photons emitted during the electron-ion interaction are detected using a broadband detection system. Preliminary experiments have concentrated on the e + C^{3+} electron impact excitation cross section. This was measured previously by Taylor et al (1977) and so provides a method of calibrating the apparatus. In the near future the broadband system will be replaced with a spectrometer to identify the satelite line arising from the stabilizing transition in dielec-tronic recombination. Ultimately, dielectronic recombination

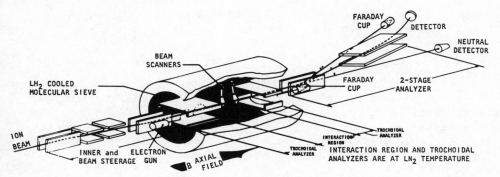

Fig. 11   Sketch of McGowan's MEIBE II apparatus for the study of
dielectronic recombination.

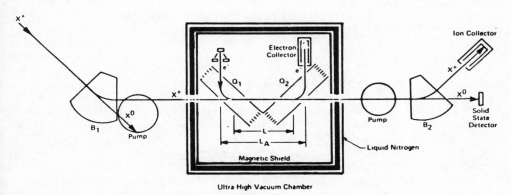

Fig. 12   Golden's merged beam apparatus for the study of
radiative and dielectronic recombination.

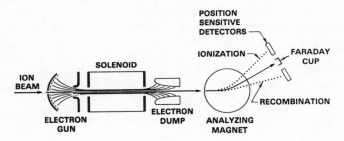

Fig. 13   Dittner's merged beam apparatus for the study of
dielectronic recombination of heavy ions.

cross sections will be measured by detecting the stabilizing photon and the recombined ion in coincidence.

D. Golden – UNIVERSITY OF OKLAHOMA

A merged beam apparatus is at present under construction for
the study of both dielectronic and radiative recombination.   The
ions are derived from a 4 MeV Van de Graaff accelerator and the
electron beam, originally travelling at right angles to the electrons is made to merge with the ions by means of a 45° parallel
plate analyzer.

The interaction region is 1 metre in length and following
the collision the electrons are removed using a second parallel
plate analyzer.   Momentum analysis of the ion beam is performed
using a magnet to separate out the recombination products which
are detected in coincidence with the photons also emitted during
the recombination process.   A diagram of the apparatus is given
in fig (12).

P. F. Dittner – OAK RIDGE NATIONAL LABORATORY

This experiment again uses the merged electron ion beam
technique for the study of multiply charged ions such as 0^{5+},
Si^{4+}, Cl^{7+}.   A diagram is given in fig. (13).   The ions are derived from the ORNL-EN Tandem accelerator which is capable of
producing a wide variety of species with multiple charge states.
The concept behind this apparatus is similar to that first used
by McGowan in early unsuccessful merged beam attempts (Cacak et
al (1971).   The electron beam is produced from a ring cathode
sitting in the fringing field of a long solenoid ($\sim$1m in
length).   The ion beam passes through the cathode and along the
axis of the solenoid.   Electrons leaving the cathode follow the
magnetic field lines and become confined to a narrow beam with
the same axis as the ion beam.   Upon leaving the solenoid, the
electron beam blows apart due to space charge repulsion and is
collected on an annular beam dump.   The ions are analyzed magnetically and the recombination/ionization products are separated
out and detected.   Electron beam modulation is used to separate
out the true signal from background effects.

As of this writing no recombination signals have been detected in this apparatus.

REFERENCES

D. Auerbach, R. Cacak, R. Caudano, T.D. Gaily, C.J. Keyser, J. Wm. McGowan, J.B.A. Mitchell and S.F.J. Wilk, J. Phys. B. 10, 3797, (1977).

J.N. Bardsley, J. Phys. B. 1, 365 (1968).

J.N. Bardsley and M.A. Biondi, Adv. At. Molec. Phys. Vol. 6 ed. D.R. Bates and I. Estermann (New York: Academic Press) p. 1, 1970.

J.N. Bardsley and B.R. Junker, Ap. J. 183, L135 (1973).

D.R. Bates and H.S.W. Massey, Proc. Roy. Soc. A192, 1 (1947).

D.R. Bates and L. Spitzer. Ap. J. 113, 441 (1951).

D.R. Bates Phys. Rev. 77, 718, 1950; 78, 492, (1950).

D.R. Bates in Case Studies in Atomic Physics, eds. M.R.C. McDowell and E.W. McDowell and E.W. McDaniel North Holland, Amsterdam, Vol. 4, 57, (1974).

D.R. Bates and E.C. Zipf, Planet Space Sci. 28, 1081, (1980).

M.A. Biondi, Comments on Atomic and Molecular Physics (1976).

M.A. Biondi and S.C. Brown Phys. Rev. 75, 1700, (1949).

C. Bottcher, J. Phys. B9, 2899 (1976).

C. Breton, C. DeMichelis, M. Finkenthal and M. Mattioli. Phys. Rev. Lett. 41, 110 (1978).

R.L. Brooks, R.U. Dalta and H.R. Griem Phys. Rev. Lett 41, 107, (1978).

N.A. Burdett and A.N. Hayhurst, Combustion and Flame, 34, 119 (1979).

A. Burgess, Ap. J. 139, 776 (1964).

A. Burgess, Ap. J. 141, 1588 (1965).

R.K. Cacak, R. Caudano, T.D. Gaily and J. Wm. McGowan VIIth ICPEAC, Amsterdam ed. J. Kistemaker (Amsterdam: North Holland) Abstracts p. 992 (1971).

J.C.Y. Chen and M.A. Mittleman, Abstr. Vth ICPEAC Leningrad. ed. I.P. Flakes and E.S. Solovyov, Nauka, Leningrad p.329, (1967).

A.J. Cunningham and R.M. Hobson, Phys. Rev. 185, 98 (1969).

A.J. Cunningham, T.F. O'Malley and R.M. Hobson, Abstr. of papers XIIth ICPEAC, Gatlinburg, USA (ed. S. Datz) 1981a, p. 484.

A.J. Cunningham, T.F. O'Malley and R.M. Hobson, J. Phys. B. 14, 773, (1981b)

A.J. Cunningham and R.M. Hobson, J. Phys. B 5, 2328 (1972b).

A.J. Cunningham and R.M. Hobson, J. Phys. B 5, 2320 (1972a).

V.S. D'Angelo. Msc Thesis University of Western Ontario 1978.

C. Derkits, J.N. Bardsley and J.M. Wadehra, J. Phys. B. 12, L529 (1979).

K.T. Dolder, Case Studies in Atomic Collision Physics Vol. 1; ed. E.W. McDaniel and M.R.C. McDowell. (Amsterdam: North-Holland) p. 249, (1969).

J. Dubau and S. Volonte, Rep. Prog. Phys. 43, 199 (1980).

R.D. Dubois, J.B. Jeffries and G.H. Dunn Phys. Rev. A17, 1314 (1978).

M. Elitzur and W.D. Watson, Ap. J. (Letters) 204, L134 (1978).
M. Elitzur and W.D. Watson, Ap. J. 236, 172 (1980).
S.R. Federman, Ap. J. 241, L109 (1980).
J.N. Fox and R.M. Hobson, Phys. Rev. Lett. 17, 161, 1966.
P.C. Frisch and M. Jura, Ap. J. 242, 560 (1980).
L. Frommhold, M.A. Biondi and F.J. Mehr, Phys. Rev. 165, 44
     (1968).
A. Giusti, J. Phys B 13, 3867 (1980).
A. Giusti-Suzor and H. Lefebre-Brion, Apt. J. 214, L101 (1977).
A.E. Glassgold and W.D. Langer, Ap. J. 186, 859 (1973).
M.F.A. Harrison, Brit. J. Appl. Phys. 17, 371 (1966).
A.N. Hayhurst and N.R. Telford, J. Chem Sec. Faraday Trans I.
     1999, 1974.
A.N Hayhurst and D.B. Kittelson, Combustion and Flame, 31, 37
     (1978).
A.N. Hayhurst and N.R. Telford, Proc. Roy. Soc. Lond. A 322, 483
     (1971).
R.A. Heppner, F.L. Walls, W.T. Armstrong and G.H. Dunn, Phys.
     Rev. A13, 1000 (1976).
J.R. Hiskes, M. Bacal and G.W. Hamilton, Lawrence Livermore
     Report UCID, 18031 (1979).
C.M. Huang, M.A. Biondi and R. Johnsen, Phys. Rev. A14, 984
     (1976).
C.M. Huang, M.A. Biondi and R. Johnsen, Phys. Rev. A11, 901
(1975).
V.L. Jacobs, J. Davis, P.C. Kepple, M. Blaha Ap. J. 211, 605
     (1977).
W.H. Kasner and M.A. Biondi, Phys. Rev. 174, 139 (1968).
W.H. Kasner, Phys. Rev. 164, 194 (1967).
C. J. Keyser, H.R. Froelich, J.B.A. Mitchell and J. Wm. McGowan
     J. Phys. E., 12, 316 (1979).
H.A. Kramers, Phil. Mag. 44, 836 (1923).
M. Krauss and P.S. Julienne, Ap. J. 183, L139 (1973).
C.M. Lee, Phys Rev, A16, 109 (1977).
M.T. Leu, M.A. Biondi and R. Johnsen, Phys. Rev. A7, 292 (1973a).
M.T. Leu, M.A. Biondi and R. Johnsen, Phys. Rev. A8, 413 (1973b).
M.T. Leu, M.A. Biondi and R. Johnsen, Phys. Rev. A, 420 (1973c).
T. Maruyama, Y. Ichikawa, R.M. Hobson, S. Teii, T. Kaneda and
 J.S. Chang, IEE Japan. Proc. Symp. "Fundamental Process in
Gas Discharges" Vol 2 P. 1 (1981).
H.S.W. Massey and D.R. Bates, Rept. Prog. Phys. 9, 62 (1943).
D. Mathur, S.U. Khan and J.B. Hasted, J. Phys. B 11, 3615 (1978).
D. Mathur, J.B. Hasted and S.U. Khan, J. Phys. B 12, 2043 (1979).
J. Wm. McGowan, P.M. Mul, V.S. D'Angelo, J.B.A. Mitchell, P.
Defrance and H.R. Froelich Phys. Rev. Lett. 42, 373 (1979).
     Erata, 42, 1186 (1979).
J. Wm. McGowan, R. Caudano and J. Keyser, Phys. Rev. Lett. 36,
     1447 (1976).
F. J. Mehr and M.A. Biondi, Phys. Rev. 181, 264, (1969).
F.J. Mehr and M.A. Biondi, Phys. Rev. 176, 322 (1968).

H.H. Michaels, R.H. Hobbs and L.A. Wright, J. Chem. Phys. 69, 5151 (1978).

H. H. Michels, Proc. 3rd Int. Conf. Atomic Phys. Boulder, 1, 73 (1972).

H.H. Michels, Air Force Cambridge Research Laboratory Report, No. AFCRL-TR-75-0509, 1975.

J.B.A. Mitchell and J. Wm. McGowan, Ap. J. 222, L77 (1978).

P. M. Mul and J. Wm. McGowan, Ap. J. 227, L157 (1979b).

P.M. Mul and J. Wm. McGowan, Ap. J. 237, 749 (1980).

P.M. Mul and J. Wm. McGowan, J. Phys. B. 12, 1591 (1979a).

P.M. Mul, J.B.A. Mitchell, V.S. D'Angelo, P. Defrance, J. Wm. McGowan and H.R. Froelich J. Phys. B. 14, 1353 (1981a).

P.M. Mul, J. Wm. McGowan, P. Defrance and W. Claeys (to be published, 1981b).

P.M. Mul, W. Claeys, V.S. D'Angelo, H.R. Froelich and J. Wm. McGowan (In preparation, 1981c).

R.S. Mulliken, Phys. Rev. 136, A962 (1964).

T.F. O'Malley (Accepted for publication, J. Phys. B. 1981).

T.F. O'Malley, A.J. Cunningham and R.M. Hobson, J. Phys. B. 5, 2126, (1972).

T.F. O'Malley, J. Chem. Phys. 51, 322, (1969).

G.L. Ogram, J.S. Chang and R.M. Hobson Phys. Rev. A21, 982, (1980).

N. Orsini, D.G. Torr, H.C. Brinton, L.H. Brace, W.B. Hanson, J.H. Hoffman, and A.O. Nier, Geophys. Res. Lett 4, 431, (1977).

H.J. Oskam, in "Case Studies in Atomic Collision Physics Vol. I", ed. E.W. McDaniel and M.R.C. McDowell, Elsevier-New York p 465 (1969).

H.J. Oskam and V.R. Mittelstadt, Phys. Rev. 132, 1445 (1963).

B. Peart, R.A. Forest and K.T. Dolder, J. Phys. B.12, 3441 (1979).

B. Peart and K.T. Dolder, J. Phys. B. 7, 1567 (1974b).

B. Peart and K.T. Dolder, J. Phys. B 8, 1570 (1975).

B. Peart and K.T. Dolder, J. Phys. B 7, 236 (1974a).

B. Peart and K.T. Dolder, J. Phys. B. 7, 1948 (1974c).

B. Peart and K.T. Dolder, J. Phys. B. 6, L359 (1973).

J.M. Peek, Phys. Rev., 52, 154 (1967).

J.M. Peek, Phys. Rev. A10, 539 (1974).

Phaneuf, D.H. Crandall and G.H. Dunn, Phys. Rev. A11, 528 (1975).

J. Phillbrick, F.J. Mehr and M.A. Biondi, Phys. Rev. 181, 271 (1969).

J. Sayers, J. Atmos. Terr. Phys. Special Suppl. 6, 212 (1956).

J. Sayers Private Communication 1943.

Y.J. Shiu and M.A. Biondi and Phys. Rev. A17, 868 (1978).

Y.J. Shiu and M.A. Biondi and Phys. Rev. A16, 1817 (1977).

Y.J. Shiu, M.A. Biondi and D.P. Sipler, Phys. Rev. A15, 494 (1977).

P. Solomon and W. Klemperer, Ap. J. 178, 389 (1972).

P.O. Taylor, D. Gregory, G.H. Dunn, R.A. Phaneuf and D.H.
    Crandall, Phys. Rev. Lett. 391, 1256 (1977).
E. Teloy and D. Gerlich, Chem. Phys. 4, 417 (1974).
M. Vogler and G.H. Dunn, Phys. Rev. A11, 1983 (1975).
M. Vogler and G.H. Dunn, Bull. Am. Phys. Soc. 15, 417, 1970.
F. Von Busch and G.H. Dunn, Phys Rev. A 5, 1726 (1972).
F.L. Walls and G.H. Dunn, Physics Today 27, 34 (1974).
F.L. Walls and G.H. Dunn, J. Geophys. Res. 79, 1911 (1974).
W.D. Watson in "CNO Isotopes in Astrophysics" ed. Jean Audouze,
    Reidel Publishing Co., Dordrecht-Holland, p 105 (1977).
C.S. Weller and M.A. Biondi Phys. Rev. 172, 198 (1968).
C.S. Weller and M.A. Biondi, Phys. Rev. Lett 19, 59 (1967).
M.A. Whitaker, M.A. Biondi and R. Johnsen, Phys. Rev. 24A, 743
    (1981).
M.A. Whitaker, M.A. Biondi and R. Johnsen, Phys. Rev. A23, 1481
    (1981a).
E.C. Zipf, J. Geophys. Res. 85, 4232 (1980a).
E.C. Zipf, Geophys. Res. Lett. 7, 645 (1980b).
E.C. Zipf, Bull. Am. Phys. Soc. 15, 418 (1970).

THEORY OF LOW ENERGY ION-ION CHARGE EXCHANGE

Antoine Salin

Laboratoire d'Astrophysique, E.R. CNRS n° 137,
Université de Bordeaux I, 33405 Talence, France

PART I. RELATION BETWEEN QUANTAL AND SEMI-CLASSICAL THEORIES

INTRODUCTION

In the semi-classical theory one treats the motion of the nuclei classically whereas the motion of the electrons is treated by quantum mechanics. This does not mean necessarily that the classical motion of the nuclei can be considered as independent of that of the electrons. It is, on the contrary, quite clear that the potential experienced by the nuclei depends on the state of the electrons. Hence the trajectory described by the nuclei depends on the evolution of the electronic state while the evolution of the electronic state can be calculated once the trajectory of the nuclei is known. This is a non trivial problem for which solutions have been proposed and tested (Riley, 1973, Green et al., 1981). We shall restrict ourselves to a more simple situation where the trajectories of the nuclei can be obtained from a known potential $V_0(R)$ depending only on the relative distance between them. This potential, by definition, will be independent of the reaction channel, hence the name "common trajectory" given to this approach (Gaussorgue et al., 1975). It is obvious that in ion-ion collisions at low energies the coulomb repulsion between the two ions plays an important role so that its effect has to be included in V_0.

We shall not discuss further the possible definitions of V_0 for the moment. Our aim here is to show the relation between the semi classical theory and the quantum distorted wave theory. This relation has also the advantage to illustrate the basic ideas collision theory.

Statement of the problem

We isolate the part of the potential V_o that we wish to treat classically. This potential is sufficiently strong so that it is worthwhile to treat it as exactly as possible. Therefore we have to start from a formulation of the full quantum theory where this potential is treated exactly (then go to the classical limit): this is the definition of the distorted wave theory. In this sense, we may say that the distorted wave theory is the quantal analogue of the semi-classical "common trajectory" method.

QUANTUM DISTORTED WAVE THEORY

Scattering from two potentials

We begin with the most simple case : the scattering of a particle of mass m by a central potential V(r) :

$$V(r) = U(r) + W(r) \tag{1}$$

We shall suppose, for simplicity, that U, W and V do not behave like coulomb potentials for $r \to \infty$ (the generalization to coulomb potentials can be done, but it involves manipulations of coulomb phases which make the derivation much more cumbersome). The hamiltonian of the system is :

$$H = T + V = T + U + W \tag{2}$$

where T is the kinetic energy operator. We first start with the solution of the eigen-equation :

$$(H_u - E)\chi^{\pm} = (T + U - E)\chi^{\pm} = 0 \tag{3}$$

with asymptotic condition

$$\chi^{\pm} \underset{r \to \infty}{\to} [e^{i\underline{k} \cdot \underline{r}} + f_u(\theta) \frac{e^{\pm ikr}}{r}] (2\pi)^{-3/2} \tag{4}$$

$\frac{1}{2m} k^2 = E$, θ is the scattering angle and $\underline{k}$ the initial momentum.

As usual the functions $\chi^{\pm}$ satisfy the Lippman-Schwinger equation :

$$\chi^{\pm} = (2\pi)^{-3/2} e^{i\underline{k} \cdot \underline{r}} + [E - T \pm i\epsilon]^{-1} U \chi^{\pm} \tag{5}$$

Note : we have introduced the factors $(2\pi)^{-3/2}$ in order to get the same normalization of our functions as used by Joachain (1975).

We now consider the solution of the problem with the full potential V :

$$(H - E)\psi^{\pm} = 0 \tag{6}$$

$$\psi^{\pm}(\underset{\sim}{r}) \underset{r\to\infty}{\longrightarrow} [e^{i\underset{\sim}{k}\cdot\underset{\sim}{r}} + f(\theta) \frac{e^{\pm ikr}}{r}] (2\pi)^{-3/2}$$

Then the transition amplitude $f(\theta)$ can be cast into the form :

$$f(\theta) = -(2\pi)^2 \, m \mathcal{C} \tag{7}$$

where the $\mathcal{C}$ matrix has the form :

$$\mathcal{C} = <\chi^- |U| e^{i\underset{\sim}{k}\cdot\underset{\sim}{r}} (2\pi^{-3/2})> + <\chi^- |W+W(E-H+i\varepsilon)^{-1}W| \chi^+>$$

The first term corresponds to the scattering by the potential U only. The second term corresponds to the <u>scattering by the potential W in the presence of U</u> - an expression which is not meaningful in the quantum theory. We note that the amplitude $f(\theta)$ can be expressed :

$$f(\theta) = f_u(\theta) + f'(\theta) \tag{8}$$

an expression which is also well known for the case when U is a coulomb potential and W a short range potential (see Joachain, 1975). It is easy to see that it is precisely the idea contained in the expression "scattering by the potential W in the presence of U" that we wish to use in the semi classical theory. If $\chi^{\pm}$ is now approximated by its classical limit, we shall get the scattering by W, the trajectory being determined by U.

Two potential formula for multichannel scattering

We shall now generalize the preceeding formula to multichannel scattering. We call i and f the initial an final channel. The state of the system in these channels can be described by an hamiltonian $H_{i,f}$. We call H the hamiltonian describing the entire system in interaction (during the collision). Then :

$$H = H_i + V_i = H_f + V_f \tag{9}$$

is defining the interactions V_i, V_f. As discussed in the introduction, we consider a potential V_o that we wish to treat as exactly as possible. V_o will be our distortion potential. We introduce the definitions :

$$V_i = V_o + W_i \qquad\qquad V_f = V_o + W_f \tag{10}$$

Finally we introduce the functions describing free motion in each channel (the equivalent to the plane wave in potential scattering) :

$$[H_{i,f} - E_{i,f}]\phi_{i,f} = 0 \tag{11}$$

We suppose that we can solve the collision problem in the presence of the potential V_o only. The solutions are the distorted wave functions :

$$\chi_{i,f}^{\pm} = \phi_{i,f} + [E - H_{i,f} \pm i\varepsilon]V_o \, \chi_{i,f}^{\pm} \tag{12}$$

Then the expression (7) can be generalized to :

$$f_{i,f} = (2\pi)^2 (M_i M_f)^{1/2} T_{i,f} \tag{13}$$

where $M_i M_f$ are the reduced masses and :

$$T_{i,f} = \langle\phi_f|\underset{\sim}{\mathcal{T}}|\phi_i\rangle = \langle\chi_f^-|V_i-W_f|\phi_i\rangle + \langle\chi_f^-|\underset{\approx}{\mathcal{G}}|\chi_i^+\rangle$$

$$\underset{\approx}{\mathcal{G}} = W_f + W_f[E-H+i\varepsilon]^{-1}W_i \tag{14}$$

The interpretation of the two terms is not so obvious as in the case of potential sacttering. Let us consider two cases :

a) rearrangement processes :

They can be defined by the condition that $H_i \neq H_f$. The particles form different aggregates in the initial and final channel. We transform the first term in (17) into :

$$\langle\chi_f^-|V_i-W_f|\chi_i^+\rangle = - i\varepsilon\langle\phi_f|\phi_i\rangle - i\varepsilon\langle\phi_f|\chi_i^+\rangle \tag{15}$$

The first term is obviously zero for $\varepsilon \to o$ since $i \neq f$. The same is true for the second term if the potential V_o by itself cannot produce the transition $i \to f$ (remember that χ_i^+ is the exact solution of the problem including the potential V_o). This is not a limitation in our case : to define a trajectory we shall use a potential that depends only on the internuclear distance. It cannot cause any electronic transition. Finally we get :

$$T_{i,f} = \langle\chi_f^-|\underset{\approx}{\mathcal{G}}|\chi_i^+\rangle \tag{16}$$

b) direct processes :

The particles are grouped in the same aggregates in the initial and final channels ($H_i=H_f$). However, ϕ_i may be different from ϕ_f (excitation of the aggregates may occur).

In the present case $V_i - W_f = V_i - W_i = V_o$, hence :

$$T_{i,f} = \langle\chi_f^-|V_o|\phi_i\rangle + \langle\chi_f^-|\underset{\approx}{\mathcal{G}}|\chi_i^+\rangle \tag{17}$$

The interpretation of the first term is the same as for potential scattering : it corresponds to the transition i → f caused under the influence of V_0 only. For the same reason as in case a), V_0 will be chosen so that it cannot cause the transition i → f by itself. Hence this term is zero except for i = f (elastic collisions).

In conclusion if the potential V_0 is chosen so that it causes by itself no transition from i to f (i ≠ f), then :

$$T_{i,f} = <\chi_f^-|V_0\phi_i>\delta_{i,f} + <\chi_f^-|\underset{\approx}{\mathcal{C}}|\chi_i^+> \tag{18}$$

In the following derivation we shall deal exclusively with the second term of (18). The first term is trivial.

CLASSICAL LIMIT OF DISTORTED WAVE EQUATIONS

We consider the case of collisions between two atoms or ions. Let R be the distance between the nuclei. We shall choose V_0 so that $V_0 = V_0(R)$. The functions $\chi^\pm$ correspond to the distortion by the potential $V_0(R)$ only, hence the electrons on one atom do not interact with the other atom. If we call collectively the electronic coordinates $\underset{\sim}{r}$, then obviously :

$$\chi^\pm(\underset{\sim}{R}, \underset{\sim}{r}) = \mathcal{F}^\pm(\underset{\sim}{R})\phi_i(\underset{\sim}{r}) \tag{19}$$

where ϕ_i is a function describing only the state of the electrons in the atom. We shall suppose that the total angular momentum of the electrons is J, M_J and we shall use instead of $\phi_i(\underset{\sim}{r})$ the notation : $|\alpha J M_J>$.

The wave functions $\mathcal{F}^\pm(\underset{\sim}{R})$ can be expanded in partial waves (with respect to the relative motion of the nuclei) :

$$\mathcal{F}_i^+ = (2\pi)^{-3/2}[\frac{4\pi}{k_iR} \sum_{L_oM_o} i^{L_o}e^{i\sigma_{L_o}} F_{L_o}^+(R) Y_{L_o}^{M_o}(\underset{\sim}{\hat{R}}) Y_{L_o}^{M_o*}(\underset{\sim}{\hat{k}_i})] \tag{20}$$

$\underset{\sim}{k_i}$ is the relative momentum before the collision. Similarly :

$$\mathcal{F}_f^- = (2\pi)^{-3/2}[\frac{4\pi}{k_fR} \sum_{L,M} i^Le^{-i L} F_L^-(R) Y_L^M(\underset{\sim}{\hat{R}}) Y_L^{M*}(\underset{\sim}{\hat{k}_f})] \tag{21}$$

Conservation of Total Angular Momentum

We shall make use of the fact that the total angular momentum is conserved in the collision. Any such property of the hamiltonian is reflected on the symmetries of $\underset{\approx}{\mathcal{C}}$.

We have started from states having a definite electronic angular momentum J, M_J and relative angular momentum of the nuclear motion

L, M_L (from now on we shall replace $Y_L^M(\hat{R})$ by $|L,M\rangle$). From this basis, it is possible to construct states having a definite total angular momentum K, M_K :

$$|\alpha JLKM_K\rangle = \sum_{M_L} \begin{pmatrix} J & L & K \\ M_J & M_L & M_K \end{pmatrix} (-)^{J+L+M_K} (2K+1)^{1/2} |\alpha LM_L JM_J\rangle \qquad (22)$$

We have introduced the notation $(\vdots \ \vdots \ \vdots)$ for Wigner's 3-J symbols. The only important thing, on physical grounds, in expression (22) is the fact that the sum runs only over M_L (that is for a given J and M_J it is possible to build a state of given K, M_K using a combinations of states corresponding to given L and only different M_L). Transformation (22) can be used either for the initial or the final state.

Now we express in mathematical terms the fact that $\underset{\sim}{\mathcal{C}}$ conserves the total angular momentum. To this end we introduce the identity operator :

$$\sum_{\alpha JLKM_K} |\alpha JLKM_K\rangle\langle\alpha JLKM_K| = 1$$

which gives :

$$\underset{\sim}{\mathcal{C}} = \sum_{\alpha'J'L'K'M_k'} \sum_{\alpha JLKM_K} |\alpha JLKM_K\rangle\langle\alpha JLKM_K| \underset{\approx}{\mathcal{C}} |\alpha'J'L'K'M_K'\rangle\langle\alpha'J'L'K'M_K'|$$

$$(23)$$

The fact that $\underset{\approx}{\mathcal{C}}$ conserves K and M_K means that :

$$\langle\alpha JLKM_K| \underset{\approx}{\mathcal{C}} |\alpha'J'L'K'M_K'\rangle = \mathcal{C}_{K,M_K}(\alpha JL, \alpha'L'L') \ \delta_{KK'} \delta_{M_K M_K'} \qquad (24)$$

The integration in (24) is over $\underset{\sim}{r}$ and $\hat{R}$. Hence $\mathcal{C}_{K,M_K}$ is only a function of R that we shall name the form factor.

We can use the above properties to simplify our initial expression (17) for the matrix elements of $\underset{\approx}{\mathcal{C}}$. We use (20), (22), (23) and (24) to show that :

$$T_{i,f}(i\neq f) = \sum_{K,M_K} T_{K,M_K} \qquad (25)$$

and :

$$T_{K,M_K} = \frac{2(2K+1)}{\pi \ k_i k_f} \sum_{L_o M_{L_o}} \sum_{L,M_L} i^{L_o-L} e^{i(\sigma_{L_o}+\sigma_L)} (-)^{J+L+J_o+L_o}$$

$$\times \begin{pmatrix} J & L & K \\ M_J & M_L & M_K \end{pmatrix} \begin{pmatrix} J_o & L_o & K \\ M_{J_o} & M_{L_o} & -M_K \end{pmatrix} Y_{L_o}^{M_{L_o}*}(\hat{\underset{\sim}{k}}_i) \ Y_L^M(\hat{\underset{\sim}{k}}_f) \ \mathcal{G}(L,L_o,K) \qquad (26)$$

with $\mathcal{G}(L,L_o,K) = \int_o^\infty dR \ F_L^{-*}(R) \mathcal{C}_{K,M_K}(\alpha JL, \alpha_o J_o L_o) \ F_{L_o}^+(R) \qquad (27)$

This expression gives us the differential scattering amplitude once the quantity $\mathcal{G}$ has been calculated. We now have to show that the quantity $\mathcal{G}$ is readily evaluated from quantities obtained in the semi classical theory (until now we have made no approximation). It can be simplified by using a quantization axis along the direction of the initial momentum $\underset{\sim}{k_i}$ so that :

$$Y_{L_o}^{M_o} (\hat{\underset{\sim}{k}}_i) = \delta_{M_{L_o},o} \sqrt{(2L_o+1)/4\pi} \tag{28}$$

Then the properties of 3J symbols impose $M_K = M_{J_o}$ hence $M_L = M_{J_o} - M_J$ so that the sum in (26) is only over L and L_o and T_{K,M_K} depends only on K.

Reduction of the radial integral $\mathcal{G}(L,L_o,K)$

We shall now introduce the classical approximation to the radial wave functions $F_L^{\pm}(R)$:

$$F_L(R) = [f(R)/k^2]^{-1/4}\sin\phi \tag{29}$$

with : $\phi = \frac{\pi}{4} + \int_{r_o}^{R} [f(r)]^{1/2} dr \qquad f(r) = k^2 - 2mV_o(r) - \frac{L(L+1)}{r^2}.$

r_o is the distance of closest approach (outer zero of $f(r)$). The only difference between F^+ and F^- is that the momentum k may be different in the ingoing and outgoing channel. We get :

$$\mathcal{G}(L,L_o,K) = -\frac{1}{4} \int_o^{\infty} dR(k_i k_f)^{1/2} [f^+ f^-]^{-1/4} [e^{i\phi^+} - e^{i\phi^-}]$$
$$[e^{i\phi^-} - e^{i\phi^+}] \mathcal{C}_{K,M_K}^{C1}(R) \tag{30}$$

We have used the classical limit of $\mathcal{C}$ (see appendix II).

Then a number of approximations are introduced (Alder et al., 1956) and we shall give only then physical content.

(i) The classically forbidden region is neglected in the evaluation of $\mathcal{G}$.

(ii) We suppose that r_o is the same in both channels. This has the consequence that energy is not conserved in the approximate theory.

(iii) We define an average angular momentum :

$$\ell = L_o + \frac{\mu}{2} = L - \frac{\mu}{2} \qquad or \qquad \mu = L - L_o \tag{31}$$

and we shall admit that $\mu \ll \ell$. Momentum will not be conserved exactly (variations of the order of μ will be neglected). For example we put :

$$L_o(L_o+1) \simeq L(L+1) \simeq (\ell+1/2)^2 = k^2\rho^2 \tag{32}$$

ρ will play, as we show later, the role of the impact parameter. Furthermore :

$$L_o(L_o+1) - L(L+1) \simeq - 2\ell\mu \tag{33}$$

(iv) in the integral (30) the terms containing $e^{i(\phi^+ +\phi^-)}$ and $e^{-i(\phi^+ +\phi^-)}$ will be neglected. Only those containing $e^{\pm i(\phi^+ -\phi^-)}$ are kept. This will be a good approximation if k is large enough (since ϕ is proportional to k).

The details of the derivation are given in the appendix. We quote the final result :

$$\mathcal{G}(L,L_o,K) = - \frac{1}{4} \frac{k}{M} \int_{-\infty}^{+\infty} dt \; e^{-i\Delta\varepsilon t} \; \mathcal{C}^{Cl}_{K,M_K}(R) \; \frac{(x+iy)^{\mu}}{R^{\mu}} \tag{34}$$

In this expression t is the time, M and k the reduced mass and relative momentum of the nuclei. R is a function of t defined by the classical law of motion in the potential V_o. x and y are the cartesian components of R in the plane of the trajectory in a con-focal system of coordinates defined in figure 1. We have transfor-med $\mathcal{G}$ into an expression involving a trajectory integral. We have now to show the relation between $\mathcal{G}$ and the solutions of a semi classical theory.

TRANSITION AMPLITUDE IN THE SEMI-CLASSICAL THEORY

Let us now start directly from the semi-classical theory. In that case we follow the evolution of the system under the influence of the time dependent field created by the motion of the nuclei. Hence the wave function describing the motion of the electrons is

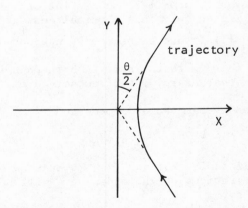

Fig. 1. Confocal system of coordinates.

solution of the time dependent Schrödinger equation :

$$[H_{el} - V_o - i \frac{d}{dt}]\psi^{\pm}_{s.c.}(\underset{\sim}{r},t)\exp\{-i\varepsilon_{i,f}t\}= 0 \qquad (35)$$

H_{el} is equal to H less the kinetic energy term corresponding to the relative motion of the nuclei.
$\varepsilon_{i,f}$ is the eigen-energy of the unperturbed system in the initial and final state respectively.
We have substracted here V_o from H for convenience in the definition of Ψ. The trajectory is defined by $\underset{\sim}{R}(t)$. As V_o depends only on R, it is clear that the fact that we use $(H-V_o)$ instead of V_o in our definition of Ψ makes only a difference of a phase factor. This phase factor may be of importance in some instances. We are here interested in making the relation with the D.W. theory and we may define our semi-classical equations in the form most suitable to make this connection.

The initial and final state satisfy an equation similar to (35)

$$[h_{i,f} - i \frac{d}{dt}]\phi^{s.c.}_{i,f}(\underset{\sim}{r},t) \exp\{-i\varepsilon_{i,f}t\} = 0 \qquad (36)$$

where the difference between $h_{i,f}$ and $H_{i,f}$ is the same as between H_{el} and H and we have the condition :

$$\psi^{\pm}_{s.c.}(\underset{\sim}{r},t) \xrightarrow[t \to \mp\infty]{} \phi^{s.c.}_{i.f.}(\underset{\sim}{r},t) \qquad (37)$$

The probability amplitude for the reaction $i \to f$ is defined by :

$$C_{i,f}(\underset{\sim}{b}) = \langle\psi^{-}_{s.c.}(\underset{\sim}{r},t) / \psi^{+}_{s.c.}(\underset{\sim}{r},t)\rangle \; e^{-i(\Delta\varepsilon)t}$$

$$= \lim_{t\to\infty} \langle\phi^{s.c.}_f / \psi^{+}_{s.c.}(\underset{\sim}{r},t)\rangle \; e^{-i(\Delta\varepsilon)t} \qquad (38)$$

where $\underset{\sim}{b}$ is the impact parameter and $\Delta\varepsilon = \varepsilon_i - \varepsilon_f$.
By using the equations (35) to (37), it is easy to show that :

$$C(\underset{\sim}{b}) = -i\int_{-\infty}^{+\infty} dt \; e^{-i(\Delta\varepsilon)t} \; \langle\phi^{s.c.}_f|W_f|\psi^{+}_{s.c.}(\underset{\sim}{r},t)\rangle \qquad (39)$$

This expression is the semi classical analogue of the T matrix for the quantal theory. The integration in the brackets of equations (38) and (39) is over the electronic coordinates, so that the resulting expression is a function of $\underset{\sim}{R}$ only. Let us introduce :

$$f(\underset{\sim}{R}) = \langle\phi^{s.c.}_f|W_f|\psi^{+}_{s.c.}(\underset{\sim}{r},t)\rangle \qquad (40)$$

We may expand $f(\underset{\sim}{R})$ in partial waves so that :

$$f(\underset{\sim}{R}) = \sum_{\lambda M_\lambda} f^{M_\lambda}_{\lambda}(R) \; Y_\lambda(\hat{\underset{\sim}{R}}) \qquad (41)$$

We choose a coordinate system with the z axis perpendicular to the plane of the trajectory. As in the semi-classical theory $\underset{\sim}{R}$ is by definition in the plane of the trajectory :

$$Y_\lambda^{M_\lambda} (\hat{\underset{\sim}{R}}) = Y_\lambda^{M_\lambda} (\frac{\pi}{2}, 0) \ e^{iM_\lambda \phi} \tag{42}$$

So that

$$C(\underset{\sim}{b}) = -i \sum_{\lambda M_\lambda} \int_{-\infty}^{+\infty} dt \ e^{-i(\Delta\varepsilon)t} \ f_{\lambda M_\lambda} (R) Y_\lambda^{M_\lambda} (\frac{\pi}{2}, 0) \ \frac{(x+iy)^{M_\lambda}}{R^{M_\lambda}} \tag{43}$$

We get in this way an expression which is very close to the expression (34). The identification can be proved in fact. As the demonstration is rather involved we quote directly the final result. Care must be exercised for the relation between the momentum variables. We shall admit that the semi classical calculations are done with a coordinate system for the electronic coordinates having its z axis perpendicular to the collision plane. This will be used in the definition of the angular momentum. E.g., M_{J_o} will be the projection of the electronic angular momentum along this z axis, etc... We define :

$$C(\underset{\sim}{b}) = \sum_{\lambda M_\lambda} C_{\lambda M_\lambda} (b) \tag{44}$$

Note that by construction our coordinate system depends on the plane of the trajectory so that $C_{\lambda M_\lambda}$ depends only on b. We make now explicit the dependence of $C_{\lambda M_\lambda}$ on the various momenta :

$$C_{\lambda M_\lambda} (b) = C_{\lambda M_\lambda} (b; \alpha J M_J; \alpha_o J_o M_{J_o}) \tag{45}$$

The relation between (equ. 34) and $C(\underset{\sim}{b})$ is :

$$\mathcal{G} = \frac{k}{4M} i^{L_o-L+1} (-)^{J+J_o} e^{i(\sigma_L-\sigma_{L_o})} C_{K,L-L_o} [b; \alpha J(K-L); \alpha_o J_o (K-L_o)] \tag{46}$$

with $b = (\ell + 1/2)/k = \rho$.

This expression is completely general. In particular it does not depend on any approximation made on the solution of the electronic part of the problem. For example C(b) can be calculated in the Born approximation, the atomic expansion or the molecular expansion, etc... From the evaluation of C(b) in the classical trajectory theory, one may evaluate the transition amplitude for scattering into the direction θ through the use of (13), (25) − (27) and (46). However a much simpler formula has been obtained recently by Dickinson (1981). The expression (26) can be transformed by using approximations consistent with the semi-classical approximation.

In fact it should be realized that though we violate momentum
conservation in a semi classical approximation (or rather the motion
of the nuclei is a "source" of momentum), the transformation of the
transition amplitude under a rotation has to be described properly.
This is implicitely contained in the expression (26).

PART II. MOLECULAR THEORY OF ATOMIC COLLISIONS

INTRODUCTION

We shall not review here all the aspects of the molecular theory
of atomic collisions. There are already many reviews on the problem
and the basic concepts are well established (Kessel et al, 1978;
Lichten, 1980). Our aim will be to discuss some essential aspects
of the theory that we shall illustrate by simple examples.

BASIC EQUATIONS AND DEFINITIONS

Consider, for simplicity, a system consisting of two nuclei A
and B and one electron. The mass of the nuclei are m_A and m_B respectively. In the center of mass frame, the hamiltonian of the system
is :

$$H = - \frac{1}{2\mu_{AB}} \nabla_R^2 - \frac{1}{2} (m_A + m_B + 1)/(m_A + m_B) \; \nabla_r^2 + V(\underset{\sim}{r}, \underset{\sim}{R}) \qquad (1)$$

The coordinates are defined in figure 2 where 0 is the center of mass
of A and B and $\mu_{AB} = m_A m_B / (m_A + m_B)$.

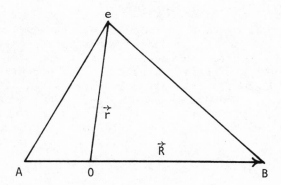

Fig. 2.   Coordinates for the OEDM.

We wish to express the solution of the Schrödinger equation

$$(H-E)\Psi(\underset{\sim}{R},\underset{\sim}{r}) = 0$$

in terms of Born-Oppenheimer states : χ_n defined by :

$$[-\frac{1}{2} \nabla^2 + V(\underset{\sim}{r},\underset{\sim}{R})] \chi_n(\underset{\sim}{r},\underset{\sim}{R}) = \varepsilon_n \chi_n(\underset{\sim}{r},\underset{\sim}{R}) \tag{2}$$

We introduce the expansion :

$$\Psi(\underset{\sim}{r},\underset{\sim}{R}) = \sum_n F_n(\underset{\sim}{R}) \chi_n(\underset{\sim}{r},\underset{\sim}{R}) \tag{3}$$

which yields the coupled equations for F_n :

$$\{-\frac{1}{2} \nabla^2 + \varepsilon_n - E\} F_n(R) = \sum_{n'} \{\frac{1}{2} <\chi_n|\nabla_R^2|\chi_{n'}>_{\underset{\sim}{r}}$$

$$+ <\chi_n|\underset{\sim}{\nabla}_R \chi_{n'}>_{\underset{\sim}{r}} \cdot \underset{\sim}{\nabla}_R\} F_{n'}(\underset{\sim}{R}) \tag{4}$$

with suitable asymptotic conditions.

In the semi-classical approximation (as discussed earlier) we have to solve the equation :

$$(H_{el} - i \frac{d}{dt})\Psi(\underset{\sim}{r},t) = 0 \tag{5}$$

with $H_{el} = -\frac{1}{2} \nabla_r^2 + V(\underset{\sim}{r},\underset{\sim}{R})$.

Expanding $\Psi(\underset{\sim}{r},t)$ onto a set of Born Oppenheimer states yields :

$$\Psi(\underset{\sim}{r},t) = \sum_n a_n(t) \chi_n(\underset{\sim}{r},\underset{\sim}{R}) \tag{6}$$

and the coupled equations :

$$i \frac{d}{dt} a_n(t) = \sum_{n'} <\chi_n|H_{el} - i \frac{d}{dt}|\chi_{n'}> a_{n'}(t) \tag{7}$$

There are two types of coupling terms in these equations.
- $<\chi_n|H_{el}|\chi_{n'}>$. If the χ_n are eigenfunctions of H_{el}, only the diagonal terms are non zero. However we shall encounter cases where it is advantageous to use basis states which are not eigenfunctions of H_{el}. Then non-diagonal terms exist and are called electronic coupling
- $<\chi_n|-i \frac{d}{dt}|\chi_{n'}>$ is the dynamical coupling. When the trajectory of the nuclei is in a plane, it can be expressed in a more convenient form. Introducing the polar coordinates of $\underset{\sim}{R}(R,\Theta)$ (see fig. 3) :

$$\frac{d}{dt} = \frac{dR}{dt}\frac{d}{dR} + \frac{d\Theta}{dt}\frac{d}{d\Theta} \tag{8}$$

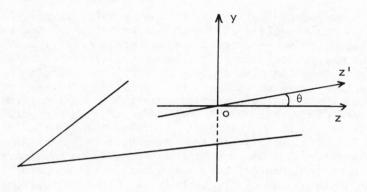

Fig. 3.   Polar coordinates.

We suppose that the wave functions χ_n are defined in a coordinate system linked with $\underset{\sim}{R}$, that is a coordinate system rotating in
the course of the collision. We call z' the axis along $\underset{\sim}{R}$ and y the
fixed axis perpendicular to the collision plane. The wave function
in the fixed system where z is along the initial direction of $\underset{\sim}{R}$ is
connected with the wave function in the rotating frame through the
relation :

$$\chi_n = e^{-iL_y\Theta}\chi_n^{fix} \tag{9}$$

where $e^{-iL_y\Theta}$ is the rotation operator. Hence :

$$\frac{d}{d\Theta}\chi_n = -iL_y\chi_n$$

Consequently :

$$<\chi_n|\frac{d}{dt}|\chi_{n'}> = \dot{R} <\chi_n|\frac{d}{dR}|\chi_{n'}> + \dot{\Theta} <\chi_n|-iL_y|\chi_{n'}> \tag{10}$$

The first term is called "radial coupling". It is multiplied
by the radial velocity. As the radial velocity is zero at the distance of closest approach, it gives no contribution for close internuclear distances. The second term is the "rotational coupling".
For a central potential, the classical trajectory is such that
$\dot{\Theta} = v\rho/R^2$ where ρ is the impact parameter. Then, this term is maximum close to the distance of closest approach. It also increases for
decreasing ρ since the minimum value of R decreases with ρ. For $\rho\rightarrow o$,
$\dot{\Theta}$ goes to a delta function.

The interpretation of the two terms is quite simple. If the velocity was vanishingly small, then the evolution of the system would be adiabatic so that the coupling terms would all be zero. Two reasons may provoke non adiabatic effects. If the change of R is too rapid, then the electronic clouds will not adjust instantaneously to the new value of R as assumed in our adiabatic evolution. Hence the electronic state of the system will no longer stay in a B.O. state. We then have to describe it with a superposition of B.O. states and this is what is described through the first term in (10). Analogously, if the internuclear axis rotates too rapidly the electronic state of the system will not follow this rotation and this is described through the second coupling term in (10). For a discussion of the methods to calculate these couplings see Macias and Riera (1978b).

Finally let us mention the selection rules for these matrix elements. Let us call Λ the projection of the electronic angular momentum along the internuclear axis. As L_z and d/dR commute, then the radial coupling is zero unless χ_n and χ_n' have the same value of Λ. Noting that $iL_y = (L^+-L^-)/2$ where L^+ and L^- are operators which increase and decrease respectively the value of Λ by one unit, the rotational coupling is zero unless the two states have a value of Λ differing by one. Note that when the trajectory of the nuclei is not in a plane the expression of the rotational coupling is more complicated and the selection rule is different.

Finally, let us mention that the molecular theory of atomic collisions suffers from basic inconsistencies which are often referred to as the "momentum transfer" problem. We shall not discuss this problem here (see detailed references in Salin, 1980 and Ponce, 1981).

ONE ELECTRON SYSTEMS

One electron systems are very good examples to illustrate the theory. Although they have some peculiarities, most of the physics of many electron systems can be illustrated by one electron systems. Hence they play the same role in atomic collision theory as the hydrogen atom for the atomic structure.

The B.O. states of a one electron diatomic molecule (OEDM) are eigenstates of the hamiltonian :

$$H_{el} = - \frac{1}{2} \nabla^2 - \frac{z_A}{r_A} - \frac{z_B}{r_B} \tag{11}$$

It is well known that the OEDM hamiltonian has an extra symmetry specific of one electron systems (see the extensive discussion in Power, 1973). It can be read in the literature that "the OEDM

violate the non crossing rule of states of the same symmetry".
Such a claim is wrong. When all the symmetries of OEDM are taken
into account, there is no violation of the non crossing rule. The
electronic energies ε_n, wavefunctions and matrix elements between
OEDM states can be calculated accurately and programs are available
(Power, 1973; Salin, 1978).

Scaling Laws

Let us introduce the scaled electronic coordinates $s=z_A r$.
Hence : $s_A = z_A r_A$, $s_B = z_A r_B$. The hamiltonian can be cast into
the form :

$$H_{el} = z_A^2 \ \{ - \frac{1}{2} \nabla_s^2 - \frac{1}{s_A} - \frac{q}{s_B} \} \tag{12}$$

where $q = z_B/z_A$. Hence all the properties of a system with charges
z_A, z_B can be derived from that of the system with $z_A=1$, $z_B=q$.
Equivalently, we can study all properties of OEDM by simply varying
q. This is what we shall do thereafter.

Correlation rules for OEDM

The correlation rules can be constructed by using the conservation of modes for OEDM wave-functions. The OEDM wave-functions can
be separated in spheroïdal coordinates $\xi = r_A+r_B/2$, $\mu = r_A-r_B/2$, ϕ.
That is :

$$\chi(\underline{R},\underline{r}) = \Xi(\xi,R)M(\mu,R)\Omega(\phi) \tag{13}$$

Let us call n_ξ, n_μ, λ the number of nodes of Ξ, M and Ω respectively. We can correlate these numbers with the number of nodes of
the functions in the united atom ($R\to o$) and separated atom limit
($R\to\infty$).

a) united atom limit :

The eigenstates in the limit go to the hydrogenic states of
an atom with charge $(1+q)$. Let us call n_r, n_ℓ, λ the number of
radial, angular and azimuthal nodes of the atomic wave function.
One has :

$$n_\xi = n_r \quad (=n-\ell-1 \text{ for hydrogenic atoms})$$

$$n_\mu = n_\ell \quad (=\ell-\lambda \text{ for hydrogenic atoms}).$$

This simple rule is used to name the OEDM states. For example the $2p\sigma$ state is the one which goes to the 2p state of the
united atom with $\lambda=o$.

b) separated atom limit :

The situation is more complex for two reasons :
(i) we have to find out on which nucleus the state is centered at infinity
(ii) for one electron systems, the atomic states, say, on center A are influenced by the Stark effect due to the coulomb field of the bare nucleus B.   Hence the molecular states do not tend to spherical states of the separated atoms but to "Stark states". The hydrogenic hamiltonian is separable in parabolic coordinates :

$$\zeta = r + z, \quad \eta = r - z, \quad \phi$$
$$\psi(\underset{\sim}{r}) = Z(\zeta)E(\eta)\Omega(\phi).$$

(14)

To the functions Z and E are associated the quantum numbers n_1 and n_2, respectively, such that $n = n_1 + n_2 + |m| + 1$.
It is important to note that n_1 and n_2 do not play the same role in the correlation rules because the axis AB is oriented : a state around B in the field of A is not invariant by reflexion through B when A is kept fixed.   It can be easily demonstrated that : $n_\zeta = n_1$. This result is independent of the nucleus on which the atomic state is centered.   The form of $M(\mu, R)$ fixes the center on which the atomic state is found at infinite separations.   The proof is more involved and is given elsewhere.   A program exists for a systematic bulding of these correlations.   Let us give an example :

Correlation rule for He^{++} - H

Name	United atom	Separated atom
$1s\sigma$	$1s$	$1s$ on He^{++}
$2s\sigma$	$2s$	$2s$-$2p$ on He^{++}
$2p\sigma$	$2p^{\circ}$	$1s$ on H^+
$3d\sigma$	$3d^{\circ}$	$2s$+$2p$ on He^{++}

Here we have expressed the parabolic states in the separated atom limit in terms of spherical states.   For example the $2s\sigma$ state does not go to the $2s$ state but to the Stark state $2s$-$2p$.

Applications

1) Resonant charge exchange for symmetrical systems (semi classical treatment) :

We neglect all inelastic reactions so that we are left with the two molecular states correlated with a ground state hydrogen atom (fig. 4) : $1s\sigma_g$ and $2p\sigma_u$, that is in (6)

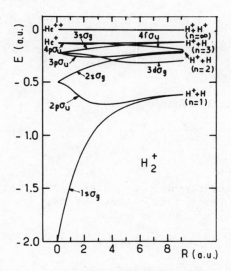

Fig. 4. Purely electronic energies of H_2^+ as a function of internuclear distance.

$$\psi(\underset{\sim}{r},t) = a^+(t)\chi_g(\underset{\sim}{r},R) + a^-(t)\chi_u(\underset{\sim}{r},R). \tag{15}$$

If we put the origin of the electronic coordinates at the
middle of the internuclear axis, the operator d/dt is invariant by
a reflexion through this point. Hence there is no coupling between
gerade and ungerade states and we get :

$$i \frac{d}{dt} a^\pm(t) = \varepsilon^\pm(R)a^\pm(t) \tag{16}$$

where $\varepsilon^+ = \varepsilon_{1s\sigma_g}$ and $\varepsilon^- = \varepsilon_{2p\sigma_u}$. The initial condition is :

$$\psi(\underset{\sim}{r},t) \xrightarrow[t\to-\infty]{} \phi_{1s}^A(\underset{\sim}{r}_A)\exp\{-i\varepsilon_o t\} \tag{17}$$

if the electron is initially on atom A (we neglect here the electron translation factors). As

$$\chi_{g,u} \xrightarrow{R\to\infty} \frac{1}{\sqrt{2}} (\phi_{1s}^A \pm \phi_{1s}^B) \tag{18}$$

we get :

$$a^\pm(t) = \frac{1}{\sqrt{2}} \exp\{-i\int_{-\infty}^{t} (\varepsilon^\pm(R) - \varepsilon(\infty))dt'\}. \tag{19}$$

The probability of finding the electron on B after the collision ($t \to \infty$) is then :

$$\mathscr{P}_{AB} = \sin^2\{ \frac{1}{2} \int_{-\infty}^{+\infty} [\varepsilon^+(R) - \varepsilon^-(R)] dt\}. \tag{20}$$

We consider the general case of the reaction :

$$Z^{n+} + (Z^{n+} + e) \rightarrow (Z^{n+} + e) + Z^{n+}, \tag{21}$$

where Z^{n+} is a bare nucleus.

As we have already shown, the molecular properties of all
these systems can be derived from those of H_2^+.
Introducing $\tilde{R} = ZR$, $\varepsilon(R) = Z^2 \tilde{\varepsilon}(\tilde{R})$, we get

$$\mathscr{P}^Z_{AB}(\rho, v) = \sin^2\{ \frac{Z^2}{2} \int_{-\infty}^{+\infty} [\varepsilon^+(\tilde{R}) - \varepsilon^-(\tilde{R})] dt\} \tag{22}$$

(i) If we use straight line trajectories : $R^2 = \rho^2 + v^2 t^2$. Introducing
$\tilde{v} = v/Z$, $\tilde{t} = Z^2 t$, $\tilde{\rho} = Z\rho$ so that $\tilde{R}^2 = \tilde{\rho}^2 + \tilde{v}^2 \tilde{t}^2$:

$$\mathscr{P}^1_{AB}(\tilde{\rho}, \tilde{v}) = \mathscr{P}^Z_{AB}(\rho, v) \tag{23}$$

so that all the results can be obtained directly from the case
$Z = 1$.

(ii) Trajectory defined from a potential $V_o(R)$. Then :

$$t = \int dR \{\frac{2}{M}[\varepsilon - V_o(R)] - \frac{v^2 \rho^2}{R^2}\}^{-1/2}$$

$$\tilde{t} = \int d\tilde{R} \{[\tilde{v}^2 - \frac{2V_o(R)}{Z^2 M}] - \frac{\tilde{v}^2 \tilde{\rho}^2}{\tilde{R}^2}\}^{1/2}. \tag{24}$$

The relation between $\tilde{t}$ and $\tilde{R}$ will be the same as between t and
R if :

$$\frac{V_o(R)}{Z^2 M} = \frac{V_o(\tilde{R})}{M_o}. \tag{25}$$

If this condition is realized, then (23) is again valid. This
is the case, for example, if $V_o(R)$ is the internuclear potential
Z^2/R and $M = 2Z$. In this case all the results can be derived from
the study of $D^+ + D$. Results for $H^+ + H$ and $He^{2+} + He^+$ are given on
figure 5 (Bates and Boyd, 1962).

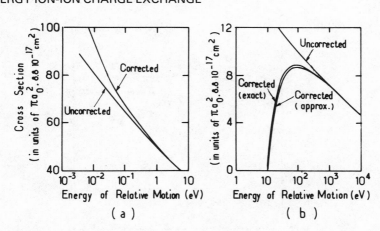

Fig. 5.   Cross-section for the reaction : (a) $H^+ + H(1s) \to$
          $H(1s) + H^+$, (b) $He^{2+} + He^+(1s) \to He^+ + He^{2+}$ calculated
          by Bates and Boyd (1962). The "corrected" curve
          takes into account the repulsion between the
          nculei which is ignored in the uncorrected curve.

Charge exchange for non symmetrical systems

     Again we need only study the problem as a function of q,
since the problem is qualitatively similar for different values
of Z_A. Extensive studies have been carried out for integer values
of q (Harel and Salin, 1977; Olson, 1979). We give in figures 6-9
the potential curves for q = 2,3,5,8. The most saliant feature
for q > 3 is the existence of pseudo crossings at large distances.
The origin of these pseudo crossings is easy to understand. If
the electron is on the atom B, $\varepsilon(R)$ is approximately proportional
to (q-1)/R for large distances, whereas if the electron is on the
nucleus A, $\varepsilon(R) \simeq -\alpha/R^4$. Note that for the sake of clarity we
have not drawn all the states for q = 8 but only those that show
the pseudo-crossing. Calculations show that the pseudo-crossings
determine the charge exchange process and furthermore that for a
given energy only one pseudo-crossing is important. A typical
behaviour of the charge exchange cross section is given in figure
10. It is clearly different from the case of symmetrical systems.

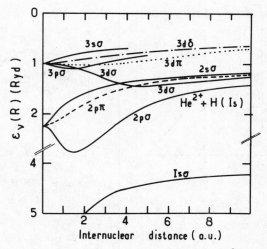

Fig. 6.   Purely electronic energies of HeH^{2+} as a function
of internuclear distance.

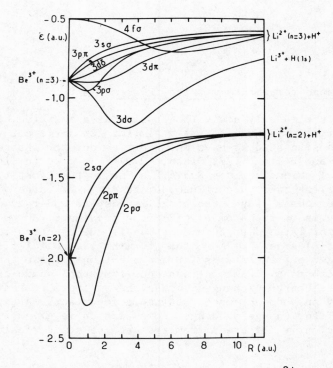

Fig. 7.   Purely electronic energies of $(LiH)^{3+}$ as a function
of internuclear distance.

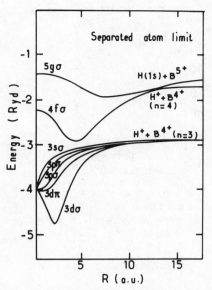

Fig. 8. Purely electronic energies of (BH)$^{5+}$ as a function of
internuclear distance.

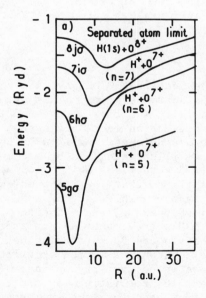

Fig. 9. Purely electronic energies of (OH)$^{8+}$ as a function of
internuclear distance.

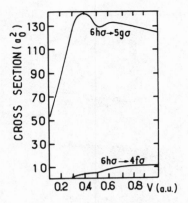

Fig. 10. Electron capture cross-section from atomic hydrogen by
          fully stripped oxygen ions as a function of relative
          velocity (Harel and Salin, 1977).

          The cases q = 2,3 are quite different since the pseudo crossings
are not so well defined as for q > 3. For a detailed analysis, see
references in Olson (1979).

MANY ELECTRON SYSTEMS

     We have stressed the fact that OEDM have a specific symmetry
not encountered in many electron systems. This symmetry is related
to the form of the potential. The first consequence for many electron systems is that even in cases where the system is very close
to an OEDM a number of crossing appearing in OEDM do not appear for
many electron systems. We illustrate this by a comparison between
Be^{4+}-H and C^{4+}-H given in figure 11. The C^{4+} ion has only two
electrons in the inner shell so that the C^{4+}-H molecule is very
close to a one-electron system. This is confirmed by the shape of
the curves in figure 11. For example the $4f\sigma$ and $3p\sigma$ cross each
other in the case of Be^{4+}-H whereas they do not cross in C^{4+}-H.
In the case of Be^{4+}-H the $4f\sigma$ and $3p\sigma$ states do not have the same
symmetry (because of the constant of motion specific of OEDM).
In the case of C^{4+}-H there is no symmetry operator (commuting with
the hamiltonian) which enables to distinguish the corresponding
states : in other terms they have the same symmetry hence they do
not cross. Note that in Be^{4+}-H the $4f\sigma$ and $3d\sigma$ states have the
same symmetry and clearly show a pseudo-crossing.

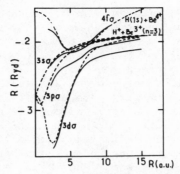

Fig. 11.   Purely electronic energies of BeH^{4+} (---) and CH^{4+} (——)
as a function of internuclear distance. The calculations
of CH^{4+} are pseudo-potential calculations of Valiron (unpublished).

Nevertheless it may still be a useful starting point to
exploit the similarity between the two-systems. The idea that the
symmetry of OEDM states arises from an observable which is nearly
a constant of motion for many electron systems is the root of the
theory of Barat-Lichten (Barat and Lichten, 1972). This theory
may be summarized as follows :
- each electron evolves independently
- the orbital correlation diagram is the same as for OEDM.
Therefore we get easily a correlation between molecular orbitals
and atomic orbitals in the limit R → o. However the problem is
not so clear for large distances and we have to look further into
this.

As we have mentioned before, the OEDM states go at larger
distances to Stark states of the separated atom. This is a special
case related to the degeneracy of the substates of given principal
quantum number in hydrogenic atoms (which causes the linear Stark
effect). For atoms with a least two electrons the states at
infinity are "spherical" states, i.e. eigenstates of L^2 - the
degeneracy between substates is removed by the electron-electron
repulsion. Consequently there is no trivial extension of OEDM
correlations to many electron systems.

Consider a special case : the molecular σ orbitals of a
nearly symmetric molecule going to the 2s and 2p orbitals in the

separated atom limit. We give on figure 12a the correlation of
OEDM and in figure 12b a tentative one for many electron systems.
The problem, first raised by Eichler et al. (1976) has been recently studied quantitatively by Falcon et al. (1981). The
authors show that no simple and general rule exists : the Stark
effect produces transitions between the 2sσ and 3dσ states so that
no one to one correlation at large distances is valid. This limitation has to be kept in mind in applications of the Barat-Lichten
rules. In fact the most significant contribution of the Barat-
Lichten theory is to give a qualitative idea on the shape of the
potential curves, whereas the one to one correlation between united
and separated atomic orbitals should not be taken too seriously.
Also the fact that these qualitative orbitals are OEDM orbitals
replaces a number of pseudo-crossings by crossings, which makes a
qualitative interpretation of experiments much easier. This is
specially significant at small internuclear distances. We give
in figure 13 the example of the 2pσ and 2sσ orbitals at small
internuclear distances. When one applies the Barat-Lichten rules
to many electron systems, difficulties or inconsistencies are often
met. We shall now discuss some of these problems.

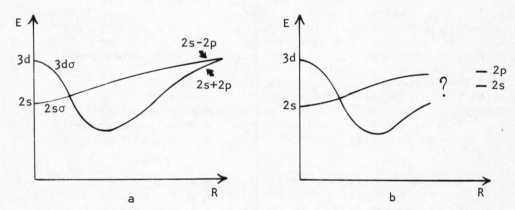

Fig. 12.   Schematic variation with internuclear distance
            of the energies of the orbitals correlated with
            the 2s and 2p atomic orbitals at infinite inter-
            nuclear distance.
            a) OEDM case - b) many-electron system.

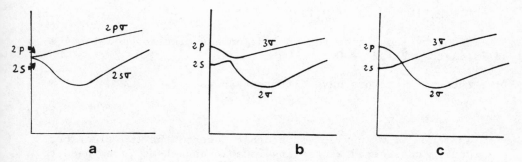

Fig. 13.   Schematic correlation rules for the 2σ and 3σ
           orbitals.
           a) OEDM case – b) Adiabatic correlation for many
           electron systems – c) Barat-Lichten correlation
           rule for many electron systems.

INDEPENDENT ELECTRON MODEL

   This name is rather ambiguous since it is often used with very
different meanings. The idea stems from the fact that we would
like to describe a many electron process as a product of processes
concerning one electron. This means that the many electron wave
function could be a product of "orbital wave function" – that is
functions depending on the space coordinate of only one electron.

$$\chi(\underset{\sim}{r}_1 \ \cdots \ \underset{\sim}{r}_N) = \underset{i}{\pi} \ \phi_i(\underset{\sim}{r}_i). \tag{26}$$

   This idea does not work in general because of the Pauli principle which means that the evolution of the state of one electron
cannot be independent of that of the other electrons. To take into
account the Pauli principle, the usual procedure is to replace the
ϕ_i's by spin-orbitals and write the N electron wave function as a
determinant :

$$\chi(\underset{\sim}{r}_1 \ \cdots \ \underset{\sim}{r}_N) = N^{-1/2} \begin{vmatrix} \phi_1(1) & \phi_2(1) \cdots \ \phi_N(1) \\ \vdots & \vdots \qquad \vdots \\ \phi_1(N) & \phi_2(N) \cdots \ \phi_N(N) \end{vmatrix} \tag{27}$$

   To trace out the consequence of the Pauli principle we consider
the case of two electrons. We write the hamiltonian :

$$H = H_1(1) + H_2(2) + V(1,2) \tag{28}$$

The first condition for an independent electron model to be valid is that $V(1,2) = 0$.

(i) the two electrons have different spins :

$$\chi_\nu = \phi_n(1)\phi_m(2). \tag{29}$$

In fact, if we neglect spin-orbit forces, the two electrons can be considered as distinguishable and there is no need to symmetrize χ_ν. Introducing (29) into (6) we obtain the coupled equations (7). Now :

$$H\phi_n\phi_m = (H_1\phi_n)\phi_m + \phi_n(H_2\phi_m) \tag{30}$$

so that

$$i\frac{d}{dt}a_{nm} = \sum_{n'm'}\{<\phi_n|H_1-i\frac{d}{dt}|\phi_{n'}> + <\phi_m|H_2-i\frac{d}{dt}|\phi_{m'}>\}a_{n'm'} \tag{31}$$

We may now introduce :

$$a_{nm} = \alpha_n\alpha_m \tag{32}$$

and show that the α satisfy the equation :

$$i\frac{d}{dt}\alpha_n = \sum_{n'}<\phi_n|H_1-i\frac{d}{dt}|\phi_{n'}>\alpha_{n'}. \tag{33}$$

This equation is simply the equation describing the evolution of the state of electron 1 and equation (32) states that the probability amplitude is a product of probability amplitudes corresponding to each particle. This is an obvious result for non-interacting distinguishable particles.

(ii) the two electrons have the same spin and form a triplet state.

$$\chi_\nu = \frac{1}{\sqrt{2}}\{\phi_n(1)\phi_m(2)-\phi_n(2)\phi_m(1)\}. \tag{34}$$

We obtain the coupled equations :

$$i\frac{d}{dt}a_{nm} = \sum_{n'm'}a_{n'm'}\{<\phi_n(1)|H_1-i\frac{d}{dt}|\phi_{n'}(1) -$$
$$<\phi_n(1)|H_1-i\frac{d}{dt}|\phi_{m'}(1)>$$

$$+ <\phi_m(2)|H_2-i\frac{d}{dt}|\phi_{m'}(2)>-<\phi_m(2)|H_2-i\frac{d}{dt}|\phi_{m'}(2)>\} \tag{35}$$

We call n_i and m_i the initial values of n and m. Introduce
the transformation :

$$a_{nm} = \alpha_n^{n_i} \alpha_m^{m_i} - \alpha_n^{m_i} \alpha_m^{n_i}. \tag{36}$$

Then the α satisfy the one-electron equations similar to (33)
with the initial condition :

$$\alpha_n^{\nu} \xrightarrow[t \to -\infty]{} \delta_{n\nu}. \tag{37}$$

The probability amplitude is no longer a simple product. However it can still be given a simple interpretation in terms of
one electron processes. We consider the process when one electron is initially in the orbital n_i and the other in m_i. At
the end, the orbitals are n and m. The first term ($\alpha_n^{n_i} \alpha_m^{m_i}$)
corresponds to the electron initially in n_i going to n whereas
the other goes from m_i to n. In the second term the electron
initially in n_i goes to m and the other from m_i to m. As the
electron are indistinguishable there is a way to distinguish
between the two processes so that the amplitudes should inter-
fere (Feynmann et al., 1970).
However the expression of the probability is more complex
since :

$$|a_{nm}|^2 = |\alpha_n^{n_i}|^2 |\alpha_m^{m_i}|^2 + |\alpha_n^{m_i}|^2 |\alpha_m^{n_i}|^2 -$$
$$2\mathscr{Re}\{\alpha_n^{n_i} \alpha_n^{m_i *} \alpha_m^{n_i} \alpha_m^{n_i *}\} \tag{38}$$

A complete discussion has been given by Reading and Ford (1980).

As a conclusion, a proper inclusion of the Pauli principle is
not incompatible with an independent electron model. However
we need the probability amplitudes rather than the probabilities. Also it is clear that an independent electron model is
not incompatible with many electron transitions, as can be seen
immediately from the expressions given above, although the
probability can be appreciable only when the probabilities of
one electron processes are already close to one.

CONFIGURATION INTERACTION

It is not my purpose here to develop the subject formally.
I want to discuss here some of the limitations of the orbital model
and show how they can be overcome.

I shall discuss the simple example of He^{++}-He collisions.

This system has been discussed by Lichten (1963) in the early development of the molecular theory of atomic collisions. It has been
studied recently in details by Harel et al. (1980) and Lopez et al.
(1978).

   We first show that the independent electron model does not
work. Suppose that we build one electron orbitals for He_2^{2+}. These
orbitals will have a gerade or ungerade symmetry by reflexion
through the middle of the internuclear axis since it is a symmetrical

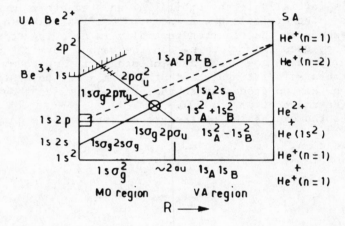

Fig. 14.   Schematic correlation diagram for He_2^{2+} (from
            Lopez et al., 1978).

system. Consider the lowest σ_g and σ_u orbitals (see fig. 14)

$$1\sigma_g \xrightarrow[R\to\infty]{} (1s_A + 1s_B)/\sqrt{2}$$
$$1\sigma_u \xrightarrow[R\to\infty]{} (1s_A - 1s_B)/\sqrt{2}$$

                                                                    (39)

where $1s_{A,B}$ designates a $1s$ atomic orbital on atom A and B respectively. If we suppose that the two electrons are in a σ_g orbital,
we obtain a $^1\Sigma$ state of configuration σ_g^2.

This configuration has the limit :

$$(\sigma_g)^2 \xrightarrow[R\to\infty]{} \frac{1}{2}(1s_A + 1s_B)^2 = \frac{1}{2}[1s_A^2 + 1s_B^2 + 2(1s_A)(1s_B)].\tag{40}$$

Similarly the $(\sigma_u)^2$ configuration gives a $^1\Sigma_g$ state with limit :

$$(\sigma_u^2) \xrightarrow[R\to\infty]{} \frac{1}{2}[1s_A^2 + 1s_B^2 - 2(1s_A)(1s_B)].\tag{41}$$

We obtain at infinity a mixture between states corresponding to He^{2+}-He and He^+-He^+. Such incorrect dissociations are always obtained when a single configuration is used to describe one molecular state, as is the case with our definition of the independent electron model. "Configuration interaction" means that we have to describe molecular states by a linear combination of different configurations. More generally, as the Pauli principle already imposes such a restriction, we have to introduce linear combinations of Slater determinants. For example in the case of He^{2+}-He, the $^1\Sigma_g$ state going at infinity to He^{2+}-He($1s^2$) may be constructed as :

$$^1\Sigma_g = [\sigma_g^2 + \sigma_u^2] \xrightarrow[R\to\infty]{} (1s_A^2 + 1s_B^2)/2\tag{42}$$

the $^1\Sigma_u$ state as :

$$^1\Sigma_u = \sigma_g\sigma_u \xrightarrow[R\to\infty]{} (1s_A^2 - 1s_B^2)/2\tag{43}$$

and the state leading to He^+(1s) - He^+(1s) by :

$$^1\Sigma_g = [\sigma_g^2 - \sigma_u^2] \xrightarrow[\to\infty]{} 1s_A 1s_B\tag{44}$$

This problem of incorrect dissociations is merely an example of the application of configuration interaction methods, which is of more general relevance. The practical aspects are discussed by Lopez et al. (1978).

DIABATIC STATES

I should first mention that there is no definition of diabatic states. Diabatic states can be introduced for convenience in the description of collisions. However the choice will depend on the system considered, the energy, the process under study. Therefore what I shall do here is to first give examples of various cases where diabatic states have been introduced. Then I shall try to make some general statements on methods that may help to construct diabatic states.

Consider a situation where two molecular curves present a pseudo-crossing, as shown in Figure 15.

Molecular curves correspond to an adiabatic approximation, i.e. if the nuclei move infinitely slowly, then when the system is in the state b for $R=R_1$, it will be in state 2 for $R=R_2$. However in a collision, the motion of the nuclei induces transitions between the two states. In some situation the transition probability may be large enough so that if the system is in state b for $R=R_1$, it will be in state (1) for $R=R_2$. In such cases one may say that the system follows the dashed curves shown on the figure. We may refer to such states as diabatic states (in a sense, a diabatic state is anything but an adiabatic state). Of course when the transition probability is close to 1/2, neither the adiabatic nor the diabatic states give an accurate description of the evolution of the system.

The concept of diabatic state is not really very useful in the simple case given above. Let us come back to the case of $He^{2+}-He$.

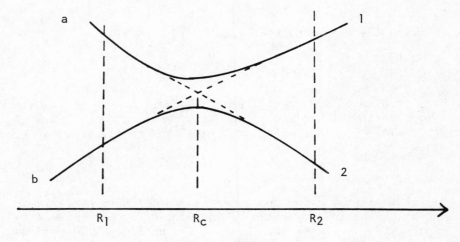

Fig. 15. Pseudo-crossing of molecular potential curves.

We show in figure 16 a diagram of adiabatic states for this system
Clearly the succession of pseudo-crossings can be interpretated
simply by the introduction of the diabatic state shown as the
dashed line. In such a situation, however, it would be desirable
to define the diabatic state beforehand so that one may avoid to
calculate the whole series of adiabatic states. Hence we shall
distinguish two situations :

Determination of diabatic states from adiabatic states

Let us take again the simple case of a pseudo-crossing between
two adiabatic states. A pseudo-crossing corresponds to a rapid
change of the wave functions which is reflected in a sharp variation
of $\varepsilon(R)$. The idea behind the diabatic states is to define a new
set of functions that vary smoothly though the pseudo-crossing
region. An elegant way to do this has been proposed by Macias and
Riera (1978a).

We shall follow the example they have given : the avoided
crossing at $R_0=0.139$ a.u. between the $2s\sigma$ and $2p\sigma$ states of BC^+.
We give in figure 17 the diagonal and non diagonal matrix elements
of the kinetic energy operator and dipole moment for these two states.

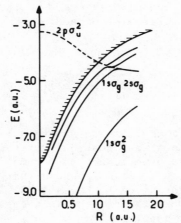

Fig. 16. Potential energy curves for the $^1\Sigma_g$ states of He_2^{2+}. Full
curves are adiabatic states and broken curves diabatic states (from
Lopez et al., 1978).

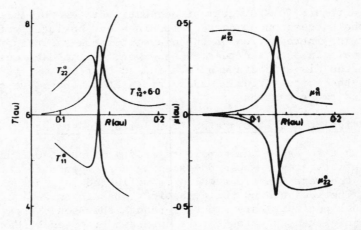

Fig. 17. Matrix elements of:(a) the kinetic energy operator
T, (b) the dipole moment μ, between the 2sσ and 2pσ
orbitals of BC$^+$ (Macias and Riera, 1978a).

Obviously these "properties" of the adiabatic states vary rapidly
through the avoided crossing region. The diabatic states can be
defined by imposing that either the kinetic or the dipole moments
vary smoothly through the crossing region. How to do this in
practice ? Let us write down the matrix elements of the operator
A (the kinetic energy or the dipole moment in the example given
above) in the diabatic representation as :

$$\underset{\approx}{A}^d = \alpha \underset{\approx}{I} + \gamma \underset{\approx}{\sigma}_1 - \Delta \underset{\approx}{\sigma}_3 \tag{45}$$

$$\alpha = 1/2(A^d_{11} + A^d_{22}), \quad \gamma = A^d_{12}, \quad \Delta = 1/2(A^d_{22} - A^d_{11})$$

$$\underset{\approx}{\sigma}_1 = \begin{pmatrix} 0 & 1 \\ 1 & 0 \end{pmatrix} \quad \underset{\approx}{\sigma}_3 = \begin{pmatrix} 1 & 0 \\ 0 & -1 \end{pmatrix} \quad \underset{\approx}{I} = \begin{pmatrix} 1 & 0 \\ 0 & 1 \end{pmatrix}.$$

By hypothesis, α, γ and Δ should vary smoothly in the vicinity of
the pseudo-crossing. The transformation from two adiabatic to the
two diabatic states is an orthogonal transformation which may be
defined by a single angle θ. If we suppose that the non diagonal
matrix element of H in the diabatic basis (H^d_{12}) is constant and
that the difference between the diagonal elements ($H^d_{11} - H^d_{12}$) varies
linearly with $X = R - R_0$, we get :

$$\theta = - \frac{1}{2} \tan^{-1} \delta/X \tag{46}$$

with $\delta(x) = 2x H^d_{12}/(H^d_{22} - H^d_{11})$. In fact we shall suppose only that
$\delta(x)$ varies slowly with x around R_0, so that :

$$\underset{\approx}{A}^a = \alpha \underset{\approx}{I} + (2\delta \frac{d\theta}{dR})^{1/2}(\Delta \underset{\approx}{\sigma}_1 + \gamma \underset{\approx}{\sigma}_3) - \frac{x}{\delta}(2\delta \frac{d\theta}{dR})^{1/2}(\gamma \underset{\approx}{\sigma}_1 - \Delta \underset{\approx}{\sigma}_3) \tag{47}$$

where $\underset{\sim}{A}^a$ is the matrix of A in the adiabatic approximation. With (47) we are able to understand the shape of the curves in figure 13 and extract the value of θ from calculations made in the adiabatic representation. The kinetic energy corresponds to the case when Δ >> γ so that :

$$A_{12}^a \simeq \Delta (2\delta \frac{d\theta}{dR})^{1/2}. \qquad (48)$$

The non diagonal matrix element shows a peak. A fit of this peak allows a determination of dθ/dR. The dipole moment corresponds to a case where Δ << γ. Then :

$$\frac{1}{2} (A_{11}^a - A_{22}^a) \simeq \gamma (2\delta \frac{d\theta}{dR})^{1/2},$$

and this quantity has now a peak.

When Δ≈γ, there is no simple behaviour. However we have a complete freedom on the choice of A which unables one to make the fit determining θ.

<u>Diabatic states and double perturbation problems</u>

Although we have taken as a first example the pseudo crossings between adiabatic states, it is not the only context in which diabatic states have been introduced. There is another class of situations that we shall describe as "double perturbation" problems. The analysis in terms of a double perturbation may help to choose the property (A) to be used in the method of the previous section or for the construction of block-diagonalisation methods to be discussed later.

Let us start from a case we have already encountered. In OEDM, due to the degeneracy of states with a given n, the molecular terms are correlated at infinity to "Stark" states (e.g. 2s+2p, 2s-2p). Consider now a collision between two ions with at least two electrons. At infinite separation, the electron-electron inter-action waives the degeneracy between n substates (if the electrons are on the same nucleus). Hence the 2s and 2p states for example do not have the same energy. When the two ions approach each other, the Stark effect increases. Ultimately it may be much stronger than the electron-electron repulsion so that the molecular orbitals will look very much like OEDM. This is illustrated in figure 18 for $Ne^{8+}-Ar^{12+}$.

Schematically we may represent this in the following way

Region I	Region II
Stark effect	Electron-electron
dominant	repulsion dominant

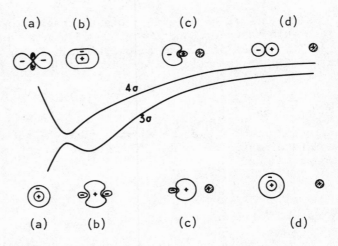

Fig. 18.   Schematic shape of the 3σ and 4σ orbitals of
(NeAr)$^{20+}$.   a) United-atom limit, b) electron
sharing region, c) Stark limit, d) separated-
atom limit.

The way to relate the terms between region I and II is dependent on
the collision parameters (energy, system considered, etc...).

Let me take another example : the role of the fine structure.
This problem has been discussed e.g. in the case of the fine structure transitions in the collision of alkalis in P states with rare
gases (Nikitin, 1974).

At infinite separations, spin-orbit forces fix the pattern of
energy levels (the $P_{3/2}$ and $P_{1/2}$ states are non degenerate).   For
smaller internuclear separations, the electrostatic forces between
the atoms are much stronger and give rise to Σ and Π molecular
states (figure 19).

Region I	Region II
Electrostatic forces dominant	Spin orbit forces dominant

In practice, the concept of diabatic states is not necessarily
useful.   In the preceeding case we were interested in transition
rates which are maximum precisely when neither the adiabatic nor
the diabatic representation are valid.   Hence discussions on the

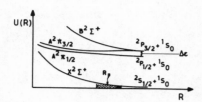

Fig. 19.   Qualitative form of the adiabatic terms of the alkali
            metal M and inert gas X system (Nikitin, 1974).

most adequate definition of diabatic states have often the same
status as those on the sex of angels in the Byzantine Empire.

A priori determination of diabatic states

        This is a more difficult problem, though we have to find some
way of solving the cases when a diabatic state crosses a whole
Rydberg series as in the case He^{2+}-He.

        Obviously, such a series of pseudo-crossings does not appear by
accident and this is what we have to find out. We can think about
this diabatic state in complete analogy with the series of autoionizing states in atoms. The autoionizing levels of helium, say
look very much like doubly excited configurations. For example the
^1S and ^1D resonances for e-He^+ scattering around 37 eV can be accounted for through the introduction of a $(2p)^2$ configuration. This
idea can be used for the determination of autoionizing states :
the autoionizing state has a p^2 "character" whereas excited states
have a 1s nl "character".

        In other terms, we need not describe the autoionizing state
by a single p^2 configuration : we would use as well a mixture of
many p^2, pd, d^2 etc... configurations. The important condition
is that, when we diagonalize the hamiltonian, we should not mix
the configurations above with those having a 1s nl character, as
shown schematically below :

This submatrix diagonal	1s nl configurations	Non zero elements	
	Non zero elements	p^2, pd, d^2, etc... configurations	This submatrix diagonal

The H matrix is diagonal only "in blocks", hence the name "block diagonalization" given to this method Lopez et al. (1978). It has been applied successfully to two electron systems like He^{++}-He (Lopez et al., 1978), He^{+}-H (Macias et al., 1981) and H^{+}-H^{-} (Borondo et al., 1981 a,b). In the case of He^{++}-He, the diabatic state is given a σ_u^2 character at small internuclear distances whereas the excited states of $(He)_2^{2+}$ have a $\sigma_g \sigma_g'$ character. At large distances, we have shown that it is important to mix the σ_u^2 configuration to the σ_g^2. This causes no difficulty. It means only that the non diagonal blocks are so important at large distances that it is preferable to switch from the diabatic to the adiabatic basis. Such a situation is not exceptional in molecular studies.

ION-ION NEUTRALIZATION

We shall discuss the reaction

$$A^{+} + B^{-} \rightarrow A + B.$$

All the theories of this process are in fact based on the theory developped for H^{+}+H^{-}. Therefore we shall discuss the details of the reaction

$$H^{+} + H^{-} \rightarrow H + H$$

which has been the subject of an extensive study by Borondo et al. (1981a,b). The main ideas traditionally exposed on the problem are as follows. One first notes the attraction in the incoming channel whereas the interaction in the final channel goes to zero much more rapidly as a function of internuclear distance. Hence one builds a diabatic picture consisting of ionic and covalent states as shown in figure 20.

Borondo et al. have first studied the characteristics of the system in the region 7-20 a.u. Their study involves various steps. In a first approach, they calculate diabatic states using the block diagonalization method described earlier. The idea is to

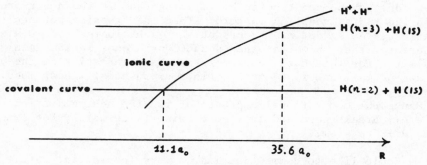

Fig. 20. Diabatic states in ($H^+ - H^-$).

calculate directly the diabatic states by imposing the ionic and
covalent characters.

     This is easy to do. As in most molecular calculations, they
use an atomic basis to expand the molecular wave function - actually six (1s) and three (2p) Gaussian orbitals. The two blocks of
the block diagonalization method are defined as follows :
- in the first block there is no product of two atomic orbitals on
  the same nucleus. The states of this block would then correspond to covalent states,
- in the second block there are only products of atomic orbitals
  centered on the same nucleus.
The results are :
- for short internuclear distances the "covalent" and "ionic"
  states so determined become linearly dependent. This means that
  the distinction between ionic and covalent curves is meaningless.
  This could be expected and shows clearly, as we have discussed
  earlier, that a single representation cannot be used for all
  internuclear distances.
- for large internuclear distances, the overlap between ionic and
  covalent states may be as large as 0.75 for distances between 10
  to 15 a.u. Furthermore this result does not seen to be dependent
  on the particular method (block diagonalization) used to make
  the calculation. This means that the distinction between ionic and
  covalent states is not significant and the conventional approach
  breaks down. The origin of this difficulty is in the diffuse-
  ness of the H^- atomic states.

     In a second step, Borondo et al. calculate adiabatic energies
and radial couplings between these adiabatic states. Diabatic

states can then be constructed by imposing that in the new representation the radial coupling in zero. Then the crossing between the
diabatic states so defined occurs at $R_0=7$ a.u. instead of 10 a.u. !
Furthermore, they show that the Landau-Zener model is completely
unrealistic for this case. However the characteristics of the crossing can be well described by Nikitin's exponential linear model.

Hence the conventional approach of ion-ion neutralization
completely breaks down and this conclusion is certainly not specific
to the H^+-H^- system.

We would like to comment at last a very interesting result
obtained by the same authors for the inverse reaction :

$$H(2s) + H(1s) \rightarrow H^+ + H^- \qquad (a)$$
$$\rightarrow H^- + H^+ \qquad (b)$$

At energies above 100 eV the two reactions can be considered
as distinct since the scattering occurs mostly in the forward
direction (anyhow, the experiment could be done with D(2s)+H(1s)).
The cross section for reaction (a) is one order of magnitude larger
than the cross section for process (b), that is the captured electron comes preferentially from the excited atom. This is a striking example of the selectivity of charge exchange processes.

MULTICHARGED IONS ON ATOMS

This is a very good example of the selectivity of charge
exchange processes at low energies. We shall not dicuss in details
these reactions since there is an extensive literature on the
subject.

Selectivity of the reaction

The selectivity of the charge exchange process for multicharged ions on atoms can be explained through the example of one
electron systems. Let us consider again figure 9. As I have
remarked earlier, transitions occur at the pseudo-crossings of
potential curves such that the behaviour is neither diabatic nor
adiabatic. Detailed calculations show that for a given energy
there is mainly one pseudo crossing that contributes to the charge
exchange reaction. This produces a selective excitation of one
excited state (Harel and Salin, 1977). This has been confirmed by
experiments. We give in figure 21 the results of measurements by
Mann et al. (1981) on very low energy collisions between Kr^{18+} and
Ne or He. Population of the n=4 state is clearly shown to be
dominant. On figure 22 we give the single electron capture cross
section for C^{4+} collisions with H(1s) (Gargaud et al., 1981).
The theory shows that one state is selectively populated.

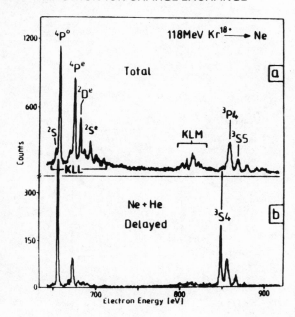

Fig. 21.

Ne K-Auger spectrum following the collision of Ne^{8+} ions with Ne and He atoms (Mann et al., 1981).

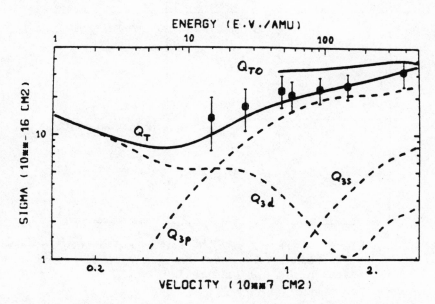

Fig. 22.   Single electron capture cross section for C^{4+} collisions with $H(1s)$ - Gargaud et al., 1981. Q_T : total single capture cross-section. Q_{nl} : contribution of capture into the state $2s^3nl$ of C^{3+}.

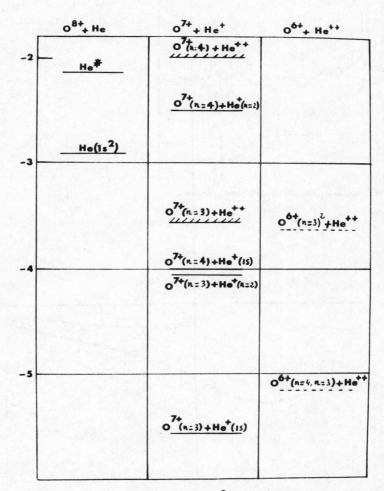

Fig. 23.   Energy levels of $(OH_2)^{8+}$ for infinite internuclear
separations. The position of the doubly excited
states of O^{6+} is qualitative.

However, which state is dominant is fixed by the collision energy.

Production of autoionizing states

We have already shown that highly excited states can be produced by charge exchange to multicharged ions. The main reason is that the ionisation energy of the electron in the atom is much less than that of the ground state of the ion formed in the reaction. However the situation is even more complex. Consider the case of the collision between O^{8+} and $He(1s^2)$. We give on figure 23 a diagram of asymptotic states of this system. It can be seen that the energy of the initial state corresponds to a continuum for the charge exchange channel. This is because the binding energy of $O^{7+}(n=3)$ for example is larger than that of the two electrons in He. Two processes may occur. Either ionisation takes place during the collision (this is the same as Penning ionisation) or autoionising states of O^{6+} are formed. A large set of experimental data show evidence of the production of autoionising states in such collisions.

APPENDIX I

DEMONSTRATION OF (25) and (26)

We start from (23). Using (22) :

$$\tilde{\zeta} = \sum_{\alpha'J'L'} \sum_{\alpha_o'J_o'L_o'} \sum_{K,M_K} |\alpha'J'L'KM_K\rangle \tilde{C}_{K,M_K}(\alpha'J'L',\alpha_o'J_o'L_o') \langle\alpha_o'J_o'L_o'KM_K|$$

$$= \sum_{\alpha'J'L'} \sum_{\alpha_o'J_o'L_o'} \sum_{K,M_K} \sum_{M_L'} \sum_{M_{Lo}'} (2K+1)(-)^{J'+L'+M'+J_o'+L_o'+M_o'}$$

$$\begin{pmatrix} J' & L' & K \\ M_{J'} & M_{L'} & -M_K \end{pmatrix} \begin{pmatrix} J_o' & L_o' & K \\ M_{J_o'} & M_{L_o'} & -M_K \end{pmatrix} |\alpha'L'M_L'J'M_J'\rangle \tilde{C}_{K,M_K}$$

$$\langle\alpha_o'L_o'M_{L_o}'J_o'M_{J_o}'| \qquad (A1)$$

To calculate the matrix elements of $\tilde{\zeta}$, we use the orthogonality condition :

$$\langle\alpha LM_L JM_J|\alpha'L'M_L'J'M_J'\rangle = \delta_{\alpha\alpha'}\delta_{LL'}\delta_{M_L M_L'}\delta_{JJ'}\delta_{M_J M_J'} \qquad (A2)$$

and we get :

$$T_{if} = (2\pi)^{-3} \sum_{K,M_K} \sum_{L_o M_o} \sum_{LM} \frac{(4\pi)^2}{k_i k_f} i^{L_o-L} e^{i(\sigma_{L_o}+\sigma_L)} (-)^{J+L+J_o+L_o}$$

$$(2K+1) \begin{pmatrix} J & L & K \\ M_J & M_L & -M_K \end{pmatrix} \begin{pmatrix} J_o & L_o & K \\ M_{J_o} & M_{L_o} & -M_K \end{pmatrix} Y_{L_o}^{M_o *}(\hat{k}_i) Y_L^M(\hat{k}_f) \mathcal{G}(L,L_o,K) \tag{A3}$$

with

$$\mathcal{G}(L,L_o,K) = \int_o^\infty dR\; F_L^{-*}(R) \widetilde{\mathcal{C}}_{K,M_K}(\alpha JL,\; \alpha_o J_o L_o) F_{L_o}^+(R). \tag{A4}$$

Hence we finally obtain expression (26).

APPENDIX II

CLASSICAL APPROXIMATION OF $\widetilde{\mathcal{C}} \chi_i^+$

An initial expression is :

$$\widetilde{\mathcal{C}} \chi_i^+ = [1+(E-H+i\epsilon)^{-1} W_f] \chi_i^+. \tag{A5}$$

Define the function :

$$\psi_i^+ = [1+(E-H+i\epsilon)^{-1} W_i] \chi_i^+. \tag{A6}$$

This function satisfies the equation :

$$(H-E)\psi_i^+ = 0 \tag{A7}$$

with implicit asymptotic conditions given by (A6). The hamiltonian can be written :

$$H = T + H_{el} \tag{A8}$$

where T is the kinetic energy operator associated with the relative motion of the nuclei and the coordinate $\underset{\sim}{R}$:

$$T = -\frac{1}{2M} \frac{1}{R} \frac{d^2}{dR^2} R + \frac{L^2}{2MR^2}. \tag{A9}$$

Let us now introduce the transformation :

$$\psi_i^+ = \mathcal{F}^+(\underset{\sim}{R})\; \Xi^+(\underset{\sim}{r},\underset{\sim}{R}) \tag{A10}$$

where $\mathcal{F}^+(\underset{\sim}{R})$ is the solution of the equation :

$$(T + V_o(R) - \frac{1}{2} \frac{k_i^2}{2M}) \mathcal{F}^+(\underset{\sim}{R}) = 0. \tag{A11}$$

$\mathcal{F}^+(\underset{\sim}{R})$ describes the scattering of the nuclei by the potential $V_o(R)$. We introduce the approximations :

(i) $\underset{\sim}{L}^2 \mathcal{F}^+ \Xi^+ \simeq \Xi^+ \underset{\sim}{L}^2 \mathcal{F}^+$ \hfill (A12)

which means that the angular momentum carried by the nuclei is much larger than the change of the electronic angular momentum during the collision.

(ii) $\mathcal{F}^+$ can be replaced by its classical limit (equations 20 and 29). We then obtain the equation for Ξ^+ :

$$-\frac{1}{2M} \mathcal{F}^+(\underset{\sim}{R}) \frac{1}{R} \frac{d^2}{dR^2} R\Xi^+(\underset{\sim}{r},\underset{\sim}{R}) - \frac{1}{M} \frac{1}{R} (\frac{d}{dR} R\mathcal{F}^+) \frac{d}{dR} \Xi^+(\underset{\sim}{r},\underset{\sim}{R})$$

$$+ \mathcal{F}^+(\underset{\sim}{R})[H_{el} - V_o - \varepsilon_i]\Xi^+(\underset{\sim}{r},\underset{\sim}{R}) = 0. \tag{A13}$$

We shall neglect the first term by comparison with the second on the basis that $d/dR \, \mathcal{F}^+ \approx k$ whereas $d/dR \, \Xi^+ \ll k$. This approximation is very close to the one done in the case of the eikonal approximation (Joachain, 1975, section 9).

Using the expression (29) it is now a trivial task to show that :

$$\frac{1}{M} \frac{1}{R} \frac{d}{dR} R\mathcal{F}^+ \frac{d}{dR} \Xi^+ \simeq i\mathcal{F}^+ \frac{d}{dt} \Xi^+ \tag{A14}$$

to first order in $1/k$. Here t is connected with R by the classical law of motion on the trajectory defined by the potential V_o (see equations A22 and A24). Hence

$$(H_{el} - V_o - \varepsilon_i - i \frac{d}{dt}) \Xi^+(\underset{\sim}{r},\underset{\sim}{R}) = 0. \tag{A15}$$

This equation is exactly the one satisfied by $\psi_{s.c}^{\pm}$ as defined in (35). A similar procedure can be carried out now for χ_i^+ which leads to equation (36). Hence (A6) can be transformed into :

$$\psi_i^+ \simeq \mathcal{F}^+(\underset{\sim}{R})[1 + (H_{el} - V_o - \varepsilon_i - i \frac{d}{dt} + i\varepsilon)^{-1} W_i]\Phi_i(\underset{\sim}{r}). \tag{A16}$$

So that we get the definition of the semi-classical $\underset{\sim}{\mathcal{G}}$ matrix :

$$\underset{\sim}{\mathcal{G}} \chi_i^+ \to \mathcal{F}^+(\underset{\sim}{R}) \underset{\sim}{\mathcal{G}}^{cl}\Phi(\underset{\sim}{r}) = \mathcal{F}^+ W_f \, \psi_{s.c.}^+$$

$$= \mathcal{F}^+ W_f[1 + (H_{el} - V_o - \varepsilon_i - i \frac{d}{dt} + i\varepsilon)^{-1} W_i] \, \Phi_i(\underset{\sim}{r}). \tag{A17}$$

APPENDIX III

DEMONSTRATION OF EQUATION (34)

Consider first the integral :

$$\mathcal{G}^1 = -\frac{1}{4}(k_i k_f)^{1/2}\int_o^\infty dR[f_{L_o}]^{-1/4}[f_L]^{-1/4}\, e^{i(\phi^+ -\phi^-)}\mathcal{C}^{c1}_{K,M_K}(R) \tag{A18}$$

Approximation (i) means that we replace o by R_o (the distance of closest approach) as the lower bound of the integral (according to (ii), it is channel independent). We make the approximation (Alder et al., 1956)

$$\phi^+ -\phi^- \simeq (k_i -k_f)\int_{R_o}^R dr\, kf_L^{-1/2} - [L_o(L_o+1)-L(L+1)]\int_{R_o}^R [f_L(r)]^{1/2}\frac{dr}{2r^2} \tag{A19}$$

We can now use some well known relations for a trajectory defined by a central potential (Goldstein, 1980). In the system of confocal coordinates defined earlier, the polar angle ϕ corresponding to a point R on the trajectory is given by :

$$\phi = \int_{R_o}^{\tilde{R}}\frac{sdr}{r^2}\left(1 - \frac{\rho^2}{r^2} - \frac{2V_o}{Mv^2}\right)^{-1/2}. \tag{A20}$$

Taking into account the relation (33), (32) and the definition of f, one obtains :

$$[L_o(L_o+1)-L(L+1)]\int_{R_o}^R [f(r)]^{-1/2}\frac{dr}{r^2} = -\mu\phi. \tag{A21}$$

Analogously the relation between position and time is given by :

$$t = M\int_{R_o}^R dr[f(r)]^{-1/2}. \qquad (t=o\ for\ R=R_o) \tag{A22}$$

Hence :

$$(k_i -k_f)k\frac{t}{M} \simeq \frac{1}{2}(k_i^2 -k_f^2)\frac{t}{M} = -\Delta\varepsilon t = -(\varepsilon_i -\varepsilon_f)t \tag{A23}$$

where $\Delta\varepsilon$ is the excitation energy of the reaction. Now we have also :

$$f_L(r)^{-1/2}dR = \frac{dt}{M} \tag{A24}$$

so that $[f_{L_o}]^{-1/4}[f_L]^{-1/4}\, dR \simeq \frac{dt}{M}$.

The final result is

$$\mathcal{G}^1 = -\frac{k}{4M}\int_0^\infty dt\ e^{i\Delta\epsilon t + i\mu\phi}\ \underset{K,M_K}{c1}(R).$$ (A25)

Similarly :

$$\mathcal{G}^{(2)} = -\frac{1}{4}(k_i k_f)^{1/2}\int_0^\infty dR[f_{L_0}]^{-1/4}[f_L]^{-1/4}e^{-i(\phi^+-\phi^-)}$$

$$= -\frac{k}{4M}\int_{-\infty}^0 dt\ e^{-i\Delta\epsilon t+i\mu\phi}\mathcal{C}_{K,M_K}(R).$$ (A25)

According to approximation (iv), $\mathcal{G} = \mathcal{G}^1 + \mathcal{G}^2$ so that we get immediately the final formula (34).

APPENDIX IV

PROOF OF EQUATION (46)

Our proof follows closely the derivation given in the appendix of Gaussorgues et al. (1975). We have to find out the relation between the angular momentum variables as expressed in (46). Using (22) :

$$\mathcal{C}_{K,M_K}(\alpha JL, \alpha_0 J_0 L_0) = \underset{M_L M_{L_0}}{\Sigma} \begin{pmatrix} J & L & K \\ M_J & M_L & -M_K \end{pmatrix}\begin{pmatrix} J_0 & L_0 & K \\ M_{J_0} & M_{L_0} & -M_K \end{pmatrix}(2K+1)$$

$$(-)^{J+L+J_0+L_0}<\alpha JM_J LM_L |\underset{\approx}{\mathcal{C}}|\alpha_0 J_0 M_{J_0} L_0 M_{L_0}>.$$ (A27)

Replacing $\underset{\approx}{\mathcal{C}}$ by its classical limit (A17) :

$$<\alpha JM_J LM_L |\underset{\approx}{\mathcal{C}}|\alpha_0 J_0 M_{J_0} L_0 M_{L_0}>=\underset{1m}{\Sigma} f_{1m}(R)\int d\hat{R}\ [Y_L^{M_L}(\hat{R})]^{*}$$

$$Y_{L_0}^{M_{L_0}}(\hat{R})Y_1^m(\hat{R})$$ (A28)

where f_{1m} has been derived in (41). This expression can be simplified with the help of the limit of 3J coefficients for large values of the momenta (see Gaussorgues et al., 1975, equation 57) :

$$\int d\hat{R}[Y_L^{M_L}(\hat{R})]^{*}\ Y_{L_0}^{M_{L_0}}(\hat{R})Y_1^m(\hat{R}) \simeq (-)^m d_{L_0-L,-m}^1(\frac{\pi}{2})Y_1^{L-L_0}(\frac{\pi}{2},\ 0)$$

$$\delta(m,\ M_L-M_{L_0}).$$ (A29)

We also transform the 3J symbols in (A27) using the same asymptotic formulae, i.e. :

$$\begin{pmatrix} J & L & K \\ M_J & M_L & -M_K \end{pmatrix} \simeq (-)^{K+M_K}(2K+1)^{-1/2}\ d_{K-L,M_J}^J$$ (A30)

so that we obtain the final result :

$$\mathscr{C}_{K,M_K} \simeq \sum_{1mM_JM_{J_O}} (-)^{m+J+J_O+L+L_O} \, d^1_{L_O-L,-m}(\tfrac{\pi}{2}) \, Y^{L-L_O}_1(\tfrac{\pi}{2},o)$$

$$d^J_{K-L,M_J}(\tfrac{\pi}{2}) \, d^{J_O}_{K-L_O,M_{J_O}}(\tfrac{\pi}{2}) \, \delta(m,M_{J_O}-M_J) f_{1m}(R). \tag{A31}$$

The coefficients d are the rotation matrices (Edmonds, 1980). We have taken into account the fact that the 3-J factors in (A27) give $M_L-M_{L_O} = M_{J_O}-M_J$. However we may drop the δ function in (A31). It is easy to check that $f_{1m}(R)$ is zero if $m \neq M_{J_O}-M_J$. Hence we may treat the variables m, M_J, M_{J_O} as independent variables. The axis of quantisation used until now is along the direction of the initial velocity of the projectile. The transformation to a quantisation axis perpendicular to the zy plane correspond to a rotation of $\pi/2$ around the y axis. It is easy to check that the rotation matrices in (A31) correspond precisely to this transformation so that :

$$\mathscr{C}_{K,M_K} \simeq (-)^{J+J_O} Y^{L-L_O}_K(\tfrac{\pi}{2},o) f_{K,L-L_O}[\alpha J(K-L), \, \alpha_o J_o(K-L_o)] \tag{A32}$$

where now the quantisation axis is perpendicular to the plane defined by the initial and final direction. Before comparing (A32) with (43), it is important to note that in (34) we have used a confocal system of coordinate, whereas in (43) ϕ is defined by reference to a fixed coordinate system. If we call θ the scattering angle the transformation from one system to the other is given by :

$$\mathscr{C}_{K,M_K} \rightarrow \mathscr{C}_{K,M_K} \exp\{i(L_o-L)(\pi-\theta)/2\}. \tag{A33}$$

Using the classical relation between scattering angle and phase shift :

$$\theta = 2 \, d\sigma_L/dL \tag{A34}$$

we may use the approximation :

$$(L_o-L) \, \theta/2 \simeq -(\sigma_L-\sigma_{L_o}). \tag{A35}$$

The final expression for $\mathscr{G}$ is then :

$$\mathscr{G}(L,L_O,K) = -\frac{1}{4}\frac{k}{M}\int_{-\infty}^{+\infty} dt \, e^{-i\Delta\varepsilon t} \, \frac{(x+iy)^\mu}{R^\mu} \, (i)^{L_O-L}(-)^{J+J_O} \tag{A36}$$

$$\exp\{i(\sigma_L-\sigma_{L_o})\} \, Y^{L-L_O}_K(\tfrac{\pi}{2},o) \, f_{K,L-L_O}[\alpha J(K-L),\alpha_o J_o(K-L_o)]$$

The comparison with (43-46) yields immediately (46). The transition amplitude takes the final form :

$$f_{if}(\theta) \simeq \frac{\sqrt{\pi}}{k} \sum_{K,L,L_0} (2K+1)(2L_0+1)^{1/2} e^{2i\sigma_L} \begin{pmatrix} J & L & K \\ M_J & -\Delta M & M_{J_0} \end{pmatrix}$$

$$\begin{pmatrix} J_0 & L_0 & K \\ M_{J_0} & 0 & -M_{J_0} \end{pmatrix} Y_L^{\Delta M}(\hat{\underline{k}}_f) C_{K,L-L_0} [b; \alpha J(K-L); \alpha_0 J_0(K-L_0)]$$

$$(A37)$$

REFERENCES

K. Alder, A. Bohr, T. Huus, B. Mottelson and A. Winther, 1956, Rev. Mod. Phys., 28, 77.

M. Barat and W. Lichten, 1972, Phys. Rev., A6, 211.

D.R. Bates and A.H. Boyd, 1962, Proc. Phys. Soc., 80, 1301.

F. Borondo, A. Macias and A. Riera, 1981, Phys. Rev. Letters, 46, 420.

F. Borondo, A. Macias and A. Riera, 1981, J. Chem. Phys. 74, 6126.

A.S. Dickinson, 1981, J. Phys. B : Atom. Molec. Phys. 14, 3685-91.

A.R. Edmonds, 1980, Angular Momentum in Quantum Mechanics, Princeton University Press.

J. Eichler, U. Wille, B. Fastrup and K. Taulbjerg, 1976, Phys. Rev., A14, 707.

C. Falcon, A. Macias, A. Riera and A. Salin, 1981, J. Phys. B : Atom. Molec. Phys., 14, 1983.

R.P. Feynmann, R.B. Leighton and M. Sands, 1970, Lectures on Physics, vol. 3, Addison-Wesley.

M. Gargaud, J. Hanssen and P. Valiron, 1981, ICPEAC XII, Book of invited lectures, Gatlinburg.

C. Gaussorgues, C. Le Sech, F. Masnou, R. Mc Carroll and A. Riera, 1975, J. Phys. B : Atom. Molec. Phys., 8, 239.

H. Goldstein, 1980, Classical Mechanics, Addison-Wesley.

T.A. Green, M.E. Riley, E.J. Shipsey and J.C. Browne, 1981, ICPEAC XII, Book of Abstracts, Gatlinburg.

C. Harel and A. Salin, 1977, J. Phys. B : Atom. Molec. Phys., 10, 3511.

C. Harel and A. Salin, 1980, J. Phys. B : Atom. Molec. Phys., 13, 785.

C.J. Joachain, 1975, Quantum Collision Theory, North Holland

Q.C. Kessel, E. Pollack and W.W. Smith, 1978, in "Collision Spectroscopy", R.G. Cooks ed., Plenum Press.

L. Landau and E. Lifchitz, 1966, Mecanique quantique, Editions Mir.

W. Lichten, 1963, Phys. Rev., 131, 229.

W. Lichten, 1980, J. of Physical Chem., 84, 2102.

V. Lopez, A. Macias, R.D. Piacentini, A. Riera and M. Yanez, 1978, J. Phys. B : Atom. Molec. Phys., 11, 2889.

A. Macias and A. Riera, 1978a, J. Phys. B : Atom. Molec. Phys., 11, L489.

A. Macias and A. Riera, 1978b, J. Phys. B : Atom. Molec. Phys.,
11, 1077.

A. Macias, A. Riera and M. Yanez, 1981, Phys. Rev., A23, 2941.

R. Mann, H.F. Beyer and F. Folkmann, 1981, Phys. Rev. Letters,
46, 646.

E.E. Nikitin, 1974, Theory of Elementary Atomic and Molecular
Processes in Gases, Oxford Clarendon Press.

R.E. Olson, 1979, ICPEAC XI, Book of Invited lectures, Kyoto.

V.H. Ponce, 1981, J. Phys. B : Atom. Molec. Phys., 14, 2823.

J.D. Power, 1973a, Phil. Trans. of the Roy. Soc., 274, 663.

J.D. Power, 1973b, Q.C.P.E., 233.

J.F. Reading and A.L. Ford, 1980, Phys. Rev. A21, 124.

M.E. Riley, 1973, Phys. Rev., A8, 742.

A. Salin, 1978, Comp. Phys. Comm., 14, 121 and 20, 462.

A. Salin, 1980, Comments in Atomic and Molec. Physics, 9, 165.

THE MEASUREMENT OF INELASTIC ION-ION

AND ELECTRON-ION COLLISIONS

K.T. Dolder

School of Physics
The University
Newcastle upon Tyne, NE1 7RU, U.K.

1.   INTRODUCTION

Dense gases or plasmas obey the laws of equilibrium thermo-
dynamics and their properties can be discussed without recourse to
atomic physics. Temperatures, for example, can be deduced from
measurements of radiated energy simply by applying Stefan's law.
But laboratory plasmas, stellar and planetary atmospheres are of
greatest practical interest and they are rarely, if ever, in full
thermodynamic equilibrium. We therefore require details of all
relevant atomic processes to construct theoretical models and deduce
plasma properties from observations.

There are other incentives to study ion-ion and electron-ion
collisions. In a thermonuclear reactor, the ion of one element may
be particularly effective at neutralizing fuel ions or inducing some
other undesirable effect and, if these elements can be identified,
it may be possible to exclude them.

More fundamental reasons also motivate the study of collisions
between charged particles. An electron can be regarded as a simple
probe with which to explore atomic and molecular ions. It might be
argued that this is an inferior technique to conventional spectroscopy because uncharged electromagnetic radiation causes much less
perturbation to a target. But electrons are not bound by the same
angular momentum selection rules and can therefore induce "forbidden"
transitions. They may also react with a target to form new, and
often unstable, particles (e.g. He^- or H^{--}). Sometimes, the charged
particles are particularly elementary structures and it is worth
noting that the simplest molecular structure (and almost the only one
for which good "ab initio" calculations are possible) is H_2^+.

We will discuss certain technical aspects of charged particle
collisions and then consider some experimental results. Since we
must be brief, it is helpful to draw attention to relevant reviews.

Electron-ion collisions were reviewed by Dolder and Peart (1976)
and this account was updated by Dunn (1980). There are other articles
by Harrison (1966, 1968, 1978), Dunn (1969, 1979) and Dolder (1969).
The theory of electron impact excitation of ions was discussed by
Seaton (1975) and Henry (1981).

Much less has been written about ion-ion collisions, but mutual
neutralization was excellently reviewed by Moseley et al (1975) and
some other aspects were recently described in a lecture by Gilbody
(1981). An extended review of ion-ion collisions is in preparation
(Dolder and Peart).

2.   GENERAL FEATURES OF EXPERIMENTS

Details of most experimental techniques are given in the reviews
and recent developments are described elsewhere in this volume. Some
"pitfalls" in experimental design will be discussed in section 5 and
so, by way of introduction, it will suffice to outline some general
principles.

Two, well-defined beams intersect to produce the required reaction. Energy transferred by these collisions is much less than the
beam energy, so collision products remain within their parent beam
until they are separated and detected. Separation is usually effec-
ted by electric or magnetic fields and sometimes by combinations of
both.

Most experimental difficulties arise either from collisions
with residual gas and metal surfaces or from the tenuous nature of
the beams. It is therefore usual to maintain ultra-high vacua ($\sim 10^{-10}$
torr) where the beams intersect but, even under these conditions,
particle densities in the beams ($\sim 10^6$ cm^{-3}) and the surrounding
"vacua" are comparable. The beam density is often limited by space
charge, which not only causes beams to spread but also gives rise to
electrostatic forces which may alter beam geometries when they cross.
Sophisticated beam modulation techniques have been developed to separate required signals from the backgrounds which are caused mainly
by collisions with residual gas. The ratio of signal to background
(SBR) differs greatly between experiments but is frequently only 10^{-2}
or 10^{-3}.

In spite of these difficulties and the need for complex apparatus
the intersecting beam technique is very attractive. The reactants
and the type of reaction can be clearly defined and the range of accessible energies spans seven or more orders of magnitude. These

properties are not possessed by complementary experiments in which
reaction rates are deduced from observations of plasmas.

Some general features of this type of experiment can be appreciated by referring to Figure 1. This illustrates an apparatus which
could be used to measure cross sections for the ionization of positive ions by electron impact, i.e.

$$e + A^+ \rightarrow A^{2+} + 2e. \tag{2.1}$$

One first produces a parallel, monoenergetic beam of A^+ ions to
serve as the target for an electron beam. This requires an ion
source (S) with an electrostatic lens (L) and deflectors (D) to focus
and align the beam before it is momentum analysed in a magnetic field
(M1). A pair of collimating slits (CS) (typically 3 mm × 1 mm) select
a beam of A^+ ions which is intersected by an electron beam flowing
between the electron gun (G) and collector (Ce). The diagram also
shows an L-shaped shutter, pierced by a narrow horizontal slit, which
can be lowered vertically through both beams to monitor their current
density distributions.

The electron impacts cause a small fraction (typically 10^{-8}) of
the A^+ ions to be ionized to the higher A^{2+} state, but ions receive

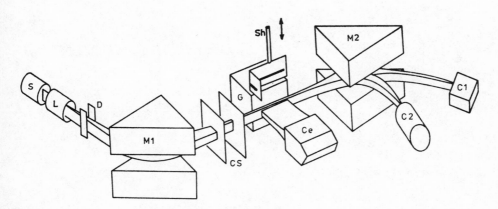

Fig. 1. Schematic diagram of an apparatus to study the ionization
         of ions by electron impact. Ions from a source (S) are
         focused by a lens (L) and deflectors (D) and momentum
         analysed in a magnetic field (M1). After collimation (CS)
         the ion beam is bombarded by electrons passing between the
         gun (G) and collector (Ce). Parent and product ions are
         separated by magnetic field M2 and collected at C1 and C2.
         A movable shutter (Sh) is used to monitor beam current
         density distribution.

very little momentum from these collisions so that the A^+ and A^{2+} ions both proceed as a well-defined beam to the second magnetic field (M2) where they are deflected according to their charge into the collectors C1 and C2. From measurements of the ion and electron currents and their spatial distributions, one can deduce the ionization cross section. Broadly similar methods can be applied to a variety of reactions involving charged particles.

3. TYPES OF REACTION INVESTIGATED WITH INTERSECTING, CHARGED BEAMS

The following is a list of the types of electron-ion collision which have been investigated.

(a) Ionization of ions by electron impact, e.g.

$$e + He^+ \rightarrow He^{2+} + 2e \tag{3.1}$$

$$e + Li^+ \rightarrow Li^{3+} + 3e \tag{3.2}$$

(b) Detachment from negative ions by electron impact, e.g.

$$e + H^- \rightarrow H + 2e \tag{3.3}$$

(c) Excitation of ions by electron impact, e.g.

$$Ca^+(4^2S_{\frac{1}{2}}) + e \rightarrow Ca^+(4^2P_{\frac{3}{2}}) + e \tag{3.4}$$

(d) Dissociative ionization, excitation and "pair production" of molecular ions, e.g.

$$e + H_2^+ \rightarrow H^+ + H^+ + 2e \tag{3.5}$$

$$\rightarrow H + H^+ + e \tag{3.6}$$

$$\rightarrow H^+ + H^- \tag{3.7}$$

(e) Dissociative recombination, e.g.

$$e + H_2^+ \rightarrow H + H. \tag{3.8}$$

Many techniques developed for these measurements have been adapted to ion-ion collisions, and the following types of reaction have so far been studied:

(a) Mutual neutralization, e.g.

$$A^+ + B^- \rightarrow A + B \tag{3.9}$$

(b) Charge transfer between positive ions, e.g.

$$A^+ + B^+ \rightarrow A^{2+} + B \tag{3.10}$$

(c) Detachment by positive ion impact, e.g.

$$A^+ + B^- \rightarrow A^+ + B + e \tag{3.11}$$

(d) Double charge transfer, e.g.

$$A^+ + B^- \rightarrow A^- + B^+ \tag{3.12}$$

(d) Ionization by positive ion impact, e.g.

$$A^+ + B^+ \rightarrow A^+ + B^{2+} + e \tag{3.13}$$

(f) Association, e.g.

$$A^+ + B^- \rightarrow AB^* \rightarrow AB^+ + e. \tag{3.14}$$

In some of these reactions the interacting systems (A and B) have been molecular. The energy ranges investigated will, of course, be determined by the nature of each reaction. Recombination processes have been investigated below 10^{-1} eV, whilst charge transfer and ionization are often most significant above 100 keV. When designing an experiment to work in a particular energy range, the choice of the angle at which beams intersect is often of paramount importance. We will therefore discuss it in some detail.

4. CHOICE OF THE ANGLE OF BEAM INTERSECTION

The four possibilities are illustrated by Figure 2. Beams may cross perpendicularly, at an acute or obtuse angle, or they may be "merged" ($\theta = 0$). Detailed discussions are contained in several reviews, e.g. Neynaber (1969), Dunn (1969, 1980), Dolder and Peart (1976) and Brouillard and Claeys in this volume.

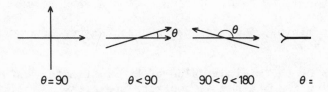

$\theta = 90$ $\theta < 90$ $90 < \theta < 180$ $\theta =$

Fig. 2. Four arrangements used with intersecting beams.

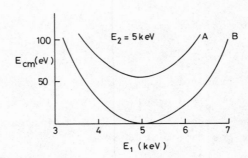

Fig. 3.   Interaction energies of H$^+$ and H$^-$ beams, intersecting at
          $8\frac{1}{2}°$ (curve B) or merged (curve A).

When beams of particles with masses and energies m_1, m_2, E_1 and
E_2 intersect at an angle θ, their interaction energy (in centre of
mass co-ordinates) is given by,

$$E_{cm} = \mu \left[\frac{E_1}{m_1} + \frac{E_2}{m_2} - 2\left(\frac{E_1 E_2}{m_1 m_2}\right)^{\frac{1}{2}} \cos \theta \right] \tag{4.1}$$

where $\mu = \dfrac{m_1 m_2}{m_1 + m_2}$.

Figure 3 illustrates the interaction energies for beams of H$^+$ and H$^-$
ions intersecting at $\theta = 8\frac{1}{2}°$ (curve B) or $\theta = 0$ (curve A). The energy
of one beam (E_2) has been held at 5 keV whilst E_1 varies as shown.

The most important conclusions are:

(a) For merged beams ($\theta = 0$), when the particle velocities (v_1, v_2)
     become equal, $E_{cm} \to 0$. This provides access to very low energies.

(b) Near the minima, $\delta E_{cm} \ll \delta E_1$, and so enhanced energy resolution
     is obtained.

(c) When $\theta \neq 0$, the minimum value of E_{cm} is finite, and enhanced
     resolution is obtained at non-zero energies.

The attainable resolution can be estimated by differentiating

equation (4.1). If θ is assumed constant (i.e. two, perfectly collimated beams),

$$\delta E_{cm} = \mu \left[\frac{1}{m_1} - \frac{1}{\sqrt{m_1 m_2}} \sqrt{\frac{E_2}{E_1}} \cos \theta \right] \delta E_1 + \mu \left[\frac{1}{m_2} - \frac{1}{\sqrt{m_1 m_2}} \sqrt{\frac{E_1}{E_2}} \cos \theta \right] \delta_2$$

(4.2)

A numerical illustration may be helpful. If $m_1 = m_2$, $E_1 = 5.0$
keV, $E_2 = 5.1$ keV, $\delta\theta = \delta E_1 = 0$, and $\delta E_2 = 10$ eV, then equations (4.1)
and (4.2) predict that, for merged beams, $E_{cm} \doteq 0.25$ eV and $\delta E_{cm} \doteq$
0.05 eV. A common error is the neglect of the effect of imperfect
collimation on the attainable resolution. This will be discussed in
section 5.4.

Inclined beams were introduced by Rundel et al (1969) who wished
to study mutual neutralization of H^+ and H^- ions,

$$H^+ + H^- \rightarrow H + H$$

(4.3)

at energies large enough for H atoms to be produced also by detachment,

$$H^+ + H^- \rightarrow H^+ + H + e.$$

(4.4)

A merged beam experiment would record H atoms formed by both reactions, but by arranging for H^+ and H^- ions to intersect at 20°, it was
possible to detect only those atoms formed in the direction of the
proton beam and it was argued that these could only be produced by
reaction (4.3); although a contrary view will be discussed in section
6. This experiment enjoyed enhanced resolution and access to low
energies characteristic of inclined beams. Inclined beams have since
been used in a number of experiments and the technique was extended
to electron-ion beams by Walton et al (1971).

Intersection at an obtuse angle (θ = 160°) was introduced by
Peart et al (1981a) who wished to study,

$$Cs^+ + Cs^+ \rightarrow Cs^{2+} + \ldots$$

(4.5)

at the highest possible energies. In their apparatus the beam energies could not conveniently exceed 14 and 80 keV but, by arranging
the beams to collide almost head on, the interaction energy was enhanced.

Perpendicular beams are relatively easy to set up and the collision geometry is well defined. Moreover, with electron-ion experiments, the width of the electron beam can be large (cf. Fig. 1) so
that bigger currents can be used to enhance signals without causing
space charge spread of the electron beam.

The relative merits of the four beam geometries can be summarized as follows:

Perpendicular beams

Well-defined geometry, larger electron currents, relatively simple to construct and monitor the spatial overlap of beams.

Inclined beams ($\theta \sim 10^{\circ}$)

Well-defined geometry, enhanced resolution (especially at non-zero values of E_{cm}), access to low energies. Ability to resolve certain competing processes (e.g. reactions (4.3) and (4.4)). Easier to set up and monitor spatial overlap than with merged beams.

Merged beams

Access to lowest energies (typically $< \sim 0.1$ eV) with good resolution. Long interaction path and hence large signal (except as velocities of two beams become equal when no collisions can occur!). Need more complicated apparatus and techniques to monitor spatial overlap of beams. There may also be some ambiguity about exact length of interaction path since extent of regions in which beams merge and "demerge" is not well defined. Consequently less well suited to absolute measurements.

Inclined beams ($\theta \sim 160^{\circ}$)

Well-defined geometry, access to higher energies.

5.   PITFALLS IN EXPERIMENTAL DESIGN

Experimental design has been described several times but a discussion of pitfalls has not previously appeared - probably for reasons that will become obvious! It is certainly an emotive and controversial topic but the writer will respond to his invitation and, where criticism is offered (our own work will not be excluded), it is hoped that it leads to fruitful discussion and improved technique.

5.1 Beam Modulation

Signals must usually be extracted from backgrounds and several modulation techniques have been developed. The three schemes described by Harrison (1968) are adequate for all types of experiment so far reported. Further details were given by Dolder (1969) and a control unit to generate the pulse trains was described by Molyneux

et al (1971). An alternative system was developed by Mitchell et al
(1977). These schemes reject backgrounds associated with particles
in either beam but <u>none</u> of them discriminate against backgrounds which
arise solely from any type of interaction <u>between</u> the two beams.
These "modulated backgrounds" can be a serious and subtle source of
error. We will therefore consider them in some detail.

5.2 <u>Modulated Backgrounds</u>

Suppose a modulated beam induces periodic pressure variations in
an apparatus, perhaps by the evolution of material occluded on a metal
surface, or by the reconversion of ions to gas. These pressure
changes may induce time-dependent variations in the background associated with the <u>other</u> beam. In this way, modulated backgrounds arise
which are partially in phase with the signal and therefore inseparable
from it.

Alternatively, space charge forces where beams intersect may
cause small geometrical changes in a beam. The modulation of one
beam then causes simultaneous changes of particle trajectories in the
other. Now, suppose that the efficiency of the device used to detect
the signal is not quite uniform over its entrance aperture. The geometrical changes in a beam path then cause spurious contributions
which are inseparable from the signal. Similar errors can arise if
the transmission efficiency of the analyser which separates signals
from their parent beam varies over its entrance aperture.

At first sight these errors may seem trivial, but in many experiments the SBR is only 10^{-3} and it follows that if the background were
modulated by only 0.01% there would be a 10% error in the signal. It
is therefore essential to be aware of the presence of any modulated
backgrounds so that steps can be taken to eliminate them.

Their presence is easily revealed if the reaction has a well
defined threshold. The ionization and excitation of ions by electron
impact are obvious examples. Figure 4 illustrates measurements by
Dance et al (1966) for,

$$e + He^{+}(1s) \rightarrow He^{+}(2s) + e. \hspace{3cm} (5.1)$$

It can be seen that the apparent cross section did not fall to zero
but increased as the electron energy was reduced below threshold.
Dance et al realized that this below-threshold signal was caused by
modulated backgrounds and, by extrapolation, they estimated its contribution at higher energies and applied the necessary corrections.
Their paper includes an instructive account of these modulations and
is commended to all experimentalists concerned with charged particle
collisions.

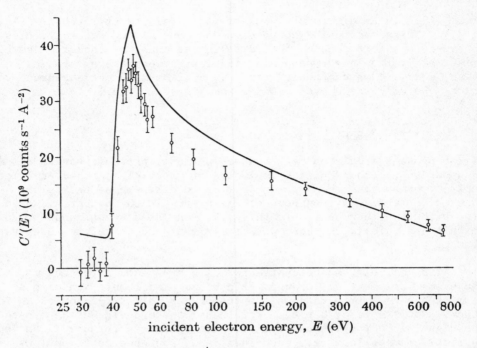

Fig. 4. Measurements for He$^+$(1s-2s) excitation by Dance et al where
measured cross section did not fall to zero at threshold.
The continuous curve shows uncorrected results. The points
denote corrected measurements.

Other examples of modulated backgrounds arose in connection with
measurements of,

$$e + H^- \rightarrow H + 2e \qquad\qquad\qquad (5.2)$$

and $e + H_2^+ \rightarrow H^+ + \ldots$ \qquad\qquad\qquad (5.3)

Both reactions have thresholds below 1 eV and slow secondary electrons can therefore collide with fast ion beams to produce H or H$^+$.
As a result, the cross sections can be overestimated, especially at
high energies, unless care is taken to sweep secondary electrons from
the ion beam path. Errors also arise if pulsed electron beams induce
outgassing from the electron collector, thereby causing synchronous
bursts of gas which can generate modulated backgrounds. The history
of these measurements and the eventual elimination of errors was
described by Dolder and Peart (1976).

This story has recently taken a new turn. Double detachment
from H$^-$,

$$e + H^- \rightarrow H^+ + 3e \tag{5.4}$$

was subsequently measured by Peart et al (1971) but the results are
larger than those recently obtained at Louvain-la-Neuve. Brouillard
suggests that slow positive H_2^+ ions might have accumulated in the
interaction space and the formation of H^+ by H^-/H_2^+ collisions would
have caused the cross section to be overestimated. This explanation
appears to require a very large cross section for the formation of H^+
by collisions between H^- and H_2^+ ions, and the problem has not yet
been resolved. But it does illustrate another possible pitfall.

It has already been mentioned that errors might arise if the
response of a particle detector is not uniform over its entrance aperture. Some interesting results were obtained by Rundel (see Dolder
1969, fig. 5.9.2) who probed Johnston and Venetian blind multipliers
with narrow ion beams. Their responses were very different and so it
is re-emphasised that modulations in the directions of particle trajectories lead to error if non-uniform detectors are used.

Similar errors would also arise if the analyser used to separate
signal and target beams had a non-uniform transmission efficiency for
particles with different angles of incidence. The effect would be
most marked in experiments with a small SBR and in which the background particles diverged appreciably as they entered the analyser.
About ten years ago, tests in this laboratory led us to conclude that
the transmission efficiency of parallel plate analysers may, under
certain conditions, depend significantly upon incident ion trajectories. The results were briefly summarized by Dolder and Peart (1976)
who reported that,"difficulties arise if the apertures in these analysers are too large or if they are situated too far from the object
and image points of an ion beam. Field penetration through the apertures (which cannot be eliminated by baffles or guard plates) may
introduce large aberrations in the beams so that not all the ions
pass through the analyser".

Figure 5 shows a grossly exaggerated view of three trajectories
as they encounter equipotentials bulging from a parallel plate analyser. If the transmission efficiencies were only slightly different
for these trajectories, and they were modulated by interaction with
another beam, an error might arise in an experiment with a small SBR.
It is, however, very strongly emphasized that there is no evidence of
such error in any electron-ion experiment or in any ion-ion experiment which used fast ions. We are, however, tempted to speculate
that this might explain the disparity between the two sets of results
for,

$$Cs^+ + Cs^+ \rightarrow Cs^{2+} + \ldots \tag{5.5}$$

illustrated by figure 16. At lower energies, where the Cs^{2+} background is more divergent, a modulated background might have arisen.

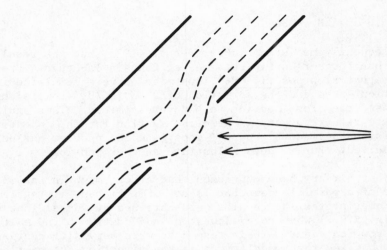

Fig. 5.   Three particle trajectories encountering equipotentials at
          the entrance of a parallel plate analyzer.

But this is an unproven and vigorously contested hypothesis. It does,
however, explain why we have always avoided this type of analyser in
ion-ion experiments.

5.4   Detection and Elimination of Modulated Backgrounds

Figure 4 illustrated that the presence of modulated backgrounds
can easily be revealed if a reaction has a well-defined threshold.
Below threshold the cross section must be zero! But detection is
much more difficult when studying ion-ion reactions or other processes without an obvious threshold. Checks devised by Peart et al
(1981a) can then be used. They have been applied to their measurements for various ion-ion reactions.

It was verified that:

(a)   the signal (corrected for a factor which accounted for spatial
      overlap of the beams) varied strictly with the product of the
      ion beam currents <u>at all energies</u>.

(b)   the measured cross sections were very insensitive to potentials
      applied to the analyser.

(c)   the cross section measured at a given interaction energy (E_{cm})
      did not depend on the choice of laboratory beam energies (E_1
      and E_2).

In addition, errors due to pressure modulation could be revealed by altering the frequency of wave trains used to pulse the beams. Effects due to space charge interactions between beams could be detected by focusing beams so that the current density profile of one beam was altered relative to the other. The measured cross section should be independent of these changes.

Even more important is the elimination of modulated backgrounds. No general rule can be given but the following precautions are suggested:

(a) Beam currents should be restricted so that space charge interactions between beams are negligible. The maximum permissible currents can be estimated from a simple theory advanced by Harrison (1966) and reviewed by Dolder (1969).

(b) When electron beams are used for reactions with low thresholds (e.g. 5.2 and 5.3), great care must be taken to sweep slow, secondary electrons away from the ion beam and to outgas collectors thoroughly. The accumulation in the interaction space of slow positive ions must also be avoided if they are likely to react appreciably with the target beam.

(c) Collectors used for gaseous ions should not be housed in the same vacuum tank as the interaction region. They should be differentially pumped.

(d) Only detectors and beam analysers with good uniformity of response over their entrance apertures should be used.

5.6   Energy Resolution in Merged Beam Experiments

This was recently discussed by Dunn (1980) who argued that the energy resolution claimed for some merged beam experiments is too high because the effects of angular divergence within the beams had been underestimated. Dunn's argument can be paraphrased as follows:

Consider electron and ion beams intersecting at an angle θ. If particle energies, masses and velocities are denoted by E, m, and v, with subscripts e and i to identify electrons and ions, we can rewrite equation (4.1),

$$E_{cm} \approx \frac{m_e}{2} (v_i^2 + v_e^2 - 2v_i v_e \cos \theta) \qquad (5.6)$$

if $m_i \gg m_e$.

Differentiating and assuming $v_i \approx v_e$ and $\sin \theta \approx \theta$, we obtain

$$E_{cm} \approx 2E_e \theta \delta \theta. \tag{5.7}$$

This enables the effect of imperfect collimation on the resolution to be estimated. We have already neglected the degradation due to spreads in laboratory beam energies (δE_1 and δE_2) given by equation (4.2), and we will also disregard contributions to $\delta \theta$ caused by imperfect mechanical collimation (typically $\delta \theta \sim 10^{-2}$), focusing fields, and space charge forces between the beams during their long interaction path. Dunn considered only the effect of lateral thermal motion of beam particles.

If $E_\perp$ represents the transverse energy of electrons relative to ions,

$$\theta \approx \left(\frac{E_\perp}{E_e}\right)^{\frac{1}{2}} \tag{5.8}$$

Writing $\theta \approx \Delta \theta$, Dunn obtained

$$\delta E_{cm} \approx 2E_\perp \tag{5.9}$$

The component of thermal energy $E_\perp \approx \frac{1}{2}kT$ and so, for a cathode operating at 1000 K, $\delta E_{cm} \approx 0.09$ eV. The claim by Auerbach et al (1977) to have used merged beams to measure dissociative recombination of H_2^+ ions at 0.01 eV must therefore be questioned. It is not easy to make a reliable estimate of the energy resolution attained in that experiment although the matter is of great interest in the theory and experimental technique of dissociative recombination. On one hand there are other factors (e.g. focusing potentials or stray fields) which might broaden the energy spread still further. On the other hand, the electron beam had to pass through circular apertures which would have rejected the more divergent electrons and so narrowed the energy distribution. It was unfortunate that the radius of these apertures was rather larger than the Larmor radius of the electrons and so their beneficial effect may not have been large.

It occurs to the writer that this apparatus might be used to explore threshold excitation of positive ions. The excitation cross section is believed to rise abruptly from threshold and so experimental results might provide a useful estimate of instrumental resolution. A further application might be threshold ionization/detachment from positive/negative ions. This is a topic of considerable technical interest which is virtually unexplored.

Another remarkable merged beam experiment was the measurement of,

$$H^+ + H^- \rightarrow H_2^* \rightarrow H_2^+ + e \qquad\qquad (5.10)$$

by Poulaert et al (1978) where results were reported at energies as low as 0.001 eV. Resolutions of this magnitude would be the envy of anyone who experiments with ion or electron beams.

For ions of equal mass ($m_1 = m_2$), equation (5.7) becomes,

$$\delta E_{cm} = E\theta\, \delta\theta \qquad\qquad (5.11)$$

and Dunn's arguments indicate that a resolution of 10^{-3} eV implies an ion temperature of only 10 K. Moreover, this estimate also neglects deleterious effects of focusing potentials or stray magnetic fields which would bend H^+ and H^- ions in opposite directions. The implied narrow resolution would only be possible if extremely tight mechanical collimation and careful magnetic shielding were applied to the beams but Poulaert et al make no mention of this, or of the ion beam energy. However, Brouillard (private communication) states that these precautions were taken, but a more detailed description of the experiment is needed. If resolutions of this magnitude could convincingly be demonstrated, it would mark a major tour-de-force in experimental physics, but published details of this experiment are far too sketchy.

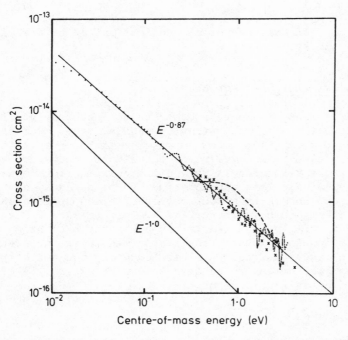

Fig. 6.   The points denote measurements of dissociative recombination of H_2^+ by Auerbach et al (1977).

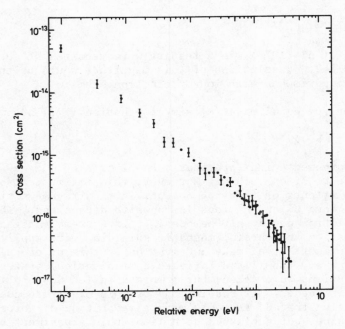

Fig. 7.   Measurements by Poulaert et al (1978) for reaction (5.10).

The results of Auerbach et al and Poulaert et al are illustrated by figures 6 and 7.

5.5   Initial Excitation of Ion Beams

We turn from one topic to another which (perhaps surprisingly) is quite controversial. When studying neutral atoms or molecules it is customary to work at or below room temperature so that initial vibrational or electronic excitation can almost always be neglected. Unfortunately the situation is very different with ion beams. Singly-charged atomic ions can sometimes be prepared by surface ion- ization or from Kunsman-type sources, and excited state populations will then be negligible. Alternatively, low-pressure PIG sources can be used and, if the anode potential is below the excitation potential of the ion, only ground-state ions are formed (e.g. Dance et al 1966). But these techniques are not feasible for every ion and Latypov et al have, for example, demonstrated (see Dolder and Peart, 1976, fig. 7) that measured cross sections for,

$$e + Xe^+ \rightarrow Xe^{2+} + 2e \qquad\qquad (5.12)$$

increased by a factor four when the ion source anode potential was raised from 35 to 50 V. Clearly, certain atomic cross sections depend quite sensitively upon the initial excitation of the ion beam beam and to ignore this effect is naive.

Theoreticians should always ask probing questions about initial ion beam excitation before they compare calculations for ground state ions with measurements. Sometimes the replies will be evasive.

The problem can be much more serious for molecular ions where vibrational excitation may enormously influence cross sections. The dissociative excitation of H_2^+, for example, varies by three or more orders of magnitude with initial vibrational level (Peek and Green, 1969).

What can the experimentalist do to combat these problems? We have already indicated that beams of unexcited, singly-charged atomic ions can be prepared for some elements but there is, as yet, no widely applicable technique for the preparation of adequate ($\gtrsim 10^{-8}$ A) beams of vibrationally de-excited molecular ions.

It is, of course, sometimes disadvantageous to measure cross sections for unexcited ions, because ions used or encountered in technical applications may have similar excitation to those which can easily be generated in the laboratory. Ground state ions may behave very differently.

On the other hand, when measurements are intended for accurate comparisons with theory, it is essential to know the initial state of excitation.

Experimentalists sometimes state that their results are insensitive to ion source conditions and this is reassuring because it suggests that similar excitation might be encountered in environments to which the results will be applied. But theoreticians who are interested in ground state ions should ask whether the results were insensitive to ion source conditions even when the anode potential was below the threshold for populating the first excited state. Even this will not prove conclusively that the ions were unexcited, but it is a question which should be asked. So beware!

6. MEASUREMENTS OF MUTUAL NEUTRALIZATION AND DOUBLE CHARGE TRANSFER

Mutual neutralization was excellently reviewed by Moseley et al (1975) and so we will merely concentrate on salient points and recent developments.

Reactions of the type,

$$A^+ + B^- \rightarrow A + B \tag{6.1}$$

Table 1. Measurements of Mutual Neutralization by Intersecting Beam
         Techniques

Reactants	Energy Range (eV)	θ	Reference
$H^+ + H^-$	0.15 – 300	0	Moseley et al, 1970
	20 – 3000	$8\frac{1}{2}$	Peart et al, 1976b
	125 – 5000	20	Rundel et al, 1969
	600 – 400	20	Gaily & Harrison, 1970a
$He^+ + H^-$	35 – 4550	$8\frac{1}{2}$	Peart et al, 1976a
	200 – 8000	20	Gaily & Harrison, 1970b
$He^+ + D^-$	20 – 350	20	Gaily & Harrison, 1970b
$O^+ + O^-$	0.1 – 25	0	Moseley et al, 1972
$N^+ + O^-$	0.1 – 100	0	Aberth & Peterson, 1970
$Na^+ + O^-$	0.1 – 7	0	Weiner et al, 1971
	0.1 – 15	0	Moseley et al, 1972
$H_2^+ + D^-$	0.15 – 50	0	Aberth et al, 1971
$O_2^+ + O_2^-$	0.1 – 50	0	Aberth & Peterson, 1970
	0.1 – 12	0	Peterson et al, 1971
$O_2^+ + O^-$	0.15 – 20	0	Moseley et al, 1972
$N_2^+ + O_2^-$	0.1 – 90	0	Aberth & Peterson, 1970
$NO^+ + O^-$	0.1 – 22	0	Moseley et al, 1972
$NO^+ + O_2^-$	0.1 – 100	0	Moseley & Peterson, 1972
$NO^+ + NO_2^-$	0.15 – 200	0	Peterson et al, 1971
$O_2^+ + NO_2^-$	0.15 – 200	0	Peterson et al, 1971
$O_2^+ + NO_3^-$	0.15 – 700	0	Moseley et al, 1975
$NO^+ + NO_3^-$	not stated	0	Moseley et al, 1975

where A and B can be atomic or molecular structures, are significant
in almost any gas which contains negative ions, and processes leading
to the removal of charge are likely to be particularly important.
Applications arise in electrical discharges, gas lasers, and planetary
and stellar atmospheres. Mutual neutralization involving hydrated
ions (e.g. $NO_2^-.H_2O$) is believed to be especially significant in the
D-region of the Earth's ionosphere.

Measurements of mutual neutralization at low energies sometimes
provide useful information about reverse processes,

$$A + B \rightarrow A^+ + B^- \qquad\qquad\qquad (6.2)$$

when A and B cannot be prepared in their ground states;although forward and reverse reactions can have very different cross sections if
the initial excitation of the reactants is not the same.

Table 1 lists measurements of mutual neutralization performed
with merged and inclined beams. The angles of intersection are noted
in the table.

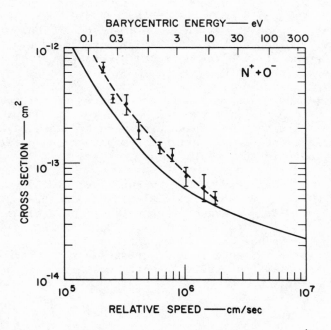

Fig. 8.   Measurements for mutual neutralization of N^+ and O^- compared
          with LZ calculation (continuous curve).

Moseley et al (l.c.) suggested the following useful technique to extrapolate merged beam data to thermal energies. At low energies LZ theory approximates to the expansion,

$$Q = Av_r^{-2} + Bv_r^{-1} + C + Dv_r \tag{6.3}$$

which relates cross section (Q) to relative velocity (v_r) and numerical constants.

Figure 8 illustrates results obtained with merged beams by Aberth and Peterson (1970) for the mutual neutralization of N^+ and O^-. The continuous curve in the figure shows the result of an LZ calculation by Olson et al (1970) and there is evidence from several measurements that this type of calculation is usually accurate within a factor two or three. But agreement is not much better even for the simplest systems.

The simplest reaction,

$$H^+ + H^- \rightarrow H + H \tag{6.4}$$

has been extensively studied experimentally and theoretically but agreement is not yet satisfactory. Figure 9 compares measurements for H^+/H^- neutralization by Peart et al (1976b) with LZ calculations by Dalgarno et al (1971), Olson et al (1970), Bates and Lewis (1955) and with a close-coupling calculation by Roy and Mukherjee (1973). Best agreement between theory and experiment occurs below 30 eV where the LZ theory of Olson et al is in remarkably close (albeit, partially fortuitous) agreement with measurements by Moseley et al (1970); this is illustrated by figure 10.

The discrepancy between theory and experiment for reaction (6.4) has recently become a live topic. A discussion of the problem and new theoretical results have just been presented by Sidis et al (1981) whilst Brouillard and Claeys (private communication) pointed to a possible ambiguity in some of the measurements.

The inclined beam experiments by Rundel et al (1969), Gaily and Harrison (1970a), and Peart et al (1976b) only detected H atoms formed from the H^+ beam. It was argued that this was sufficient to discriminate between reactions (4.3) and (4.4). But Brouillard's group have recently detected in coincidence both H atoms formed by reaction (4.3). Preliminary results indicate smaller cross sections without structure and they suggest that the earlier experiments measured the sum of cross sections for,

$$H^+ + H^- \rightarrow H + H \tag{4.3}$$

and

$$H^+ + H^- \rightarrow H + H^+ + e \tag{6.5}$$

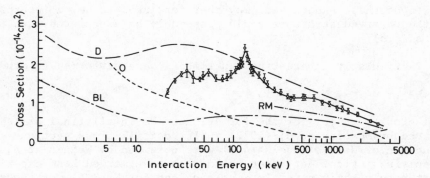

Fig. 9.   Measurements of mutual neutralization of H$^+$ and H$^-$ ions
           compared with LZ calculations by Bates and Lewis (BL),
           Dalgarno et al (D), Olson et al (O) and a close coupling
           calculation by Roy and Mukherjee (RM).

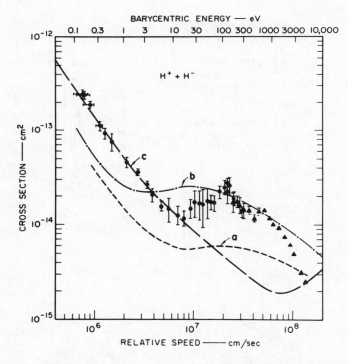

Fig. 10.   Measurements for mutual neutralization of H$^+$ and H$^-$ ions
           by Moseley et al (●) and Rundel et al (▲) compared with LZ
           calculations by Bates and Lewis (a), Dalgarno et al (b),
           and Olson et al (c).

Results of these experiments may therefore need to be re-interpreted
and the observed structure could presumably be associated with reaction (6.5).

Particularly interesting experiments were performed to study,

$$Na^+ + O^- \rightarrow Na + O. \tag{6.6}$$

Total cross sections for the formation of Na in all final states were
measured by Moseley et al (1972) whilst partial cross sections for
production of $Na(3^2D)$ and $Na(4^2P)$ were obtained by Weiner et al (1971)
by applying optical detection techniques to a merged beam experiment.
Unfortunately their <u>partial</u> cross sections for $Na(3^2D)$ were about
three times larger than the measured <u>total</u> cross section! But it was
still a worthwhile experiment because information, even about the relative population of final states, is useful. Throughout collision
physics there is so little experimental information about state population that the application of good modern optical calibration tech-

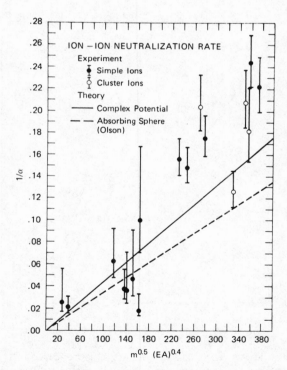

Fig. 11.   Scaling laws of Hickman (continuous line) and Olson (broken
           line) compared with experimental reaction rates which can
           be identified in Table 2.

Table 2. Values of experimental neutralization rates (α) and the product $\mu^{0.5} \chi^{0.4}$ for various ion pairs.

Ions	$m^{0.5}(E.A.)^{0.4}$	α	Reference
$H^+ + H^-$	27	39 ± 21	13
$H_2^+ + D^-$	38	47 ± 15	17
$N_2^+ + O_2^-$	118	16 ± 5	14
$H^+ + O^-$	137	26 ± 8	13
$O^+ + O^-$	142	27 ± 13	13
$Na^+ + O^-$	153	21 ± 10	15
$NO^+ + O^-$	162	49 ± 20	15
$O_2^+ + O^-$	164	10 ± 4	15
$NO^+ + NO_2^-$	232	6.4 ± 0.7	16
$NH_4^+ + Cl^-$	247	6.7 ± 0.7	16
$NH_4^+ \cdot (NH_3)_2 + NO_2^-$	270	4.9 ± 0.6	16
$NO^+ + NO_3^-$	278	5.7 ± 0.6	16
$NH_4^+ \cdot (NH_3)_2 + Cl^-$	329	7.9 ± 1.0	16
$H_3O^+ \cdot (H_2O)_3 + Cl^-$	350	4.8 ± 0.6	16
$H_3O^+ \cdot (H_2O)_3 + NO_3^-$	356	5.5 ± 1.0	16
$CClF_2^+ + Cl^-$	359	4.1 ± 0.4	16
$CCl_3^+ + Cl^-$	375	4.5 ± 0.5	16

niques (e.g. Taylor, 1972, or Taylor et al, 1980) would be extremely rewarding.

It is not possible to produce beams of clustered ions [e.g. $NO^+.H_2O$ or $(H_3O)^+.(H_2O)_3$] with the same internal energies as those encountered in nature, but the effects of internal energy and clustering was considered theoretically by Moseley et al (1975) and Hickman (1979). This is clearly a complicated subject but some tentative conclusions have emerged.

Electronic excitation usually provides access to more final states and should therefore tend to increase cross sections. Vibrational excitation, on the other hand, should only be important if it increases the effective vibrational overlap with final states. Even rotational excitation was found to be significant in the case of H^+/H^- neutralization (Browne and Victor, see Moseley et al, 1975).

The effect of clustering was explored theoretically by Hickman

with special reference to NO^+ and $NO^+.H_2O$. He concluded that there
may be a small decrease in recombination rates as the cluster size
increases, due to the change of the relative mass of the motion, but
this may be offset by the possibility of enhanced internal excitation.
Preliminary results suggest that recombination rates for small clusters should not differ greatly from those of bare ions.

There is now a fair amount of good experimental data for mutual
neutralization and this encouraged Olson (1972) and Hickman to derive
scaling laws, which was a commendable enterprise. Olson considered
only electron transfer between the reactants but Hickman elaborated
the model by including internal excitation of the short-lived complex
formed during the collision. It transpired that electron transfer is
usually dominant for small ions and clusters.

Both models led to scaling laws which are compared with experimental results by figure 11 and table 2. The figure shows the reciprocal of the recombination rates at 300 K (α^{-1}) plotted against
$\mu^{0.5} \chi^{0.4}$, where μ denotes the reduced mass and χ is the detachment
energy of the negative ion. Both theories predict linear relations.
The points denote experimental results (scaled to 300 K) for simple
and cluster ions which can be identified from the table. This seems
to offer a powerful technique for predicting recombination rates, but
it has been pointed out (D. Smith, private communication) that fig. 11
includes a selection of recombination rates obtained by afterglow and
beam experiments. In some cases (e.g. NO^+/NO_2^-) the two techniques
give results which differ by an order of magnitude and Smith suggested
that if all experimental results had been included, or if the selection had been made differently, the accord between experiment and
theory in fig. 11 would be far less impressive.

Double charge transfer between H^+ and H^-, i.e.

$$H^+ + H^- \rightarrow H^- + H^+ \tag{6.7}$$

has been studied for energies between 30 and 200 eV by Brouillard et
al (1979) and between 44 and 570 eV by Peart and Forrest (1979).
Results of both experiments are illustrated by figure 12 and they
indicate maxima at 50, 75 and 200 eV. The broken curve illustrates
measurements for the sum of reactions (4.3) and (6.5) but there is
no obvious correlation between the two results.

7.   MEASUREMENTS OF THE INTERACTIONS BETWEEN PROTONS AND OTHER
     POSITIVE IONS

When protons react with another positive ion (X^+), the dominant
reactions are usually ionization,

$$H^+ + X^+ \rightarrow H^+ + X^{2+} + e \tag{7.1}$$

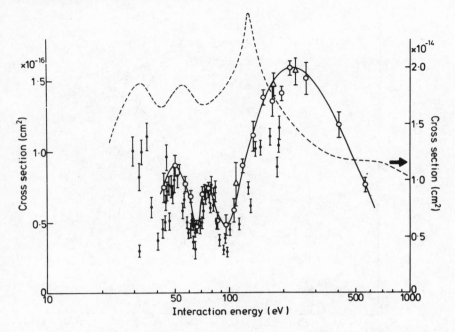

Fig. 12.   Cross sections for double charge transfer between H^+ and H^-
           ions measured by Brouillard et al ($\bullet$) and by Peart and
           Forrest (o, Δ).   The broken curve shows measurements for
           single charge transfer with different ordinate scale.

and charge transfer,

$$H^+ + X^+ \rightarrow H + X^{2+}. \tag{7.2}$$

Experiments have been reported (see Gilbody, 1981) on the interaction
of protons with He^+, Li^+, Mg^+, C^+, N^+, Ti^+ and Fe^+.   Most of these
ions are probable impurities in fusion plasmas whilst Mg^+, and possibly Li^+, are of interest in astrophysics.   The relevance to fusion
can be appreciated by considering reactions (7.1) and (7.2).   Charge
transfer allows fuel ions of H^+ (or D^+ in a reactor) to escape from
the containing fields and sputter further impurities into the plasma
when they strike the reactor walls.   Both reactions enhance the charge
on the impurity ions and so increase the energy lost by bremsstrahlung, although the net radiation loss might not increase because line
radiation is usually smaller from more highly ionized impurities.

When dealing with charge transfer from neutral atoms or molecules, it is usual to classify a process as "resonant" or "non-
resonant".   A resonant reaction, e.g.

$$H^+ + H \rightarrow H + H^+ \tag{7.3}$$

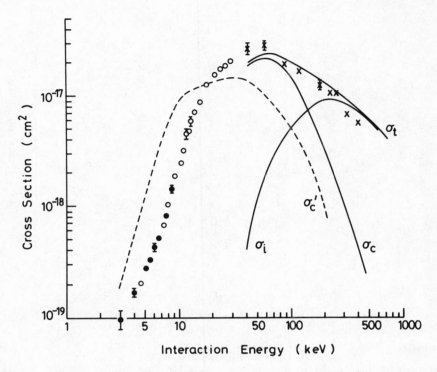

Fig. 13.   Cross sections for He^{2+} production by H$^+$/He$^+$ collisions
measured by Peart et al (●) and Angel et al (×). Continuous
curves denote results of scaled classical calculations by
Banks et al whilst broken curves illustrate a Born calculation for ionization and a close-coupling calculation for
the reverse charge transfer reaction (see text).

occurs when the commuting electron neither gains nor loses energy
($\Delta E = 0$). This concept cannot properly be applied to reactions between positive ions because a reaction is only strictly resonant at
zero interaction energy, but at low energies ion-ion reactions are
dominated by Coulomb repulsion. Nevertheless, it is still true that
larger cross sections ($\sim 10^{-15}$ cm^2) have been observed for "near resonant" reactions of protons with Ti$^+$ ($\Delta E = 0.025$ eV) and Mg$^+$ ($\Delta E =
1.4$ eV), whilst much smaller cross sections ($\sim 10^{-17}$ cm^2) were reported
for He$^+$ ($\Delta E = 40$ eV) or Li$^+$ ($\Delta E = 60$ eV).

Most experiments measured the production of doubly-charged ions
and so gave "total" cross sections (σ_T) which (if we neglect multiple
ionization and other improbable processes) are the sum of cross
sections for charge transfer (σ_c) and ionization (σ_i). We will
briefly review some results.

Helium is an inevitable impurity in DD and DT reactors because it is a fusion product. Mitchell et al (1977), Angel et al (1978) and Peart et al (1977a) measured σ_T for H^+/He^+ collisions. Figure 13 compares the two latter measurements with five theoretical predictions. The three continuous curves labelled σ_T, σ_i and σ_c illustrate results of classical calculations for H by Banks et al (1976) which have been scaled for He^+. The broken curve (σ_i') is obtained from a Born calculation for H by Bates and Griffing (1953) which has also been multiplied by classical scaling factors. Rapp (1974) performed an eleven state close-coupling calculation for,

$$He^{2+} + H(1s) \rightarrow He^+(1s) + H^+. \tag{7.4}$$

This result is illustrated by the broken curve, σ_c', but it should only provide a lower limit since, in reverse, it includes only H atoms produced in the H(1s) state. At low energies the agreement with experiment is not good.

Comparison of experiment and theory indicates that the classical theory of Banks et al is very satisfactory, at least at the higher energies, and that ionization dominates below about 100 keV.

An example of accidental near resonance ($\Delta E = 1.4$ eV) is,

$$H^+ + Mg^+ \rightarrow Mg^{2+} + H. \tag{7.5}$$

Measurements of σ_T for H^+/Mg^+ (Peart et al, 1977b) are illustrated by open circles in figure 14. An interesting comparison is afforded by the semi-classical, two-state calculation for

$$H + Mg^{2+} \rightarrow H^+ + Mg^+ \tag{7.6}$$

by Bates et al (1964) and corresponding measurements by McCullough et al (1979). The agreement is remarkably close, indicating that charge transfer dominates at these energies and, as one might expect for a near-resonant process, transitions occur mainly between unexcited states. Curves B and C are estimates of σ_T and σ_i obtained by classically scaling the calculations for H^+/H by Banks et al (1976) to H^+/Mg^+. Curve D denotes electron impact ionization cross sections for Mg^+ (Martin et al, 1968) scaled to the same projectile speed.

Very recently, Dickinson and Hardie (private communication) found that a two-state close-coupling calculation for reaction (7.5) agreed well with experiment above 5 keV, although at lower energies it was less accurate than the calculation by Bates et al. This difference is not yet understood.

The paper by McCullough et al is noteworthy because not only does it present measurements for reaction (7.6) at energies between 0.8 and 40 keV, but it provides similar results for Ba^{2+}, Tl^{2+}, Cd^{2+}, Zn^{2+},

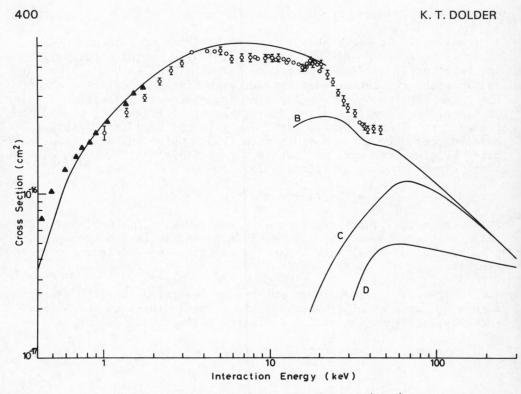

Fig. 14.   Measurements of production of Mg^{2+} by H^+/Mg^+ collisions (●)
           compared with measurements (▲) and calculations (curve A)
           of charge transfer between H and Mg^{2+}.   Curves B and C are
           classical estimates of σ_T and σ_i (see text) and curve D
           shows measured electron impact ionization cross sections of
           Mg^+ scaled to same projectile speed.

Kr^{2+} and B^{2+} which will offer interesting comparisons when the reverse processes are studied.

      An example of remarkably close accidental resonance ($\Delta E = 0.025$
eV) is charge transfer between H^+ and Ti^+ and this may have practical
implications because Ti is a constituent of many alloy steels.
Measurements of,

$$H^+ + Ti^+ \rightarrow Ti^{+} + \dots \tag{7.7}$$

by Hobbis et al (see Gilbody, 1981) are illustrated by figure 15
which includes results of two-state and three-state close-coupling
calculations by Dickinson and Hardie.   The two-state calculation
employed "ab initio" adiabatic potentials, whilst the three-state
calculation used model potentials and made some allowance for trans-

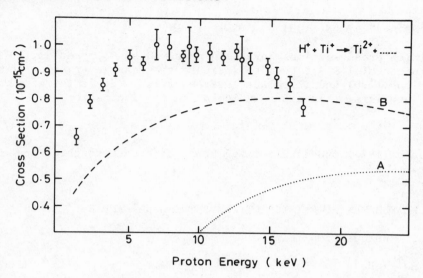

Fig. 15. Measurements of Ti^{2+} production by H^+/Ti^+ collisions by
          Hobbis et al compared with two-state (A) and three-state
          (B) close-coupling calculations by Dickinson and Hardie.

itions by the 3d electrons. The latter result agrees better with
experiment. Comparable agreement was achieved between a three-state
model potential calculation for,

$$H^+ + Fe^+ \rightarrow Fe^{2+} + \ldots \qquad (7.8)$$

and measurements by Hobbis et al (l.c.) in the energy range 2 – 13 keV.

Two other measurements relevant to magnetically confined fusion
plasmas were recently performed by Neill et al (see Gilbody, 1981) who
investigated

$$H^+ + C^+ \rightarrow C^{2+} + \ldots \qquad (7.9)$$
and
$$H^+ + N^+ \rightarrow N^{2+} + \ldots \qquad (7.10)$$

Comparisons with the reverse charge transfer process,

$$C^{2+} + H \rightarrow C^+ + H^+ \rightarrow \quad (\Delta E = 10.8 \text{ eV}) \qquad (7.11)$$

and

$$N^{2+} + H \rightarrow N^+ + H^+ \qquad (\Delta E = 16 \text{ eV}) \qquad\qquad (7.12)$$

for which cross sections have been measured by Goffe et al (1979) and
Phaneuf et al (1979), seem to indicate that charge transfer dominates
reactions (7.9) and (7.10) at lower energies. It is therefore somewhat surprising that a Born calculation for ionization,

$$H^+ + N^+ \rightarrow H^+ + N^{2+} + e \qquad\qquad (7.13)$$

by Peach (1971) predicts a large cross section throughout the energy
range.

8.   MEASUREMENTS OF INTERACTIONS BETWEEN He^+ AND He^{2+} IONS

The unique simplicity of,

$$He^+ + He^{2+} \rightarrow He^{2+} + He^+ \qquad\qquad (8.1)$$

prompted observations of He^+/He^{2+} interactions by Jognaux et al (1978)
and Peart and Dolder (1979), but neither experiment was entirely satisfactory. Jognaux et al used merged beams to explore the energy
range 0.01 - 1.7 keV whilst Peart and Dolder employed inclined beams
for energies between 0.1 and 20 keV. Owing to mutual repulsion between the collision products, it was not possible, in either experiment, to collect all the He^{2+} ions formed. Corrections derived from
differential scattering theory (e.g. Dickinson and Hardie, 1979) were
therefore applied to the measurements which then agreed with theoretical predictions within rather broad limits of error. Better experiments are needed.

9.   EXPERIMENTS RELEVANT TO "HEAVY ION FUSION" (HIF)

Design studies are being made into the feasibility of producing
controlled fusion by bombarding small DT pellets with intense ($\sim 10^{14}$
watts), energetic (~ 10 GeV) pulses of heavy ions. Problems would
arise during the acceleration and storage of such intense ion pulses
from ion-ion interactions, e.g.

$$Cs^+ + Cs^+ \begin{array}{l} \nearrow \quad Cs^+ + Cs^{2+} + e \qquad\qquad (9.1) \\ \searrow \quad Cs + Cs^{2+} \qquad\qquad\quad (9.2) \end{array}$$

which would cause ions to be lost from the accelerators and lead to
damage due to sputtering. The probable cost of the accelerating and
ion storage systems is of order $\$10^9$ and so it is essential to have
reliable data for relevant ion-ion reactions to optimize accelerator
designs.

Cross sections for,

$$Cs^+ + Cs^+ \quad Cs^{2+} + \ldots \tag{9.3}$$

were reported by Dunn et al (1979) and Peart et al (1981a). Results are illustrated by figure 16 and disagreement below 100 keV is evident. Peart et al performed a number of developments on their apparatus and applied detailed checks which were summarized in sections 5.3 and 5.4 which also include a possible explanation of the discrepancy.

Since HIF requires ions with relatively small cross sections for ionization and charge transfer, closed shell systems (e.g. Cs^+) are obvious candidates. Theoretical work is proceeding in parallel with experiment but severe problems are encountered, especially for the heavier ions. Peart et al (1981b) therefore measured cross sections for reactions analogous to (9.3) for each of the alkali ions, Li^+, Na^+, K^+, Rb^+ and Cs^+. Theoreticians are therefore invited to apply their approximations to systems of progressively increasing complexity. Results are illustrated by figure 17 but they reveal no obvious correlation between cross sections and atomic mass or ionization

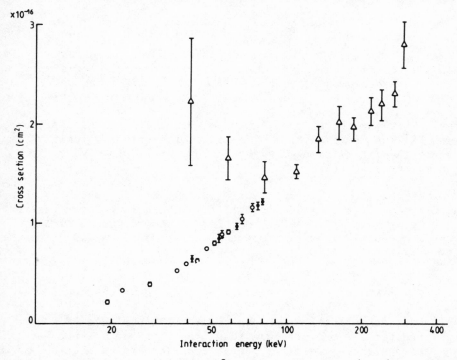

Fig. 16. Cross sections for Cs^{2+} production by Cs^+/Cs^+ collisions measured by Peart et al (o•) and Dunn et al (Δ).

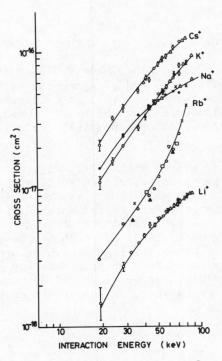

Fig. 17.   Cross sections for $Cs^+ + Cs^+ \rightarrow Cs^{2+} + \ldots$ and similar
measurements for Li^+/Li^+, Na^+/Na^+, K^+/K^+ and Rb^+/Rb^+.

energy.   This may not be surprising because the large mass difference
between Li^+ and Cs^+ implies a wide range of particle velocities and
different theoretical models might be appropriate for the various ions
at these energies.

Other reactions relevant to HIF include Xe^+/Xe^+ which has been
studied by Angel et al (1980) and Tl^+/Tl^+ which is being investigated
in this laboratory.   Cross sections for the formation of Tl^{2+} by collisions between Tl^+ ions rise monotonically from 1.4×10^{-17} cm^2 at
33 keV (E_{cm}) to 6.7×10^{-17} cm^2 at 92 keV.

10.   MEASUREMENTS OF CHARGE TRANSFER BETWEEN POSITIVE IONS

We have seen that for near-resonant reactions (e.g. H^+/Mg^+)
charge transfer is more probable than ionization at moderate energies.
Cross sections exclusively for charge transfer can be obtained if an
experiment is designed to detect the neutral atoms formed, for example
by reaction (9.2).   In this way, Peart et al (1981c) obtained cross
sections for charge transfer between Cs^+ ions and showed that, at

energies between 28 and 68 keV, it is an order of magnitude less probable than ionization. But the SBR in their experiment was poor and
random errors were typically ±50%.

It is therefore better to arrange for coincident detection of the
ion and atom formed by charge transfer and this technique was applied
to H^+/Li^+ by Sewell et al (1980) and to H^+/He^+ by Angel et al (1978).
Figure 18 illustrates their measurements for,

$$H^+ + He^+ \rightarrow H + He^{2+} \tag{10.1}$$

and shows excellent agreement with Olson's (see Angel et al, 1978)
classical Monte-Carlo calculation. A lower theoretical estimate is
afforded by Rapp's (1974) calculation for the reverse process,

$$He^{2+} + H(1s) \rightarrow He^+(1s) + H^+. \tag{10.2}$$

This calculation is not expected to agree with the measurements
because the experiment detects H atoms in all excited states. By comparing measurements for charge transfer (σ_c) with earlier results for
σ_T ($\sigma_T = \sigma_i + \sigma_c$), the ionization cross sections (σ_i) can be deduced
and figure 18 shows that they are relatively unimportant below 100 keV.

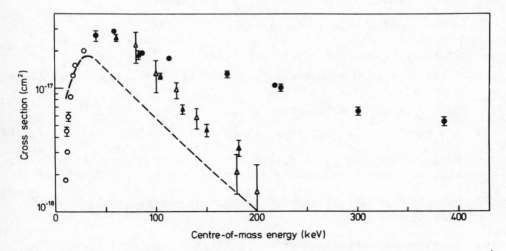

Fig. 18. Measurements by Angel et al of charge transfer between H^+
          and He^+ ions (▲) compared with calculations by Olson (Δ).
          The hollow and closed circles show measurements for the sum
          of ionization (σ_i) and charge transfer (σ_c) and the broken
          curve represents Rapp's calculation for reaction (10.2).

For H^+/Li^+ reactions, it was found that $\sigma_c = \sigma_i = 10^{-17}$ cm^2 at about 150 keV and that charge transfer dominates at lower energies. Sewell et al obtained detailed measurements for σ_c and σ_T between 62 and 350 keV and values of σ_i deduced from these results agree well with cross sections for the ionization of Li^+ by fast electrons ($v > 2.10^9$ cm s^{-1}) scaled to the same projectile speed.

11.   A BRIEF COMMENT ON MEASUREMENTS OF DISSOCIATIVE RECOMBINATION OF MOLECULAR IONS

Detailed discussions of this topic appear elsewhere in this volume and so we will be brief. Dissociative recombination is important because it governs the behaviour of planetary atmospheres, flames, combustion processes, etc. There was particular interest in

$$H_2^+ + e \rightarrow H + H \tag{11.1}$$

because this (with the possible exception of He_2^+) was the only ion for which "ab initio" calculations were possible and good measurements were needed to allow theoretical approximations to be developed confidently.

Until quite recently most information was obtained from afterglow experiments (Bardsley and Biondi, 1970) but these cannot be applied to H_2^+ because it is chemically reactive and rapidly transforms to H_3^+, H_5^+ etc. Beam methods were therefore developed.

Four techniques have been reported:

(a)   Inclined beams (e.g. Peart and Dolder, 1973a)

(b)   Static ion trap with energy-resolved electron beams (e.g. Walls and Dunn, 1974)

(c)   Merged beams (e.g. Auerbach et al, 1977)

(d)   Ion beams in electron space charge trap (e.g. Mathur et al, 1978).

Details can readily be obtained from the references cited but we note that any experiment should endeavour to achieve the following:

(a)   Unambiguous definition of the chemical nature of the reacting ions (e.g. H_2^+ or H_n^+ ...?).

(b)   Unambiguous definition of the initial vibrational state of an ion. Vibrational excitation (and to a lesser extent, rotational excitation) influence recombination cross sections enormously.

(c)   Good energy resolution. Recombination is of greatest interest at energies below a few eV and so good resolution ($\Delta E \ll 1$ eV)

is clearly important. Moreover, there is usually structure in
the recombination cross section which can give important insights
into physical processes.

(d)   Considerable interest was directed at the temperature or energy-
      dependence of dissociative recombination. Results over an energy
      range of several eV are therefore valuable and it is important to
      distinguish between kinetic temperature and vibrational or rota-
      tional temperature.

(e)   The production of absolute cross sections without recourse to
      normalization to other experimental or theoretical results.

We note in table 3 the extent to which the available techniques
meet these requirements.

There is also an important limitation in some experimental techniques which theoreticians may overlook. If non-absolute measurements
are made, they are sometimes normalized to absolute measurements or
theory. But dissociative recombination depends sensitively upon
initial vibrational excitation and, if this excitation is different
for the two cross sections being compared (this generally is the case)
the normalization is invalid!

12.   DETACHMENT FROM NEGATIVE IONS AND SEARCH FOR DOUBLY-NEGATIVE IONS

The reaction,

$$e + H^- \rightarrow H + 2e \tag{12.1}$$

was once of astrophysical interest because it was believed that it
might influence the concentration of H^- in the solar photosphere and
hence affect its opacity to visible radiation. It transpired, however,
that associative attachment,

$$H + H^- \rightarrow H_2 + e \tag{12.2}$$

is the dominant process (Schmeltekopf et al, 1967); but investigations
of (12.1) revealed two structures in the cross section illustrated by
figure 19 (Peart and Dolder, 1973b). These were interpreted (Taylor
and Thomas, 1972; Thomas, 1974) as resonances due to the formation of
short-lived ($\sim 10^{-15}$ s) states of $H^{--}(2s^2 2p)$ and $H^{--}(2p^3)$. Detachment
from C^-, O^- and F^- (Peart et al, 1979a,b,c) was therefore studied to
find evidence of similar structure. Only in the case of O^- was structure observed and this is illustrated by figure 20.

No theoretical interpretation has yet been advanced and no other
measurements of these structures has been reported.

Table 3. Comparison of five techniques used to measure dissociative recombination.

Requirement / Technique	Define Chemical Nature of Ion	Define Vibrational State	Good Energy Resolution ΔE ~ 0.1 eV	Wide Energy Range	Absolute Cross Section Without Normalization
Afterglow	Not for reactive ions, e.g. H_2^+	Defined by thermal equilibrium of plasma	Measure reaction rates over range of apparatus temperatures T~200–500 K	No T~200–500 K	Yes
Inclined beam	Yes	In general, vibrational states not defined but quite good definition for H_2^+, D_2^+ and H_3^+	Typically ΔE ≈ 0.1 eV	0.2 – 5 eV	Yes
Merged beam	Yes		Good resolution but see section 5.4	0.1 – 5 eV but see section 5.4	Difficult, but possible in principle
Electron-trapped ion beam	Yes	No – not in present form	The advantages of merged beams partially offset by local potentials	0.3 – ~1 eV	No
Static ion trap	Not for reactive ions, e.g. H_2^+	Ions pre-cooled to v=0 state. Excellent definition	Uses energy resolved electrons. ΔE 50 meV	Typically 0.05 – 6 eV	No

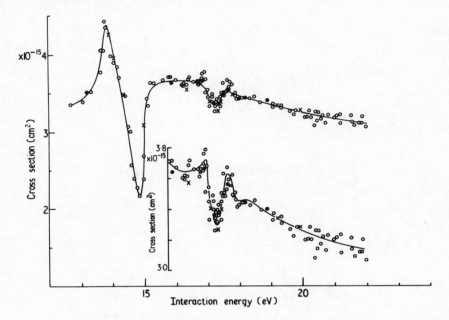

Fig. 19. Measurements of cross sections for detachment from H⁻ by
          electron impacts. The insert shows the smaller structure
          with an enlarged ordinate scale.

The availability of electron impact detachment cross sections
for H⁻, C⁻, O⁻ and F⁻ prompted a search for scaling laws similar to
those which have proved useful for positive ions. Within a factor
of four, the results can be scaled to a single curve by the classical
law,

$$\frac{\sigma_1}{\sigma_2} = \frac{n_1}{n_2}\left(\frac{\chi_1}{\chi_2}\right)^2 \tag{12.3}$$

where σ_1 and σ_2 represent cross sections for ions with detachment
energies χ_1, χ_2 and n is the "effective" number of electrons in the
outer shell. When s and p electrons co-existed in a shell, only the
more loosely-bound p electrons were included, so, for H⁻, C⁻, O⁻ and
F⁻, n took the values 2, 3, 5 and 6, respectively.

It is, of course, paradoxical to apply classical scaling to such
non-classical structures as negative ions, and Esaulov (1980) noted
that measurements of the stripping of negative ions by collisions with
neutral gas are best explained by a model in which only one electron
is considered to be loosely bound. This implies the scaling law,

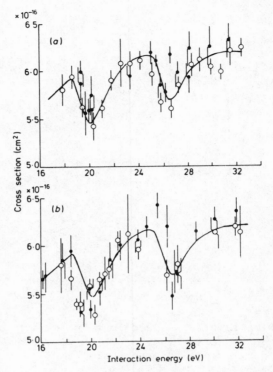

Fig. 20.   Measurements of cross sections for detachment from O^- by
           electron impact. Figures 20(a) and 20(b) show results
           with ion beam energies of 40 and 50 keV, respectively.

$$\frac{\sigma_1}{\sigma_H} = \left(\frac{\chi_1}{0.75}\right)^2 \tag{12.4}$$

which gives a two-fold improvement on equation (12.3). Here, σ_H and
0.75 (eV) represent the cross section and detachment energy of H^-

REFERENCES

Aberth, W., and Peterson, J.R., 1970, Phys. Rev. A, 1:158.
Aberth, W., Moseley, J.T., and Peterson, J.R., 1971, Air Force Camb-
      ridge Report, AFCRL-71-0481.
Angel, G.C., Sewell, G.C., Dunn, K.F., and Gilbody, H.B., 1978, J.
      Phys. B, 11:L297.

Angel, G.C., Dunn, K.F., Neill, P.A., and Gilbody, H.B., 1980, ibid, 13:L391.

Auerbach, D., Cacak, R., Caudano, R., Gailey, T.D., Keyser, C.J., McGowan, J.W., Mitchell, J.B.A., and Wilk, S.F.J., 1977, ibid, 10:3797.

Banks, D., Barnes, K.S., and Wilson, J.McB., 1976, ibid, 9:L141.

Bardsley, J.N., and Biondi, M.A., 1970, Adv. Atom. Mol. Phys., 6:1.

Bates, D.R., and Griffing, G., 1953, Proc. Phys. Soc. A, 66:961.

Bates, D.R., and Lewis, J.T., 1955, ibid, 68:173.

Bates, D.R., Johnston, H.C., and Stewart, I., 1964, ibid, 84:514.

Brouillard, F., Claeys, W., Poulaert, G., Rahman, G., and van Wassenhove, G., 1979, J. Phys. B, 12:1253.

Dalgarno, A., Victor, G.A., and Blanchard, P., 1971, Air Force Cambridge Report, AFCRL-71-0342.

Dance, D.F., Harrison, M.F.A., and Smith, A.C.H., 1966, Proc. Roy. Soc. A, 290:74.

Dickinson, A.S., and Hardie, D.J.W., 1979, J. Phys. B, 12:4147.

Dolder, K., 1969, Experiments in Colliding Charged Particle Beams in "Case Studies in Atomic Collision Physics", North Holland, Amsterdam, 1:249.

Dolder, K., and Peart, B., 1976, Rept. Prog. Phys., 39:639.

Dunn, G.H., 1969, Colliding Beams in "Atomic Physics", Plenum Press, New York, 1:417.

Dunn, G.H., 1979, Atomic Processes in Fusion Plasmas, Proc. Nagoya Seminar, Nagoya Univ. Japan Rept. IPPJ-AM13.

Dunn, G.H., 1980, Physics of Ionized Gases, Invited lecture SPIG-80 Boris Kidrić Inst. of Nuclear Science, Belgrade. Edited by M. Matić.

Dunn, K.F., Angel, G.C., and Gilbody, H.B., 1979, J. Phys. B, 12:L623.

Esaulov, V.A., 1980, ibid, 13:1625.

Gailey, T.D., and Harrison, M.F.A., 1970a, ibid, 3:L25.

Gailey, T.D., and Harrison, M.F.A., 1970b, ibid, 3:1098.

Gilbody, H.B., 1981, Invited paper, Proc. 12th Int. Conf. Phys. Elec. and At. Collns., Gatlinburg, U.S.A.

Goffe, T.V., Shah, M.B., and Gilbody, H.B., 1979, J. Phys. B, 12:3763.

Harrison, M.F.A., 1966, Brit. J. A. P., 17:371.

Harrison, M.F.A., 1968, Electron Impact Ionization and Excitation of Positive Ions in "Methods of Experimental Physics", Academic Press, New York, 7B:95.

Harrison, M.F.A., 1978, "Colliding Beam Studies of Atomic Collision Processes", Inst. Phys. Conf. Series No. 38, ch.4:190.

Henry, R.J.W., 1981, Phys. Repts., 68:1.

Hickman, A.P., 1979, J. Chem. Phys., 70:4872.

Jognaux, A., Brouillard, F., and Szucs, S., 1978, J. Phys. B, 11:L669.

McCullough, R.W., Nutt, W.L., and Gilbody, H.B., 1979, ibid, 12:4159.

Martin, S.O., Peart, B., and Dolder, K., 1968, ibid, 1:537.

Mathur, D., Khan, S.U., and Hasted, J.B., 1978, ibid, 11:3615.

Mitchell, J.B.A., Dunn, K.F., Angel, G.C., Browning, R., and Gilbody, H.B., 1977, ibid, 10:1897.

Molyneux, L., Dolder, K., and Peart, B., 1971, J. Phys. E, 4:149.

Moseley, J.T., Aberth, W., and Peterson, J.R., 1970, Phys. Rev. Letts., 24:435.

Moseley, J.T., Aberth, W., and Peterson, J.R., 1972, J. Geophys. Res., 77:255.

Moseley, J.T., and Peterson, J.R., 1972, Bull. Amer. Phys. Soc., 17: 1136.

Moseley, J.T., Olson, R., and Peterson, J.R., 1975, Ion-Ion Mutual Neutralization in "Case Studies in Atomic Collision Physics", North Holland, Amsterdam, 5:1.

Neynaber, R.H., 1969, Experiments in Merging Beams in "Advances in At. and Molec. Phys.", Academic Press, New York, 5:57.

Olson, R., 1978, private communication to Angel et al, 1978.

Olson, R., 1972, J. Chem. Phys., 56:2979,

Olson, R., Peterson, J.R., and Moseley, J.T., 1970, J. Chem. Phys., 53:3391.

Peach, G., 1971, J. Phys. B, 4:1670.

Peart, B., and Dolder, K., 1973a, ibid, 6:2409.

Peart, B., and Dolder, K., 1973b, ibid, 6:1497.

Peart, B., and Dolder, K., 1979, ibid, 12:4155.

Peart, B., Grey, R., and Dolder, K., 1976a, ibid, 9:L373.

Peart, B., Grey, R., and Dolder, K., 1976b, ibid, 9:L369.

Peart, B., Grey, R., and Dolder, K., 1977a, ibid, 10:2675.

Peart, B., Gee, D., and Dolder, K., 1977b, ibid, 10:2683.

Peart, B., and Forrest, R.A., 1979, ibid, 12:L23.

Peart, B., Forrest, R.A., and Dolder, K., 1979a, ibid, 12:849.

Peart, B., Forrest, R.A., and Dolder, K., 1979b, ibid, 12:L115.

Peart, B., Forrest, R.A., and Dolder, K., 1979c, ibid, 12:2735.

Peart, B., Forrest, R.A., and Dolder, K., 1981a, ibid, 14:1655.

Peart, B., Forrest, R.A., and Dolder, K., 1981b, in course of publication, ibid.

Peart, B., Forrest, R.A., and Dolder, K., 1981c, ibid, 14:L383.

Peek, J.M., and Green, T.A., 1969, Phys. Rev., 183:202.

Peterson, J.R., Aberth, W., Moseley, J.T., and Sheridan, J.R., 1971, Phys. Rev. A, 3:1651.

Phaneuf, R.A., Meyer, F.W., and McKnight, R.H., 1979, ibid, 17:534.

Poulaert, G., Brouillard, F., Claey, W., McGowan, J.W., and van Wassenhove, G., 1978, J. Phys. B, 11:L671.

Rapp, D., 1974, J. Chem. Phys., 61:3777.

Rundel, R.D., Aitken, K.L., and Harrison, M.F.A., 1969, J. Phys. B, 2:954.

Seaton, M.J., 1975, Electron Impact Excitation of Positive Ions in "Atomic and Molecular Physics", Academic Press, New York, 11:83.

Sewell, E.C., Angel, G.C., Dunn, K.F., and Gilbody, H.B., 1980. J. Phys. B, 13:2269.

Schmeltekopf, A.L., Fehsenfeld, F.C., and Ferguson, E.E., 1967, Astrophys. J., 148:L155.

Sidis, V., Kubach, C., and Fussen, D., 1981, Phys. Rev. Letts., 47: 1280.

Taylor, H.S., and Thomas, L.D., 1972, ibid, 28:1091.

Taylor, P.O., 1972, Ph.D. Thesis, University of Colorado, University

Microfilms, Ann. Arbor., Mich.
Taylor, P.O., Phaneuf, R.A., and Dunn, G.H., 1980, Phys. Rev. A, 22: 435.
Thomas, L.D., 1974, J. Phys. B, 7:L97.
Walls, F.L., and Dunn, G.H., 1974, Phys. Today, 27:30.
Walton, D.S., Peart, B., and Dolder, K., 1971, J. Phys. B, 4:1343.
Weiner, J., Peatman, W.B., and Berry, R.S., 1971, Phys. Rev. A, 4:1825.

ON THE MEASUREMENT OF ION(ATOM) - ION(ATOM) CHARGE EXCHANGE

F. Brouillard and W. Claeys

Institut de Physique, Université Catholique de Louvain
Chemin du cyclotron, 2   B-1348 Louvain-la-Neuve
Belgium

A. KINEMATICS OF BEAM-BEAM INTERACTIONS

In this section we derive some useful kinematic relations for the measurement of cross sections by means of intersecting beams. These include the formulation of center-of-mass energy of the collidants, laboratory-energies and angles of the products formed in elastic or inelastic processes.
Kinematics provides a simple way to reach extremely low collision energies with energetic beams (merged beams) or reversely to convert tiny CM-inelastic losses or gains into observable laboratory effects.
As a preliminary step, the concept of cross section is briefly reviewed.

CROSS SECTIONS

The concept of cross section is introduced as an observable related to the probability of a particular process taking place in a binary collision.

General definition

Let us consider one fixed particle of type 1 (target) and a second particle of type 2 (projectile) fired on it, repeatedly with a specified velocity $\vec{v}$ but at all possible values of the impact parameter b. Let us say we have an "event" each time the particular process considered is taking place (see fig. 1).
If ν is the number of trials (number of particles 2 fired) per unit area of the plane orthogonal to $\vec{v}$ (impact parameter plane),

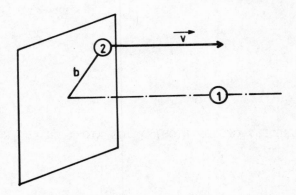

Fig. 1   Initial conditions for a binary collision.

and N is the number of events, then the cross section σ of this particular process is simply defined as

$$\sigma \equiv \frac{N}{\nu} \ .$$

Particularizing to the number dN of events observed during a given time dt,

$$\sigma = \frac{(dN/dt)}{(d\nu/dt)} = \frac{(dN/dt)}{j} \tag{1}$$

where j is the flux of particles 2 :

$$\vec{j} = n_2 \vec{v} \qquad\qquad n_2 : \text{density of particles 2.}$$

The cross section of a collisional process is thus defined as the number of events observed per unit time when a unit flux of projectiles is fired on a single target. CGS units are used here so that the cross section is expressed in cm^2.

Specification of process (differential, total, channel-specified
                        cross sections)

The particular process to which the cross section relates is completely described when :
1) the nature and quantum state,
2) the velocities $(\vec{v}_i)$, (see fig.2),
of all products are specified.

In that case, the cross section is said to be a differential cross section. It gives the number of events observed with one target, per unit time, unit incident flux and unit volume in the

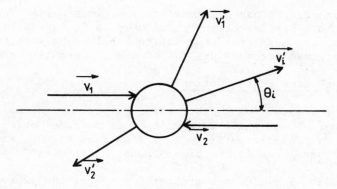

Fig. 2   Initial and final velocities for a binary collision.

3p-dimensional velocity space of the p products.
We write it I :

$$I \equiv \frac{dN}{dt\,dv'_{1x}\,dv'_{1y}\,dv'_{1z}\,dv'_{2x}\ldots dv'_{ix}\,dv'_{iy}\,dv'_{iz}\ldots}$$

It is customary to use polar rather than cartesian coordinates
to express the differential cross section and to use the kinetic
energies ε_i rather than the velocities v_i

$$I \equiv \frac{dN}{dt\,d\varepsilon_1\,d\Omega_1\,d\varepsilon_2\,d\Omega_2\ldots d\varepsilon_i\,d\Omega_i\ldots}$$

where $d\Omega_i = \sin\theta_i\,d\theta_i\,d\phi_i$
where θ_i and ϕ_i are the polar and asimuthal angles of $\vec{v}_i$.

The differential cross section appears to depend on 3p varia-
bles $(\varepsilon_i\theta_i\phi_i)$ but energy and momentum conservation during the collision imposes 4 relations between these variables so that the cross
section actually depends on (3p-4) independent variables.
For example, if 2 products only emerge out of the collision, the
differential cross section is a two-dimensional quantity.

If the velocities are left unspecified, the cross section is
the total cross section. The total cross section is of course the
integral of the differential cross section

$$\sigma = \int I(\varepsilon_i\Omega_i)\,d\varepsilon_i\,d\Omega_i.$$

Practically, measurements (or calculations) are not always
capable of specifying the quantum state of the products : the
channel is not specified and the cross section is an "all-channel-
cross section" in contrast with a "channel-specified-cross section".

Channel-specified cross sections are sometimes called "partial cross sections" but this expression is unlucky as it logically leads to call "total cross section" the "all-channel cross section", a term already reserved for the integral of the differential cross section.

Alternative forms of definition

If the interaction of the target particle with the incident flux $\vec{j} = n_2\vec{v}$ of particles 2 is watched in a frame moving with the particles 2, a single particle 1 is then seen to move with velocity $(-\vec{v})$ across a density n_2 of stationary particles 2 (fig. 3).

Formula (1) yields the probability dP for the process taking place on a distance dx :

$$dP = \sigma n_2 v dt = \sigma n_2 dx.$$

On a finite distance x, the probability P is either $P=\sigma n_2 x$ or $P=(1-e^{-\sigma n_2 x})$ depending on the non-destructive or destructive character of the process.

Finally, considering a flux $\vec{j}_1=n_1\vec{v}_1$ of particles 1 moving through a flux $\vec{j}_2=n_2\vec{v}_2$ of particles 2, we derive the expression of the number of events per second in the elementary volume dV

$$dN = \sigma n_1 n_2 v dV \tag{2}$$

where v is the relative velocity $v = |\vec{v}_2-\vec{v}_1|$.

It might be noted that the cross section is not a relativistic invariant. Formula (2) is therefore an ambiguous definition of σ as long as the referential frame is not specified. The most natural frame to define the cross section is the frame where the center of mass of the collidants is at rest (CM-frame).

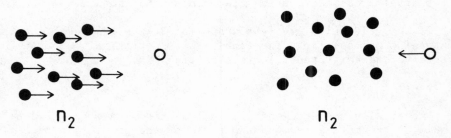

Fig. 3.   Two alternative ways of watching the collision of two species.

Measurement with a beam and a stationary target

The simplest method to measure a cross section is to use a monokinetic beam and a stationary target (gas cell, foil).

In a beam, the velocity of the particles can be regarded as constant in direction as well as in magnitude. We choose this direction as the z axis and write for the flux :

$$j_2(xyz) = n_2(xyz)v.$$

Integrating (2) over the volume of interaction we derive the number of events per second :

$$N = \int \frac{dN}{dV} \, dV = \sigma \int n_1(xyz) j_2(xyz) dV.$$

The target density is almost always independent of x and y

$$N = \sigma \int n_1(z) dz \int j_2(xyz) dxdy.$$

The second integral is just the total intensity I_2 of the beam and is z independent if the target is thin enough

$$N = \sigma I_2 \int n_1(z) dz.$$

The quantity $\int n_1(z) dz$ is called "thickness" E of the target. The cross section is derived from the measurement of N, I_2 and E.

Measurement with two beams

When two intersecting beams are used to measure the cross section, the number of events produced per second is found as follows (see fig. 4).

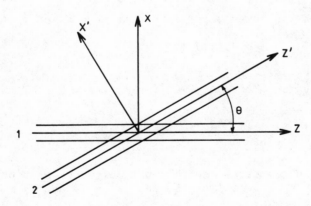

Fig. 4.   Coordinates for intersecting beams.

Let us take the z axis along the direction of beam (1) and z' along beam (2); y perpendicular to z and z'; x and x' in the plane of (zz') respectively perpendicular to z and z'.
Let θ be the angle of intersection.

From formula (2) we get the number of events per second :

$$N = \frac{dN}{dV} \, dV = \sigma v \int n_1 n_2 dV.$$

In beam (1), the density n_1 and the flux j_1 are independent of z

$$n_1 = \frac{j_1(xy)}{v_1} \, .$$

In beam (2), n_2 and j_2 are independent of z'

$$n_2 = \frac{j_2(x'y)}{v_2} \, .$$

The volume element, in zz'y coordinates is

$$dV = dydzdz'\sin\theta.$$

Therefore,

$$N = \frac{\sigma v}{v_1 v_2} \sin\theta \int dy F_1(y) F_2(y)$$

with

$$F_1(y) = \int j_1(xy) dz' = \frac{1}{\sin\theta} \int j_1(xy) dx = \frac{J_1(y)}{\sin\theta}$$

$J_1(y)$ is the one-dimensional density along y in beam 1 : $J_1 = dI_1/dy$

$$F_2(y) = \int j_2(x'y) dz = \frac{1}{\sin\theta} \int j_2(x'y) dx' = \frac{J_2(y)}{\sin\theta}$$

$J_2(y)$ is the one-dimensional density along y in beam 2 : $J_2 = dI_2/dy$.

The number of events per second is thus given by the relation :

$$N = \frac{\sigma v}{v_1 v_2 \sin\theta} \int dy J_1(y) J_2(y).$$

If we express the relative velocity v in terms of the velocities v_1 and v_2 in the beams, we finally get :

$$N = \frac{\sigma \sqrt{v_1^2 + v_2^2 - 2v_1 v_2 \cos\theta}}{v_1 v_2 \sin\theta} \int J_1(y) J_2(y) dy. \tag{3}$$

To obtain the cross section, not only the reaction rate N but also the overlap integral of the densities J_1 and J_2 must be measured.

Formula (3) is sometimes written :

$$N = \frac{\sigma\sqrt{v_1^2+v_2^2-2v_1v_2\cos\theta}}{v_1v_2\sin\theta} I_1 I_2 F$$

with a so-called "form factor" F :

$$F = \frac{\int J_1(y)J_2(y)dy}{\int J_1(y)dy \quad \int J_2(y)dy)} .$$

F has the dimension of an inverse length.

COLLISION ENERGY

The physically significant kinetic energy in the collision of 2 particles is the center-of-mass energy E_{CM}

$$E_{CM} = \frac{1}{2} \mu v^2 = \frac{1}{2} \mu(v_1^2+v_2^2 - 2v_1v_2\cos\theta) \qquad (4)$$

where $\mu = m_1m_2/m_1+m_2$ is the reduced mass and $v = |\vec{v}_2-\vec{v}_1|$ is the relative velocity.

In the case where the collisions are produced at the interaction of a beam with a stationary target (gas, foil) the relative velocity is simply the velocity of the projectiles and we have therefore :

$$E_{CM} = \mu\left(\frac{E_1}{m_1}\right)$$

where E_1 is the kinetic energy of the projectiles.

In the case of collisions between two beams intersecting at an angle θ the more general expression is

$$E_{CM} = \mu\left(\frac{E_1}{m_1} + \frac{E_2}{m_2} - 2\sqrt{\frac{E_1}{m_1}\frac{E_2}{m_2}}\cos\theta\right). \qquad (5)$$

Merged beams

An interesting case arises when the two beams travel with almost the same velocity and intersect at a vanishingly small angle ($\theta \sim 0$). This is the so-called "merged" or "superimposed beams". They provide a means to reach extremely small collision energies with energetic beams.

In this case, we write :

$$v_2 = \bar{v} + \frac{\Delta v}{2}$$

$$v_1 = \bar{v} - \frac{\Delta v}{2}$$

$$\Delta v \ll \bar{v}$$

$$E_2 \equiv \frac{m_2}{2} v^2 \simeq \frac{m_2}{2} \bar{v}^2 + \frac{m_2}{2} \bar{v}\Delta v = \bar{E}_2 + \frac{m_2}{2} \bar{v}\Delta v$$

$$E_1 \equiv \frac{m_1}{2} v_1^2 \simeq \frac{m_1}{2} \bar{v}^2 - \frac{m_1}{2} \bar{v}\Delta v = \bar{E}_1 - \frac{m_1}{2} \bar{v}\Delta v.$$

Note that $\bar{E}_1$ and $\bar{E}_2$ are the energies of the beams when they travel with the average velocity $\bar{v}$.

$$(E_2 - E_1) = (\bar{E}_2 - \bar{E}_1) + (\frac{m_1 + m_2}{2})\bar{v}\Delta v.$$

Let us write ΔE for the quantity :

$$\Delta E \equiv (E_2 - E_1) - (\bar{E}_2 - \bar{E}_1)$$

$$= (E_2 - \bar{E}_2) + (\bar{E}_1 - E_1),$$

which is the sum of the absolute displacements of the beam energies relatively to their velocity-average values, and that we call the "energy displacement". Then

$$\Delta v = \frac{2\Delta E}{(m_1 + m_2)\bar{v}} .$$

Furthermore if θ is small, we may replace $\cos\theta$ in formula (4) by $1 - \theta^2/2$:

$$E_{CM} = \frac{1}{2} \mu\{(\bar{v} - \frac{\Delta v}{2})^2 + (\bar{v} + \frac{\Delta v}{2})^2 - 2(\bar{v}^2 - (\frac{\Delta v}{2})^2)(1 - \frac{\theta^2}{2})\}$$

$$= \frac{1}{2} \mu\{(\Delta v)^2 + \bar{v}^2\theta^2)\}$$

$$= \frac{1}{2} \mu\{\frac{4(\Delta E)^2}{(m_1 + m_2)(m_1\bar{v}^2 + m_2 v^2)} + \frac{(m_1\bar{v}^2 + m_2\bar{v}^2)}{(m_1 + m_2)} \theta^2\}$$

$$= \frac{1}{2} \mu\{\frac{2(\Delta E)^2}{(m_1 + m_2)(\bar{E}_1 + \bar{E}_2)} + \frac{2(\bar{E}_1 + \bar{E}_2)}{(m_1 + m_2)} \theta^2\}$$

$$E_{CM} = \frac{\mu}{M} \frac{(\Delta E)^2}{(\bar{E}_1 + \bar{E}_2)} + \frac{\mu}{M} (\bar{E}_1 + \bar{E}_2)\theta^2. \qquad (6)$$

In merged beams, θ is zero in principle and the center-of-mass energy then appears to be a quadratic function of ΔE, the energy displacement in the laboratory. Because of that, the CM energy is easily made much smaller than the energy displacement.

Let us take, as an example, the case of 2 equal masses.

$$m_1 = m_2$$

$$\frac{\mu}{M} = \frac{1}{4}$$

$$\bar{E}_1 = \bar{E}_2 = E$$

$$E_{CM} = \frac{(\Delta E)^2}{8E}.$$

If the fast beam is 10 keV and the slow beam 9.9 keV,

$$E_{CM} = \frac{(0.1)^2}{8 \times 9.95} = 0.000125 \text{ keV} = 0.125 \text{ eV !}$$

How accurately can the CM energy be settled with merged beams ? What energy resolution can be obtained ?

The energy resolution with merged beams depends on 2 factors :
- the energy spreads δE_1 and δE_2,
- the divergence $\delta\theta$,
of the beams.

The contribution of the energy spreads is obtained by differentiating the first term in formula (6) :

$$(\delta E_{CM})_1 = \frac{2\mu}{M} \frac{\Delta E}{(\bar{E}_1 + \bar{E}_2)} \delta E_1$$

$$(\delta E_{CM})_2 = \frac{2\mu}{M} \frac{\Delta E}{(\bar{E}_1 + \bar{E}_2)} \delta E_2.$$

The contribution of the divergence is similarly obtained by differentiating the second term, although it must be done up to the second order as we want to consider the case where $\theta = 0$

$$(\delta E_{CM})_\theta = \frac{2\mu}{M} (\bar{E}_1 + \bar{E}_2)\theta\delta\theta + \frac{\mu}{M} (\bar{E}_1 + \bar{E}_2)(\delta\theta)^2.$$

For the relative resolution $(\delta E_{CM}/E_{CM})$, we get, with merged beams :

$$(\frac{\delta E_{CM}}{E_{CM}})_1 + (\frac{\delta E_{CM}}{E_{CM}})_2 = 2(\frac{\delta E_1 + \delta E_2}{\Delta E}) = 2\sqrt{\frac{\mu}{M}} \frac{(\delta E_1 + \delta E_2)}{\sqrt{E_{CM}}(\bar{E}_1 + \bar{E}_2)}$$

$$(\frac{\delta E_{CM}}{E_{CM}})_\theta = (\frac{\bar{E}_1 + \bar{E}_2}{\Delta E})^2 (\delta\theta)^2 = \frac{\mu}{M} (\frac{\bar{E}_1 + \bar{E}_2}{E_{CM}})(\delta\theta)^2.$$

We note that the relative energy resolution :
- is deteriorating when E_{CM} decreases,
- is better when working with more energetic beams if the energy

spread in the beams is large but is becoming worse if the effect
of the divergence dominates. Optimum resolution is therefore
obtained at some intermediate energy.

We also see, in the formula giving $(\delta E_{CM})_\theta$ that merged beams
give a better energy resolution than inclined beams (θ small but
not zero), as far as divergence is concerned.

As an example of very low energy work performed with merged
beams, a measurement of the cross section for associative ionisation in collisions of H^+ and H^- ions is illustrated below(fig. 5).

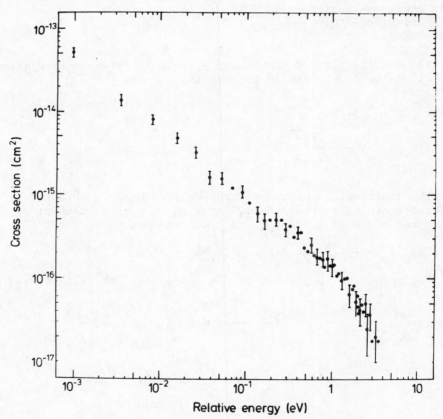

Fig. 5. Cross section for the associative ionisation reaction :
$H^+ + H^- \rightarrow H_2^+ + e$.

ENERGIES OF PRODUCTS

We consider here the general case of two particles having arbitrary velocities $\vec{v}_1$ and $\vec{v}_2$ and scattered in an arbitrary way in an inelastic collision. We want to find how the inelastic energy released or absorbed in the collision is shared between the emerging particules.

General case

Let us call $\vec{v}_1$ and $\vec{v}_2$ the incident velocities in the laboratory frame, $\vec{f}$ the velocity of their center of mass, $\vec{s}_1$ and $\vec{s}_2$ the velocities in the frame attached to the center of mass and Θ_0 the angle between $\vec{s}_1$ and $\vec{f}$ (fig. 6).

After the collision, the particles emerge with velocities $\vec{s}_1'$ and $\vec{s}_2'$, in the CM frame and $\vec{v}_1'$ and $\vec{v}_2'$ in the laboratory frame.

Note that $\vec{s}_1'$ and $\vec{s}_2'$ are generally not in the plane of $\vec{f}$ and $\vec{s}_1$, $\vec{s}_2$. Θ is now the angle between $\vec{s}_1'$ and $\vec{f}$.

Momenta $\vec{p}_1$ and $\vec{p}_2$ in the CM frame are opposite vectors and so are $\vec{p}_1'$ and $\vec{p}_2'$ also :

$$P_1 = P_2 = p = \mu v$$

$$P_1' = P_2' = p' = \mu v'$$

where v and v' are the relative velocities before and after collision.

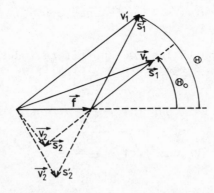

Fig. 6.   Laboratory and center of mass parameters in a binary
          collision.

The center of mass energies E_{CM} and E'_{CM} are :

$$E_{CM} = \frac{p^2}{2\mu} \qquad\qquad E'_{CM} = \frac{p'^2}{2\mu}$$

so that we have, between the inelastic effect ΔE_{CM} of the collision and the change Δp of the CM-momentum following relations :

$$p'^2-p^2 = 2\mu\Delta E_{CM} \tag{7}$$

$$(\Delta p)^2+2p\Delta p-2\mu\Delta E_{CM} = 0.$$

When the last relation is solved for Δp, we get :

$$\Delta p = p \left[\sqrt{1 + \frac{\Delta E_{CM}}{E_{CM}}} - 1\right]. \tag{8}$$

Now let us calculate the change of laboratory kinetic energies

$$\Delta E_1 \equiv E'_1-E_1 = \frac{m_1}{2} (v_1'^2 - v_1^2)$$

but

$$v_1'^2 = f^2+s_1'^2+2fs_1'\cos\Theta = f^2+ \frac{p'^2}{m_1^2} + 2f \frac{p'}{m_1} \cos\Theta$$

$$v_1^2 = f^2+s_1^2+2fs_1\cos\Theta_0 = f^2+ \frac{p^2}{m_1^2} + 2f \frac{p}{m_1} \cos\Theta_0,$$

thus

$$\Delta E_1 = \frac{p'^2-p^2}{2m_1} + fp(\cos\Theta-\cos\Theta_0) + f\Delta p \cos\Theta.$$

Or, using (7) and (8) :

$$\Delta E_1 = \frac{\mu}{m_1} \Delta E_{CM}+fp(\cos\Theta-\cos\Theta_0) + fp\cos\Theta\left(\sqrt{1 + \frac{\Delta E_{CM}}{E_{CM}}} -1\right)$$

$$= \frac{\mu}{m_1} \Delta E_{CM} + \mu fv\left(\sqrt{1 + \frac{\Delta E_{CM}}{E_{CM}}} \cos\Theta-\cos\Theta_0\right).$$

For the second particles, we substitute index 2 to index 1 and change the signs of the cosinus. Finally we get :

$$\Delta E_1 = \frac{\mu}{m_1} \Delta E_{CM} + \mu vf\left(\cos\Theta \sqrt{1 + \frac{\Delta E_{CM}}{E_{CM}}} - \cos\Theta_0\right)$$

$$\Delta E_2 = \frac{\mu}{m_2} \Delta E_{CM} - \mu vf\left(\cos\Theta \sqrt{1 + \frac{\Delta E_{CM}}{E_{CM}}} - \cos\Theta_0\right). \tag{9}$$

Stationary target

When one of the particles is at rest (v_2=o), the velocity $\vec{f}$ of the center of mass and the CM velocity $\vec{s}_1$ of the other particle (projectile) are parallel so that Θ_o is zero. Further we have :

$$v = v_1$$

$$f = \frac{m_1 v_1}{M} \quad \text{where} \quad M = m_1 + m_2$$

$$E_{CM} = \frac{\mu v_1^2}{2} \; .$$

If the deflection angle θ is small and $\Delta E_{CM} \ll E_{CM}$, the formulae (9) become :

$$\Delta E_1 = \frac{\mu}{m_1} \Delta E_{CM} + \frac{\mu}{M} \frac{m_1 v_1^2}{2} (\frac{\Delta E_{CM}}{E_{CM}} - \theta^2)$$

$$\Delta E_2 = \frac{\mu}{m_2} \Delta E_{CM} - \frac{\mu}{M} \frac{m_1 v_1^2}{2} (\frac{\Delta E_{CM}}{E_{CM}} - \theta^2) .$$

$$\Delta E_1 = \Delta E_{CM} - \frac{\mu}{M} E_1 \theta^2$$

$$\Delta E_2 = + \frac{\mu}{M} E_1 \theta^2 \tag{10}$$

The term in θ^2 is the elastic contribution.
The inelastic part is present in ΔE_1 only : the projectile carries away the entirety of inelastic gain or loss of the collision.

This feature allows an identification of the inelastic process from an analysis of the kinetic energy of the projectile after collision. It is the basis of the so-called translational spectroscopy.

In-flight dissociations

When molecules or molecular ions are involved, dissociation in flight leads to a considerable amplification in the energy spectrum of the fragments, as can be seen from formulae (9) again. In this case, we regard the two fragments as emerging from a zero-energy collision (half-collision)

$$E_{CM} = 0$$

$$\Delta E_1 = \frac{\mu}{m_1} \Delta E_{CM} + \cos\Theta \sqrt{2\mu f^2 \Delta E_{CM}}$$

$$\Delta E_2 = \frac{\mu}{m_2} \Delta E_{CM} - \cos\Theta \sqrt{2\mu f^2 \Delta E_{CM}}$$

$$\Delta E_1 = \frac{\mu}{m_1} \Delta E_{CM} + \cos\theta \sqrt{\frac{4\mu}{M} E_o \Delta E_{CM}}$$

$$\Delta E_2 = \frac{\mu}{m_2} \Delta E_{CM} - \cos\theta \sqrt{\frac{4\mu}{M} E_o \Delta E_{CM}}$$

(11)

where $E_o = Mf^2/2$ is the kinetic energy of the molecule or molecular ion.

The inelastic energy released in the dissociation is shared between the two fragments, proportionnally to the inverse of the masses but an additional kinematic effect increases the energy of the fragment ejected in the forward direction and decreases the energy of the other fragment by the same amount.

This additional energy is roughly a geometrical average of ΔE_{CM} and E_o. It can be several orders of magnitude larger than ΔE_{CM} itself and thus provides an interesting amplification of ΔE_{CM} that allows a fine analysis of the excitation of the molecule or molecular ion. This amplification is essentially the same kinematic effect as used with merged beams.

B. GENERAL APPROACH TO BEAM-BEAM CROSS-SECTION MEASUREMENTS

This section is devoted to some specific problems related to absolute cross-section measurements using a beam-beam collision arrangement. The part of the interaction geometry in the determination of the cross section is here prominent and will be studied first. In a second part the importance of background contribution is studied and methods to measure and to reduce it are analysed.

PROBLEMS RELATED TO THE INTERACTION GEOMETRY

When measuring cross sections with two interacting beams, the relation between cross section and actually measured quantities implies geometrical factors. We discuss them for the case of inclined and merged beams. An alternative method which circumvents the geometrical factors problem is also presented.

Inclined beams

When two beams intersect at an angle θ (figure 7) the number of events per unit time N can be written (see relation 3)

$$N = \sigma \frac{v}{v_1 v_2 \sin\theta} \int_{-\infty}^{+\infty} J_1(y) J_2(y) \, dy$$

where $J_i(y) = \int_{-\infty}^{+\infty} j_i(x,y) \, dx$.

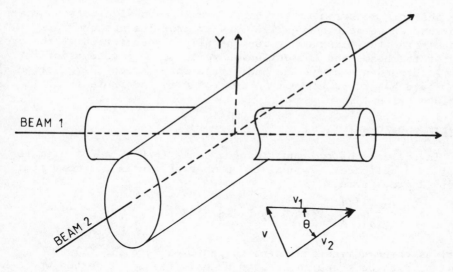

Fig. 7.   Intersecting beams.

In order to derive absolute values for σ it is necessary to measure the density profiles of both beams simultaneously along the y direction. This is generally achieved by moving a slit across both beams and measuring the product of the transmitted currents (see fig. 8). This measurement is not performed simultaneously

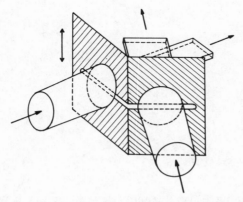

Fig. 8.   Vertical density profile measurement of two intersecting beams.

with the main measurement and therefore ignores fluctuations that
might happen between successive density profiles measurements.
Furthermore, the measurement is generally not performed at the crossing of the beams and corrections have to be made accounting for
spatial dependence along the beam trajectory (space charge effects).
Great care has to be taken in order to derive accurate (± 10%)
absolute values.

Superimposed beams

    In the case of two superimposed beams ($\theta=0$), the problem is
still more delicate. Indeed, the number of events per unit time
(derived from formula 2)

$$N = \sigma \frac{v}{v_1 v_2} \iiint j_1(x,y,z) j_2(x,y,z) \, dx \, dy \, dz$$

involves three-dimensional density profiles (see fig. 9).
These have to be measured in both beams. The ideal method would be
to have an x-y position sensitive detector and to move it along
the z direction.

    A simpler method consists in moving a small aperture across
both beams and measure the transmitted currents at different x,y,
z locations. This method needs an elaborate sampling and computing
system.

    An easier method applicable only to beams of charged parti-
cles is to keep the small aperture fixed and to scan both beams
across by means of two sets of deflecting plates like shown in
figure 10 for the case of superimposed beams of H^+ and H^- ions.
The transmitted currents are collected as a function of the
deflecting voltages.

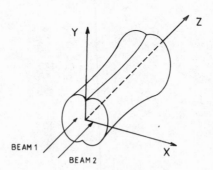

Fig. 9. The reaction rate in a superimposed beams experiment is
        related to the three-dimensional density profiles of the
        beams.

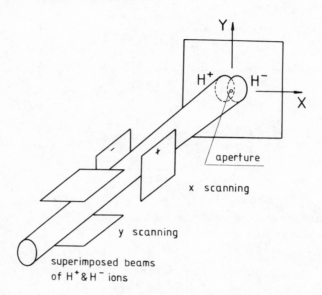

Fig. 10. Method to measure the x-y overlap integral at a given z
          location in a superimposed beams experiment.

Some simplification can be introduced in the experimental set-
up if some assumptions concerning the beam profiles are made. If
for instance the beams are supposed to be Gaussian (at a given z
position) :

$$j_1(x,y) = j_1 \exp - (\frac{x^2}{k_1^2} + \frac{y^2}{k_2^2})$$

$$j_2(x,y) = j_2 \exp - (\frac{(x-x_o)^2}{k_3^2} - \frac{(y-y_o)^2}{k_4^2})$$

then the overlap integral can be written

$$\int j_1(x,y) j_2(x,y) dxdy = \frac{I_1 I_2}{\pi \sqrt{(k_1^2+k_3^2)(k_2^2+k_4^2)}} \exp(\frac{-x_o^2}{(k_1^2+k_3^2)} + \frac{-y_o^2}{(k_2^2+k_4^2)})$$

the parameters x_o, y_o, k_1, k_2, k_3 and k_4 are easy to be obtained.

A vertical slit is moved in the x direction, the transmitted
currents are recorded as a function of x. As an example, figure 11
shows the profiles of the two beams along x and how to obtain the
parameters x_o, k_1, k_3. By doing the same with an horizontal slit

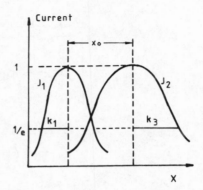

Fig. 11.   Intensity profiles of two superimposed beams obtained by
          moving a vertical slit across both beams.

along the y direction, the parameters y_0, k_2 and k_4 are obtained.
The overlap integral can readily be calculated. The method can
easily be adapted for direct computation. All these methods need
appropriate equipment (stepping motors) and have to be treated with
great care.

     A way to circumvent determining the overlap integral is to
work with homogeneous coaxial beams. This can be achieved by
passing the superimposed beams through a tiny diaphram just before
the interaction region. Figure 12 shows schematically how by
using only the core of the beams, homogeneity can be achieved.
Small inhomogeneities do not greatly affect the accuracy of the
cross section determination as exemplified in the following case.

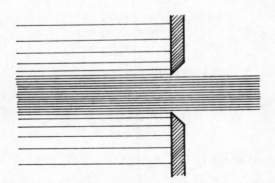

Fig. 12.   Passing superimposed beams through a tiny collimator
          is a way to produce homogeneous coaxial beams.

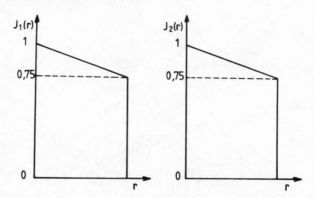

Fig. 13. Example of radial density distribution in superimposed beams.

Consider two beams where the density profiles (see fig. 13) are such that they linearly drop from the center to the edge by 25 per cent. The error made on the density profiles overlap integral by assuming both beams being homogeneous is only 6%. Thus, when the profiles are smooth the beams can be regarded as homogeneous over the diaphragm area. As far as the z dependence is concerned, it can be neglected to a first approximation when the interaction region is short. Further, corrections can be made by measuring the overall divergence of the beams.

This method is certainly the most appropriate for merged beams. However, the resulting loss of intensity is a limitation to its applicability.

Circumventing beam profile problems : sweeping one beam

In crossed beams the event rate is related to the product of beam densities along y, this is a peculiarity arizing from the absence of motion in the y direction of both beams. One slice at position y of one beam can only cross the corresponding y slice of the other beam. Inhomogeneties in the y direction directly reflect upon the counting rate. The extreme case being alternated layers of beam 1 and 2 so that collisions are avoided (see for illustration fig. 14) the same cannot happen in the other directions. This is illustrated in figure 15. Any portion of beam 1 (see for instance shaded area) crosses any portion of beam 2 (see also shaded area). This problem is still more acute in the case of superimposed beams where there is absence of motion in the xy plane.

A way to remove the peculiarity associated with the y direction is to sweep one beam vertically through the other while keeping it parallel to its initial axis (see fig. 16).

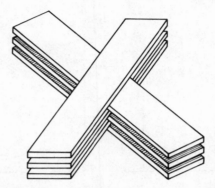

Fig. 14. Extreme situation of vertical inhomogeneity in a crossed
          beams arrangement.

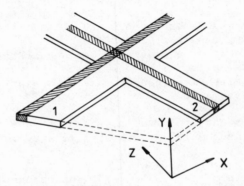

Fig. 15. Inclined beams experiments : each fraction of beam 1
          crosses each fraction of beam 2 at a given y location.

If the sweeping is done at constant speed u = dy/dt and if $J_1(y)$
and $J_2(y)$ are respectively the profiles of beam 1 and 2 at time
t=o, then at a given time t the slice $J_2(y)$ of beam 2 will cross
slice $J_1(y-ut)$ of beam 1 (see illustration of this on fig. 17).
The reaction rate at time t can be written as before :

$$N(t) = \sigma \frac{v}{v_1 v_2 \sin\theta} \int J_1(y-ut) J_2(y) dy.$$

     If the displacement is large enough that, at the extreme
positions, the beams no longer overlap, then the number of events
K during one sweep can be written

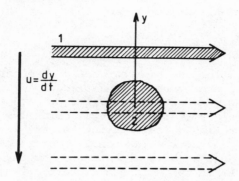

Fig. 16.   Illustration of a crossed beams experiment where beam 1
            is swept across beam 2.

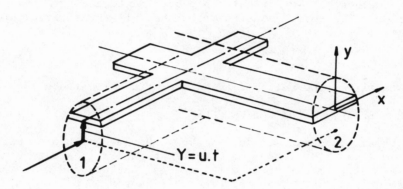

Fig. 17.   Collision of one slice of beam 2 with the corresponding
            slice in beam 1 at a given time t in a situation where
            the first beam is swept across the other one at constant
            speed u.

$$K = \int_{-\infty}^{+\infty} N(t)dt = \int_{-\infty}^{+\infty} N(t) \frac{dY}{u} \qquad \text{if} \qquad Y = ut$$

$$= \sigma \frac{v}{v_1 v_2 \sin\theta} \int_{-\infty}^{+\infty} \int_{-\infty}^{+\infty} J_1(y-Y) J_2(y) dy \frac{dY}{u}$$

$$= \frac{\sigma}{v_1} \frac{v}{v_2 u \sin\theta} \int_{-\infty}^{+\infty} J_2(y) dy \int_{-\infty}^{+\infty} J_1(y-Y) dY$$

and as $\quad \int_{-\infty}^{+\infty} J_2(y) dy = I_2 \quad , \quad \int_{-\infty}^{+\infty} J_1(y-Y) dY = I_1$

$$K = \frac{\sigma v I_1 I_2}{v_1 v_2 u \sin\theta}.$$

(12)

From this relation, we see that the determination of the cumbersome integral over y is no longer required but we need, instead, a measurement of the sweep speed u. It is generally easy to measure accurately a sweep speed. Also, as the sweeping can be achieved by a deflecting voltage acting on one of the beams, the corresponding sweep speed remains constant over the entire measuring time and does not require elaborate measuring equipment. Care has however to be taken to perform a distortionless translation of the beam. The counting rate has to be recorded as a function of time and in phase with the beam displacement.

The method has so far only been applied to electron impact ionisation. The results obtained for the well investigated process

$$e^- + He^+(1s) \rightarrow He^{++} + 2e^-$$

have established the reliability of the method (Defrance 1981).

PROBLEMS RELATED TO THE BACKGROUNDS

Nothing has been said so far about the backgrounds arising in beam-beam experiments. Besides reactions occuring between the two beams each of them can react with the residual gas and with the surfaces on its way. These are the main causes of background in beam-beam experiments. Other sources of background may show up in particular situations, for instance photon emission, field ionisation of excited states, bremstralung ... To determine absolute cross sections the background contribution has to be assessed. Furthermore, if the signal is too small, methods to reduce the background to an acceptable level have to be developped.

Methods to measure the background contribution

All methods to measure the background contribution involve some kind of modulation of the signal.

Beam intensity modulation

In a situation where two beams interact and only the product particles of the second beam are detected, an easy way to determine the background contribution is to modulate the intensity of the first beam. Figure 18 shows how a square wave intensity modulation of beam 1 reflects on the counting rate N of the product particles of beam 2. The signal is extracted from the difference in counts between the "on" and "off" periods of beam 1.

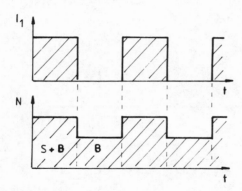

Fig. 18. Beam intensity modulation and corresponding count rate.

        The situation becomes different when the products of both
beams are detected together and cannot be identified separately
(using pulse height analysis, coincidence technique, ...). Intensity modulation of both beams is then needed to extract the background contribution.

        Figure 19 examplifies such a situation. A square wave modu-
lation is applied to the intensity of beam 1. In phase with the
modulation of beam 1, a half period intensity modulation is applied
to beam 2. Four different levels show up in the counting rate.
They are listed from (1) to (4) in figure 19. Using following
symbols :

        S   : signal
        B1   : background from beam 1
        B2   : background from beam 2
        BC   : instrumental background.

The count rate sources in the four situations are then

        (1) → S + B1 + B2 + BC
        (2) → B1 + BC
        (3) → B2 + BC
        (4) → BC

The signal can then be obtained from [(1)+(4)]-[(2)+(3)] using
appropriate in phase counting.

Beam position modulation

        In a previous paragraph the method of sweeping one beam across
the other has been described. Besides circumventing the problem

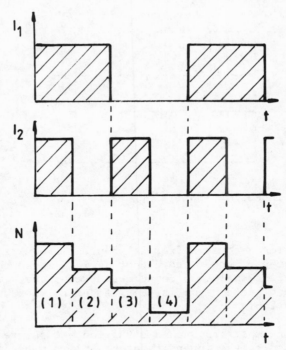

Fig. 19. Two beam intensity modulation and corresponding count rate.

of determining geometrical factors the method produces an elegant
way to measure background contribution. Figure 20a shows the position of one beam with respect to the other one as a function of
time. One beam is swept back and forth through the other at constant speed and with a large enough amplitude that at the extreme
positions, the beams practically no longer overlap. Figure 20b
shows the corresponding reaction rate on the same time scale. When
the beams do not overlap, only background is observed (flat parts).
When the beams cross one another then a signal peak is produced.
The number of events during one sweep is obtained from the peak
area when the background has been substracted.

Problems related to modulation

    When intensity modulation is used, attention must be paid to
the following phenomena that might invalidate the results of the
measurements :
- pressure modulation : this is likely to occur when the modulated
  beam strikes a surface and increases the local residual pressure.
  Its influence can be reduced by increasing the modulation
  frequency so that the corresponding pressure modulation is
  smoothed (due to built up and evacuation response time of
  residual gas).

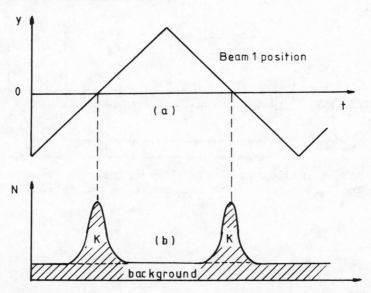

Fig. 20.   Example of count rate in a crossed beam arrangement where
           one beam is swept across the other one along y.

- space charge interaction : the space charge of one beam can modify the configuration of the other beam which in turn can affect
  the reaction rate and the collection efficiency of the reaction
  products.
- secondary electrons : when a modulated energetic beam strikes a
  surface, secondary electrons can be ejected and eventually interact with the other beam. A manner to avoid this is to bias
  negatively the interaction region.
- charged particle trapping : electrons and ions formed by the
  ionisation of the residual gas can be trapped in the beams.
  When their concentration in the interaction region becomes
  large enough, they contribute to the signal. As an illustration of the question of ion-trapping, we report some observations made during a recent investigation of the process :

$$e^- + H^- \rightarrow 3e^- + H^+.$$

A 90° crossed beam set up was used. Figure 21 shows a schema
of the electron gun. A cathode (K), an anode (A) and a pair of
plates (B) acting as a lens are used to form the electron beam.
The electrons are fired at right angle at the ion beam (I).   The
deflector plates B are also used to sweep the electron beam
across the ion beam. A set of wires (W) is used to measure the

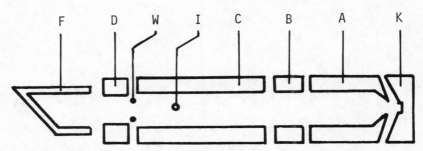

Fig. 21.   Schematic diagram of the electron gun used in a crossed
           beam experiment where beam position modulation is used.

sweep speed. The electron current is collected on a Faraday cup
(F) preceeded by a pair of suppressing plates (D). A pair of equi-
potential plates C is to bias the interaction region.

     Let us consider what happens with positive ions formed by
electron impact with residual gas. The concentration of these
ions in the interaction region depends upon the electron density
and upon the voltage distribution along the electron beam path.
The top part of figure 22 shows schematically the electron gun.
The full curve in the lower part shows the corresponding voltage
distribution along the electron beam axis in a situation where one
tries to avoid secondary electrons to move back from the Faraday
cup into the interaction region (Voltage is increasingly positive
from interaction region to Faraday cup). In our experimental set-
up plates (B) in figure 22 are usually kept positive. This voltage
distribution concentrates in the interaction region the positive
ions trapped in the electron beam. On the other had, when a negative voltage is applied on the wires W (dashed line in lower
part of fig. 22), trapped ions are removed from the interaction
region and secondary electrons are retained by the Faraday cup.
Varying the voltage of wires (W) is a good way to explore the
relative importance of this ion trapping effect on a cross section
measurement. For H^+ formation from H^- by electron impact, measurements of apparent cross sections have been made at 330 eV for
different wire voltages. The results are shown in figure 23.
The apparent cross section drops by a factor 12 when one passes
from a positive wire voltage (positive ions extracted from interaction region, corresponding to full line in fig. 22) to a
negative wire voltage. This shows how positive ions can dras-
tically influence results.

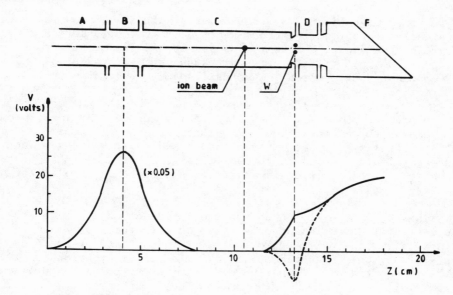

Fig. 22.   Voltage distribution in electron gun.

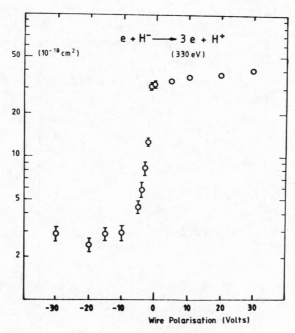

Fig. 23.   Apparent cross section for reaction H⁻+e→H⁺+3e at 330eV
           for different wire voltages in the electron gun of fig.22.

Another significant way to show this influence is by varying
the electron density. The cross section of proton formation from
H⁻ by electron impact has been measured at 330 eV for different
electron beam intensities in two different situations : +15V and
-15V on electron gun wires. The results are shown in figure 24.
The electron current is varied from 2.4 mA down to 50 µA. When a
positive voltage is applied to wires (open circles), the cross
section drops rapidly for intensities below 100 µA. At this intensity, the depth of the potential well of the electron beam is too
small to still significantly trap the formed positive ions. When a
negative voltage is applied to the wires (full circles in figure 24)
no change in cross section is observed as a function of electron
intensity.

Finally, a good way to detect the influence of positive ion
trapping in electron impact cross section measurements, is to
measure below threshold. Unfortunately, in the case of reaction
H⁻+e → H⁺+3e, the threshold (14.35 eV) lies close or below the threshold for positive ion formation by electron impact on residual gas
(i.e. : 15.4 eV for H_2). As a consequence, proton formation from
H⁻ by electron impact and positive ion formation from electron
impact on residual gas vanish for almost the same electron energy.

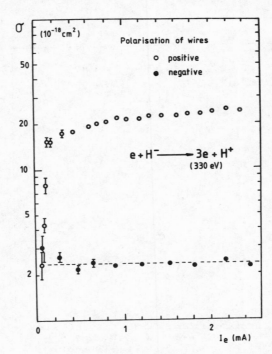

Fig. 24.   Cross section for reaction H⁻+e→H⁺+3e at 330eV for two
           different wire voltages in the electron gun as a function
           of electron beam intensity.

Methods to reduce or eliminate backgrounds

Some techniques are available to reduce or eliminate background contributions to a level where some measurements become possible or easier to perform. We list here some of these techniques.

Energy analysis of products

In some situations parasitical reactions can be energetically distinguished from the investigated one. For instance, in the crossed beam study of the ionisation :

$$e^- + He^+ \rightarrow He^{++} + 2e^-$$

the ionisation on residual gas :

$$x + He^+ \rightarrow He^{++} + x + e^-$$

yields He^{++} ions with a kinetic energy smaller by 54 eV than the kinetic energy of the ions produced by electron impact.

Energy analysis can be achieved by the use of electrostatic or magnetic selectors but also by chopping the beam and making a time-of-flight measurement. In the case of very energetic beams direct energy analysis can be done by a pulse height study on a solid state detector.

Coincidence technique

Coincident detection can be used when two interacting beams produce pairs of particles which can be separately detected. The time interval separating the pulses of two such particles is equal to their time of flight difference from the reaction point to the detectors. This time correlation is used in the coincidence technique. The pulses of one of the detectors are fed into the start input of a time-to-amplitude-converter (TAC) while those of the other action the stop input. Time correlated pairs of particles will produce a fixed amplitude output pulse. Particles formed in reactions of each beam with residual gas will reach the detectors randomly and produce random heights for the output of the TAC. A pulse height analysis of the TAC output will produce a spectrum like show in figure 25. The width of the coincidence peak is related to the spread in time of flight of the pairs of particles (interaction length, energy spread) and to the detection electronics. The remaining background is only due to random coincidences producing pulses with a height falling within the coincidence peak pulse height width τ. This background can become extremely small if τ can be sufficiently reduced.

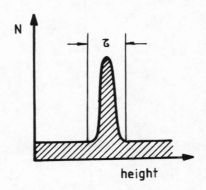

Fig. 25.   Pulse height distribution obtained from a time-to-amplitude
           converter used in a coincident detection.

        The special case of auto-coincidence is worth mentioning.   It
shows up when both particles of a pair reach the same detector
within a time interval sufficient to produce two output pulses.
These pulses will fed the start and stop of the TAC, of course,
the pulses directed to the start input have to be delayed.   If
the two particles are not time resolved, then pulse height analysis may sometimes be able to detect a coincidence (solid state
detector).

Biasing interaction region

        If one or both beams have to travel over long distances
where it is impossible to maintain low pressures, it is usefull
to bias the interaction region so that charge exchange with residual gas in this region yields particles with displaced kinetic
energy in contrast with those created outside.   On the other hand,
for charge exchange reactions studied with a superimposed beams
arrangement, such a biased region can also serve to define accurately the interaction length because the same kinetic energy
displacement felt by particles formed from reactions with residual
gas will apply to particles formed in charge exchange between
both beam.

        As an illustration we describe an experimental method using
superimposed beams, which was used to measure the symmetric resonnant charge exchange :

$$He^+ + He^{++} \rightarrow He^{++} + He^+.$$

        A beam of He^+ is formed in a source and accelerated to an
energy E_1, a beam of He^{++} is formed in another source and accelerated to an energy E_2.   The two beams are brought parallel and

superimposed using a set of electrostatic deflectors (Burniaux 1977).
It is interesting to note that the superimposed beams can be
obtained directly from one source. Extracting both He^+ and He^{++}
from a discharge source produces superimposed beams and as the
acceleration voltage acts on different charges they travel at different energies. This very straight method is nevertheless not
suitable for very low center of mass energies.

Biasing (voltage V) a given region along the beams path will
define the interaction length (see fig. 26). Inside this region
(interaction region) He^+ and He^{++} will have respectively E_1-V and
E_2-2V for energy. The products of a charge exchange will leave
the interaction region with energy E_1+V for He^{++} and E_2+V for He^+.
Charge exchange (even with residual gas) outside the interaction
region produces He^{++} of energy E_1 and He^+ of energy E_2. Subsequent
energy analysis will allow to collect only ions formed by charge
exchange inside the interaction region. The interaction length is
then well defined and the background contribution is limited to
charge exchange inside the interaction region where ultra high
vacuum can be achieved.

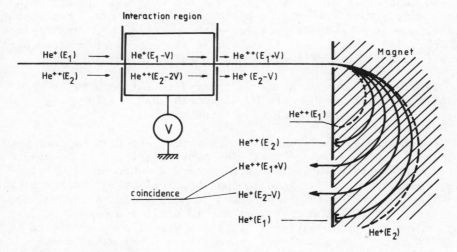

Fig. 26. Experimental method used to measure the symmetric
resonnant charge exchange cross section of reaction :
$He^+ + He^{++} \rightarrow He^{++} + He^+$.

SPECIAL METHODS FOR DEALING WITH EXCITED STATES

INTRODUCTION

In the problem of charge exchange, the question to know in which quantum state the collidants are before or after the collision is not a secondary question.
Initial excitation indeed increases the cross section considerably (roughly as n^4).
Excitation in the final state, on the other hand, is very common. It is a rule in the case of electron capture by multiply (z) charged ions where a simple argument of internal energy resonance shows at once that capture takes place dominantly to states with principal quantum number n=Z. Highly charged ions therefore always lead to the formation of highly excited products.

Even when ground state capture is dominant, the small amount of highly excited states created in the collision can play a decisive part in the subsequent processes, owing to their long life times and to their large collisional cross sections. Their ability to radiate also makes them responsible for energy loss and cooling in thermonuclear plasmas.

Studies on charge exchange involving excited states in the initial or final channel are therefore important. Much work has been devoted to them in the last years, both theoretically and experimentally but a clear, general picture has not yet been obtained. Available general theoretical predictions are approximations, not valid at low energies and not even well justified at high energy. More elaborate theories exist, of course, but they require numerical calculations for each specific case and their validity also is still questionable, as the number of experimental results to which predictions can be compared is not large.

Theory of charge exchange is treated in this course by A. SALIN and does not need to be discussed here. A few very general points, however, can be mentionned.

For instance, it is useful to know that electron capture into excited states obeys a n^{-3} law. The cross section for capture into a state with principal quantum number n (whatever the substate) decreases as n^{-3}. This law has been well verified for high values of n (n > 10) and is believed to hold to quite low values of n, provided that the energy is not to small and one is well away from resonant situations. This law was already predicted by BRINKMAN and KRAMERS (1930), using the approximation associated with their names and obtaining the cross section for capture into the S states.
Capture into all substates of a given n has been calculated by

MAY (1964) and OMIDVAR (1967), also in the B-K approximation. The cross section for capture into state n' on an ion of charge Z' from an hydrogenic ion of charge z in state n is given, in that approximation by :

$$\sigma_{nn'} = \frac{1.12 \times 10^{-15}(ZZ')^5}{n'^3 n^5 \frac{E}{M} [\frac{Z}{n^2} + \frac{E}{M} - \frac{1}{2}(\frac{Z^2}{n^2} - \frac{Z'^2}{n'^2}) + \frac{M}{16E}(\frac{Z^2}{n^2} - \frac{Z'^2}{n'^2})^2]^5} \tag{13}$$

where M and E are the mass and the impact energy of the projectile, expressed in proton mass and 100 keV units respectively, the cross section being expressed in cm^2.

Figure 27 illustrates this formula, showing the dependence on (Z'/n') of the cross section for capture from ground state hydrogen.

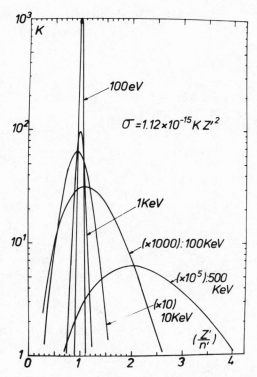

Fig. 27.   (Z'/n') dependance of the cross section for capture from ground state hydrogen at different collision energies.

Formula (13) is useful to predict the distribution of the excitation but, being the results of a B-K approximation, it largely overestimates the cross section. Correction factors have been recently given by CHAN and EICHLER (1979) for capture from 1s state.

An empirical rule to predict the most probable value of n' in the case of capture by multicharge ions has been given by RYUFUKU and WATANABE (1979)

$$n'_{max} = (Z')^{0.77}.$$

Its physical origin can be understood from very simple considerations. We state that the most probable value of n' is the one giving a resonance for :

$$H + A^{Z'+} \rightarrow H^+ + A^{(Z'-1)+}(n')$$

but we take the repulsion of the products into account. At infinite separation, resonance is obtained when n'=Z'.

But, if charge exchange is taking place at a finite distance R^*, then :

$$2 \frac{(Z'-1)}{R^*} = (\frac{Z'}{n'})^2 - 1.$$

We now state that charge exchange takes place when the electron undergoes an attraction from the projectile (Z') comparable to the attraction by the proton.
Equal attraction is achieved when :

$$\frac{Z'}{(R^*-1)^2} = 1$$

it is, when

$$R^* = Z'^{1/2} + 1.$$

Charge exchange already takes place at distances somewhat larger but not too much larger. It is therefore sensible to take :

$$R^* = 2(Z'^{1/2} + 1)$$

as the typical separation for charge exchange. Doing this, we find :

$$\frac{2(Z'-1)}{2(Z'^{1/2}+1)} = (\frac{Z'}{n'})^2 - 1$$

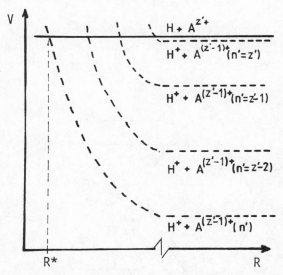

Fig. 28. Potential curves of the system formed of a proton and
          an excited multiply charged ion.

and thus

$$n' = (z')^{3/4}.$$

A question of importance is also : what is the distribution
of capture among the substates ,
To this question, the B-K calculation of OMIDVAR (1967), for
example, provides an answer in terms of a parameter A, depending
on Z/n, Z'/n' and E/m. Figures 29 and 30 below show an application to capture into the first excited states from a ground state
hydrogenic atom or ion.

More recent predictions are, for instance, the unitarized
distorded wave calculation of RYUFUKU and WATANABE (1979) or the
classical trajectory Monte Carlo calculation of SALOP (1979),
which of course is not expected to be valid for low excitation.

The agreement between the different theoretical results is
not too bad but cannot be considered as satisfactory. Unfortunately, experimental data are scarce and are of little help to
check the validity of theoretical predictions.

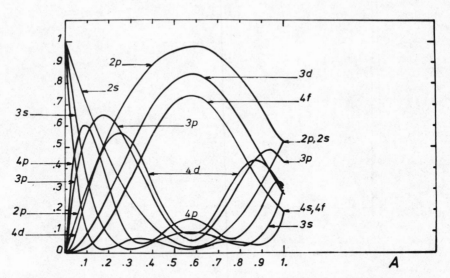

Fig. 29. Distribution among substates for capture into the first
         excited states from a ground state hydrogenic atom or
         ion as a function of parameter A (defined in figure 30).

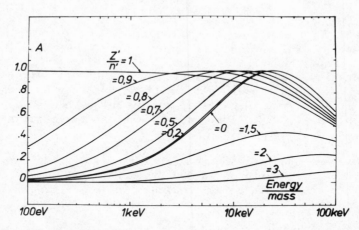

Fig. 30. Relation of parameter A of fig. 29 to (Z'/n') and (E/m).

At least, one point seems clear : capture populates high angular momentum states, up to very high energy, and the old statement of BRINKMAN-KRAMERS that "at high energy, capture takes place in the S states only" has no pratical validity as the "high energy" ment is above 19 or perhaps 100 MeV/amu.

In the following, we consider the experimental aspects of electron capture with excited states involved. We treat separatedly capture into excited states (excitation in the exit channel) and capture for excited states (excitation in the incident channel). The two processes are not fundamentally different but they involve different experimental approaches.

CAPTURE INTO EXCITED STATES

When experimenting on electron capture into a definite excited state, one must obviously have a technique to selectively detect the product atom or ion in that state.

The most straightforward technique is the observation of the deexcitation light. It is a very general, all purpose technique. Another current technique consists in analysing the kinetic energy of the excited projectile after collision. It is the "translational spectroscopy" mentioned in the chapter on kinematics. It is only feasible if the projectile is a charged particle and if the inelastic effect is large. It is therefore mainly restricted to electron capture by multiply charged ions, at not too high energy. Field ionisation is also an attractive technique, but its applicability is limited to highly excited states.

In this lecture, we briefly discuss the main features of the first technique. Field ionisation will be described in the last section as this technique is also used for the study of electron capture from excited states. Analysis of kinetic energy of the products is a rather simple method and will not be discussed here. Good examples of its application can be found, for instance, in a paper by PANOV (1979).

Identification of the states

Radiation from excited atoms or ions is mostly UV- or visible light. Wavelength analysis by monochromators thus provides an easy way to separate the contribution of a particular state. In case of hydrogenic ions, however, the l-degeneracy of the substates is an obstacle. It hinders even a global measurement of the population of a state, as a same number of photons can be produced by a large population of a long lived substate or a small population of a short lived one.

p-states are always unambiguously identified from the Lyman spectrum. For other states, the lifetime must be observed. If the excited species are travelling in a beam, this can be done by observing the spatial decay of the radiation outside the collision region.

Quantitative measurements

Absolute measurements of cross sections for capture into a specified excited state based on the observation of the deexcitation light are extremely delicate. Many sources of error exist, that cannot always be completely eliminated : efficiency of the collection and detection of photons, cascades, doppler effects, quenching by electric fields, ...

The main problem is clearly the calibration of the instrumentation for light measurement, it is to determine the ratio of the observed photon counting rate to the actual number of deexcitations taking place per second. In this ratio, many factors are involved : the efficiency of the detector, the transmission of the optical elements such as lenses, mirrors and filters, the geometrical factors related to the size of the region viewed and the solid angle accepted by the detector.

If the experiment is performed in a beam-gas arrangement, it is often possible to reach an equilibrium, it is a situation where excitation and deexcitation rates are equal. Then, only a short region of the beam needs to be observed.

Merged beams offer the same possibility, although it is difficult to keep merged beams parallel over a long distance. When crossed beams are used, there is evidently no such equilibrium. Photons have to be collected over a length extending from the crossing region to a point situated several lifetimes beyond the crossing. Capture into long-lived excited states in therefore more difficult to measure. S-states, however, can be quenched by electric fields so that their deexcitation can be forced at a well localized position.

Another question arises in relation to the angular distribution of the deexcitation light. In most experiments, light is observed in one particular direction only. Can we relate the number of photons observed (per unit solid angle) in that direction to the total number of photons emitted ? The relation is unfortunately ambiguous, as the angular distribution of the light is a priori not isotropic and is a combination of two different possibilities :

$$P_o(\theta)d\Omega = \frac{3}{8\pi} \sin^2\theta \ d\Omega \qquad \text{if} \quad \Delta m = 0,$$

$$P_1(\theta)d\Omega = \frac{3}{16\pi} (1+\cos^2\theta)d\Omega \quad \text{if} \quad \Delta m = \pm 1.$$

Here, $P(\theta)d\Omega$ is the probability for the photon to be emitted in a solid angle $d\Omega$ around a particular direction, at angle θ with the reference axis. This reference axis is normally the line joining the colliding atoms. It coincides with the beam axis in the case of beam-gas or merged beams arrangements. Δm is the change of the component on this axis of the angular momentum associated with the radiative transition.

Measurements are most often done with $\theta=90°$. The ratio of the probabilities P_0 and P_1 to the probability P_{iso} corresponding to an isotropic distribution :

$$P_{iso}d\Omega = \frac{d\Omega}{4\pi} ,$$

is then :

$$\frac{P_0}{P_{iso}} = \frac{3}{2} \quad \text{and} \quad \frac{P_1}{P_{iso}} = \frac{3}{4}.$$

One sees that the error made when no attention is given to the angular distribution (and thus isotropic distribution is assumed) ranges from -25% to 50%.

An easy way to avoid the problem is to work at the so-called "magic angle" $\theta=54.74°$ where $P_0 = P_1 = P_{iso}$. But the knowledge of the angular distribution or, what is equivalent, of the polarisation of the deexcitation light is a significant quantity by itself, that contains useful information regarding the electron capture process. To illustrate this, we consider the example of the neutralisation :

$$H^+ + H^- \rightarrow H(2p) + H(1s).$$

This reaction can proceed through radial or rotational coupling. In the incident channel, the system (H^+-H^-) is in a Σ state. In the case of radial coupling, it will remain a Σ state in the exit channel and the 2p state of the hydrogen atom will therefore be a $2p_0$. Deexcitation to (1s) will be of the type : $\Delta m = 0$.

On the other hand, rotational coupling will bring the system to a π state and the 2p atomic state will then be a $(2p)_{\pm 1}$. Deexcitation light will be of the type $\Delta m = \pm 1$. A measurement of the angular distribution of the deexcitation light thus provides information regarding the nature of the coupling.

Cascades, when present, are a serious complication. The observed state is then populated not only by the electron capture process but by the deexcitation of higher excited states and the radiation output is no longer a direct measurement of the cross section. Of course, simultaneous observation of the different lines present in the light spectrum can, in principle, solve the problem but, practically, the analysis becomes soon inextricable. Cascades also modify the apparent lifetime of the observed state. This can be used, positively, to check the absence of cascades in a measurement.

Electric fields also strongly modify the lifetime of excited states by producing a mixing of the substates. Even very weak fields, as those created on poorly conducting materia located in the vicinity of the beam, must be eliminated. Electric fields, on the other hand, are also used positively to quench metastable states.

Attention must also be paid to possible Doppler effects when energetic beams are used. If the region viewed by the detector is long, light emitted from the edges can be considerably Doppler shifted and fall outside the acceptance window of the monochromator.

Positively, Doppler shift is sometimes used to distinguish the photons emitted by the projectile from those radiated by the target. This is specially useful in the case of symmetric charge exchange, where electron capture by the projectile and target excitation give the same kind of radiation.

In beam-beam experiments, finally the problem arises of the background light produced by the interaction of the beams with the residual gas. The light yield of a beam-beam interaction is always very small and can be difficult to extract out of the background. Beam modulation and photon-ion coincidence techniques are then necessary.

CAPTURE FOR EXCITED STATES

The major difficulty with the measurement of cross sections for collisions of atoms or ions in excited states is the preparation of the excited species.

Ideally, one would desire, for instance, a beam consisting of the reactant excited in one state only. In practice, several states will coexist and the contribution of a particular one to the investigated process must be separated out. This is obtained by changing the population in a controlable way. The population of the investigated state must, of course, be large enough to produce an observable effect. For short lived states, an additional

difficulty arises with the necessity that the distance between
excitation and reaction points be short and quite accurately
controlled.

Laser excitation

Laser excitation, when feasible, is an attractive approach.
As a rule, the laser must be tunable and frequency stabilized as
the width of atomic levels is quite narrow.

Direct excitation from ground state requires U.V. light in
most cases and can therefore not be achieved, at the present time,
as tunable UV-lasers are not available. Frequency doubling, on
the other hand, reduces very much the laser power and is not
operational excepted for some limited regions in the near UV.

Laser excitation is therefore often used in conjunction with
other excitation methods that provide an important population of
some long lived (metastable) lower excited state.

For instance, PESNELLE et al (1981) produced a beam of $He(3^1P)$
for the study of associative ionisation in collisions with Ne
atoms, by exciting by means of 5015 Å-Coumarine-dye laser light a
beam of $He(2^1S)$ effusing out of an electron bombardment source.

As another example, CORNET et al (1981) populate the 3p state
of H by laser excitation at 6450 Å of H(2s), itself produced by
charge exchange of protons in Caesium.

The efficiency of laser excitation of a beam depends on many
factors. For a beam of atoms or ions, travelling along the z axis
with velocity v and density n(xy), intersecting a laser beam
travelling along the x axis and with a photon flux j(y,z), the
number of excitations occuring in a second is :

$$\int j(yz)\sigma n(xy)dxdydz$$

where σ is the excitation cross section.
The number of atoms crossing the interaction region in one second
is :

$$\int n(xy)vdxdy.$$

The efficiency ε of the excitation is thus :

$$\varepsilon = \frac{\sigma\int dyJ(y)T(y)}{\int T(y)dy} . \tag{14}$$

Where we use the notations :

$$J(y) \equiv \int j(yz)dz, \text{ for the 1-dimensional flux along y in the photon beam,}$$

$$T(y) \equiv \int n(xy)dx, \text{ for the "thickness" of the atom beam as seen by the photons.}$$

A general feature of crossed beams, discussed earlier, is met again : the efficiency depends on the density profiles of the beams along y and not only on the photon flux. If the flux is assumed constant across the photon beam, i.e. if :

$$J(y) = \frac{I}{H} \qquad \text{for} \qquad -\frac{H}{2} < y < \frac{H}{2}$$

$$J(y) = 0 \qquad \text{elsewhere,}$$

and if the photon beam completely overlaps the atomic beam, then :

$$\varepsilon = \frac{\sigma I}{vH}$$

where I is the total intensity of the photon beam (number of photon/sec) and H is its height.

When fast atoms or ion beams are laser excited, the efficiency is considerably reduced by Doppler effects.

In the formula (14) above, the use of a cross section is only valid if the frequency of the light is exactly matched to the transition frequency. This cannot be so for all the atoms of a real beam because the velocity dispersion in the atomic beam and the non-zero divergence of both the atom and the laser beam produce Doppler shifts.

Consider the effect of the divergence in the folloqing example.

v = velocity of atoms = 10^8 cm/sec.
α = divergence of atom beam = 10^{-3} rad.
ν = 5 x 10^{14} (orange light).

If the frequency of the light is matched to excite the atoms tra-velling on the axis, it will be unmatched for the atoms at the edge by an amount :

$$\Delta\nu = \nu \frac{v}{c} \sin\alpha \simeq 3.3 \times 10^{-6}\nu = 1.6 \times 10^9 \text{ Herz.}$$

Such a shift is likely to be larger than the frequency width ($\delta\nu$) of an atomic transition and only a fraction of the atoms in

the beam will be available to excitation. The efficiency will be
reduced, roughly by a factor $(\delta\nu/\Delta\nu)$. The divergence of the laser
beam must not be forgotten and it must be remembered that the
narrower the beam waist is, the larger is the divergence. It should
be clear also that the efficiency of excitation cannot be calculated accurately and that it has thus to be measured.

Finally, let us mention that Doppler shift can be used positively to tune the frequency of the light seen by the atoms. This
is best achieved with merged atom (ion) and light beams.

Field ionisation

An interesting method to modify the population of excited
states is field ionisation. In this case, the excitation itself
is created by some collisional process. For instance, in the
case of hydrogen, a beam of protons is passed through some gaseous
target and is partially neutralized. The hydrogen atoms so formed
are in all possible states, distributed following the n^{-3} law for
the principal quantum number n.

Field ionisation is used to remove all the excited states
above a certain value of n that can be changed just by changing
the field. The method developped by RIVIERE and SWEETMAN (1963)
is based on the fact that the ionisation rate of an excited atom
in an electric field increases so rapidly with the field that
one can assign to each value of the quantum number n, a critical
value of the field above which the atom is ionised practically
instantaneously and below which it is not ionised. The critical
field, defined to give an ionisation rate of 5×10^7 (sec^{-1}) is
proportional to n^{-4}. For H, it is :

$$E = \frac{6 \times 10^8}{n^4} \quad \text{volt/cm.}$$

If 5×10^5 volt/cm is taken as the higher limit for a feasible
field, hydrogenic states down to n=6 can be investigated by this
method. The critical field also somewhat depends on substate
numbers, so that the values of n for which a given field becomes
critical is unprecise. Adjacent values of n can therefore not be
resolved.

The method has been used by KOCH and BAYFIELD (1975) to
measure the cross section of the process :

$$H(n) + H^+ \xrightarrow[\sigma]{} \left\{ \begin{array}{l} H^+ + H \\ H^+ + e + H^+ \end{array} \right.$$

in the energy range 0.4-61 eV and for 44 < n < 50.
It has also been applied by BURNIAUX et al (1977) to the study of

$$H(n) + He^{++} \rightarrow H^+ + He^+(n')$$

in the range 0.2–470 eV and for $8 < n < 25$.

The main feature is the constancy of the cross section below 100 eV. This is in rather good agreement with classical calculations as those of PERCIVAL and RICHARDS (1975) and OLSON (1979).

The last paper presents a scaling law for n and for Z, in the case of a capture by a bare nucleus of charge Z

$$\sigma = kn^4 Z.$$

Such a prediction can be derived directly from simple classical arguments. A molecular calculation has also been done by SALIN (1977), applying the Landau–Zener model to the multiple crossing of ionic and covalent curves. It predicts a cross section of the form :

$$\sigma = \frac{1}{2} \pi \alpha^2 n^4$$

where α is almost velocity independent at large values of n.

REFERENCES

Dz. Belkic, R. Gayet and A. Salin, 1979, Physics Reports, Vol. 56 n° 6.

E.G. Berezhko and N.M. Kabachnik, 1980, Soviet Phys. JETP 52 (2), p. 205.

H.C. Brinkman and H.A. Kramers, 1930, Proc. Acad. Sci. Amsterdam, 33, p. 973.

M. Burniaux, F. Brouillard, A. Jognaux, T. Govers and S. Szücs, 1977, J. Phys. B, 10, p. 2421.

A. Cornet, 1981, private communication.

F.T. Chan and J. Eichler, 1979, Phys. Rev. A, Vol. 20, p. 1841.

P. Defrance, F. Brouillard, W. Claeys and G. Van Wassenhove, 1981, J. Phys. B, Vol. 14, p. 103.

T.M. El Sherbini, A. Salop, E. Bloemen and F.J. de Heer, 1979, J. Phys. B, Vol. 12, L579.

J.D. Jackson and H. Schiff, 1953), Phys. Rev. 89, p. 359.

R.K. Janev and L.P. Presnyakov, 1981, Phys. Reports, Vol. 70, n° 1.

P.M. Koch and J.E. Bayfield, 1975, Phys. Rev. Letters, 34, p. 448.

R.M. May, 1964, Phys. Rev., 136A, p. 669.

R.E. Olson, 1980, J. Phys. B, 13, p. 483.

K. Omidvar, 1967, Phys. Rev., 153A, p. 121.

J.R. Oppenheimer, 1928, Phys. Rev. 31, p. 349.

M.N. Panov, 1979, XI ICPEAC, Invited papers, p. 437.

I.C. Percival and D. Richards, 1975, Adv. in At. and Mol. Physics, n° 11, p. 1.

A. Pesnelle, S. Runge, D. Sevin, N. Wolffer and G. Water, 1981,
    J. Phys. B, Vol. 14, p. 1827.
A.C. Riviere and D.R. Sweetman, 1963, Proc. VI Conf. Ion Phen. in
    gases, Paris, Vol. 1, p. 105.
M. Rødbro and F.D. Andersen, 1979, J. Phys. B, 12, p. 2883.
H. Ryufuku and T. Watanabe, 1979, Phys. Rev. A, Vol. 20, p. 1828.
A. Salin, 1977, X ICPEAC, Paris, Abstracts of Papers, p. 452.
A. Salop, 1979, J. Phys. B, Vol. 12, p. 919.

CONFINEMENT OF IONS FOR COLLISION STUDIES

J.B. Hasted

Birkbeck College
(University of London)
Malet Street
London WC1E 7HX, U.K.

1. INTRODUCTION

All collision experiments in which individual impacting particles make more than a single trajectory across a significant part of the collision region are in effect confinement situations and will form the subject of this article. This definition excludes collisional orbiting, and also excludes plasma and afterglow studies of collision processes, provided that the mean free path is smaller than the 'significant part' of the collision region which we choose to consider. We shall discuss confinement by electric fields, magnetic confinement, the use of steady magnetic and electric fields in conjunction, and confinement in high frequency electromagnetic fields.

The purposes of confinement, as distinct from the various geometries of impacting electron and ion beams, are threefold:

1.   To avoid the presence of long-lived excited states of the ions, which can be allowed to decay radiatively or collisionally during the period the ions spend confined in the trap.

2.   To avoid the competition between collisions of the electrons or ions with background gas, and those with the relatively low densities of beam ions in the collision region. A magnetic-electric ion trap can readily offer ion number densities of $\simeq 10^6$ cm^{-3}, and a space-charge trap number densities of $\simeq 10^9$ cm^{-3}, contrasting with ion beam number densities of order 10^7 cm^{-3}.

3.   To offer relatively small and inexpensive collision facilities, avoiding the necessity of building ion beam hardware.

The disadvantages of confinement experiments will become apparent during our discussion of the various designs. The principal disadvantage is that absolute cross-sections are not obtained; the accurate relative cross-sections which become available may be normalized to bench-mark experiments, so that the confined ions are used as a target for monochromated beams, thus enabling structure in the cross-section energy variation to be elucidated directly by scansion of the beam energy, as in collision spectroscopy. The trapped ion targets can also be used for studies in optical spectroscopy (Schuessler 1979).

2. ELECTRICAL CONFINEMENT

To confine positive ions purely by static electric fields, the ions must be surrounded on all sides by positive potential, a situation which is impossible to achieve by the use of shaped electrodes only. A central negative potential must be provided by electron space charge, continuously replenished. Much is known about the electric potential configurations so obtained; we shall discuss them in various geometries relevant to successful confinement situations. The relevance of this to crossed-beam experiments will not be lost upon those who measure absolute cross-sections using crossed and merged beams of charged particles.

2.1 Cylindrical Electron Beams

We first consider the charge density ρ_c and hence the potential configuration, within a cylindrical electron beam of radius a, carrying a total current I_e. Such an ideal beam can only be maintained artificially, since it would expand under its own space-charge unless held by externally applied fields. These might either be (i) uniform axial magnetic fields confining the electrons to helical paths or (ii) electric fields intended to resist the expansive tendency of the electron trajectories; in particular, fields applied by the 67.5° conical guard electrodes proposed by Pierce (1954).

$$\rho_c = I_e / \pi a^2 \sqrt{2\eta V} \tag{1}$$

with $\eta = |e|/m$ and V the potential with respect to the cathode, which is assumed to produce electrons of zero velocity (cool electron approximation). The Poisson equation is written

$$\frac{1}{r} \frac{\partial}{\partial r} \left(r \frac{\partial V}{\partial r} \right) = \frac{I_e}{\pi \varepsilon_0 a^2 \sqrt{2\eta V}} \tag{2}$$

This equation has two solutions, the first of which can be perturbed catastrophically into the second by the presence of sufficient positive ions (Pierce 1954). There is an axial $V(0) = 0$ virtual cathode solution

$$V = \left(\frac{9}{16\pi\varepsilon_0\sqrt{2\eta}}\right)^{2/3} \left(\frac{r}{a}\right)^{4/3} I_e^{2/3} \qquad (3)$$

There is also a second solution with axial potential $V(0) \neq 0$. This potential V_0 may be taken to be that governing the electron energy of the axial electrons.

We define a reduced potential

$$\phi = V/V_0 \qquad (4)$$

and a reduced radius variable, unfortunately denoted

$$\rho = r\left[I_e/V_0^{3/2}\varepsilon_0\pi a^2\sqrt{2\eta}\right]^{\frac{1}{2}} \qquad (5)$$

so that equation (2) becomes

$$\frac{1}{\rho}\frac{\partial}{\partial\rho}\left(\frac{\partial\phi}{\partial\rho}\right) = \phi^{-\frac{1}{2}} \qquad (6)$$

which must be solved with boundary conditions $\phi = 1$ and $\partial\phi/\partial\rho = 0$ at $\rho = 0$. Numerical solutions have been obtained (Pierce 1954), but for $\rho \leqslant 0.3$, the relatively weak microampere beams encountered in many electron guns, an approximate solution may be written

$$\rho^2 \simeq 4(\phi - 1) \qquad (7)$$

so that

$$V - V_0 \simeq \frac{I_e r^2}{4V_0^{\frac{1}{2}}\varepsilon_0\pi a^2\sqrt{2\eta}} = K_c r^2 \qquad (8)$$

This represents a radial parabolic potential well which can readily be made sufficiently deep to confine thermal positive ions. For $I_e = 200$ μA, $V_0 \simeq 30$ V, $a = 0.5$ mm, $K_c = 2021$ V/cm^2, so that cylindrical potential wells in excess of 0.4 V are readily obtained; these are adequate for radial trapping of laboratory temperature positive ions of mean energy $\frac{3}{2}kT \simeq 0.04$ eV.

However for the axial trapping which is also necessary for full confinement, additional understanding of the space-charge configuration is necessary. The solution of the finite length cylindrical beam problem is non-analytical (see section 2.2); but an analytical solution is available (Lloyd 1965) for the axial potential distribution $V(z)$ along a one-dimensional beam of current I_e passing between two equipotentials separated by a distance ℓ and each held at potential V_ℓ:

$$\frac{I_e z(z - \ell)}{\varepsilon_0 \sqrt{2\eta}} = 4\left[\left(V_\ell^{\frac{1}{2}} - V_m^{\frac{1}{2}}\right)V_m^{\frac{1}{2}} - \left(V^{\frac{1}{2}} - V_m^{\frac{1}{2}}\right)V_m^{\frac{1}{2}}\right.$$
$$\left. + \frac{1}{3}\left(V_\ell^{\frac{1}{2}} - V_m^{\frac{1}{2}}\right)^{3/2} - \frac{1}{3}\left(V^{\frac{1}{2}} - V_m^{\frac{1}{2}}\right)^{3/2}\right] \tag{9}$$

At $z = \ell/2$ there is a potential minimum of depth $V_\ell - V_m$. For a cylindrical beam, the minima are less deep than those predicted by this equation; but even if an extreme factor of 100 difference is assumed, the axial trap depth is as much as 1 V for $\ell = 1$ cm, $I_e = 50$ μA, $V_\ell = 30$ V. Thus axial trapping of ions is also possible provided that the beam terminations do not counteract the effect. These analytical solutions are of value for the calculation of potential distributions in interacting beam experiments.

In the cathode-anode region the field is such as to accelerate ions out of the confinement region; it is necessary that the beam passes through a small orifice in the anode into the confinement region, so that the field penetration into this region is minimized. At the beam collection end no such problem arises. Either an excess of negative space charge is produced by secondary electrons, thus affecting the trap depth; alternatively, and more effective, an extra field gradient may be applied (Redhead 1969), penetrating through an orifice in the beam collecting electrode.

Confinement of ions within cylindrical electron beams has enabled early studies of electron-ion collision cross-section functions to be made (Baker and Hasted 1966). The importance of confinement was realized as a result of the appearance of doubly charged ions just above the potential corresponding to the difference between the first and second ionization potentials, in abundances proportional to the square of the electron current.

As a confinement technique, the cylindrical electron beam has been superseded by systems which do not suffer from the disadvantage that the same electron beam is responsible for the trapping, for the formation of the original ions, and for further collisions with them. However, molecular ion thermodynamic information can be obtained with nothing more than a commercial mass-spectrometer, and a number of such studies have been made, under the title of 'sequential mass-spectrometry' (Daly and Powell 1967, Cuthbert et al 1966). Perhaps the most interesting example of exploratory work was that carried out by Daly and Powell (1967) on the resonances lying below the n=3 level of He^+.

From the point of view of ion confinement studies, the effect of the trapped ion dynamics upon the potential configuration is of importance. A purely radial motion treatment (Baker and Hasted 1966) in terms of oscillations within the $V(r) = K_c r^2$ potential well (equation 8) is as follows. An ion formed at $r = r_0$, will possess radial velocity v_r given by

$$(r_0 - r)\eta_+ \frac{dv}{dr} = \frac{v_r^2}{2} \tag{10}$$

where η_+ is the ion charge/mass ratio. The ion radial density function is taken to be the inverse of the radial velocity distribution function

$$\frac{n_+}{\bar{n}_+} = \frac{1}{a} \int_r^a r_0 dr_0 / v_r \tag{11}$$

which may be combined with the positive ion Poisson equation

$$\frac{1}{r} \frac{\partial}{\partial r}\left(r \frac{\partial V_+}{\partial r}\right) = \frac{n_+ e}{\varepsilon_0} \tag{12}$$

Using as boundary condition $dV/dr = -k$ at $r = a/2$, a solution is obtained as follows:

$$-V = 3C(2r^5 - 3ar^3) + A\ln r + B \tag{13}$$

where

$$C = e\bar{n}_+/3\varepsilon_0 a\sqrt{\eta_+ k} \tag{14}$$

$$A = -\frac{ka}{2} + 3Ca\{(a/2)^{3/2} - \frac{2}{5}(a/2)^{5/2}\} \tag{15}$$

$$B = 3C(3a^4 - 2a^5) - A \ln a \tag{16}$$

We see from the form of this solution, which is illustrated in figure 1, that V is infinite at $r = 0$, and zero at $r = a$. Although it is not rigorous to treat the potential distribution due to ion space charge and that due to electron space charge as separate and additive functions, we nevertheless note that in figure 1, where the two functions are added, the following features emerge:

1. A region exists close to the axis in which slow electrons can be trapped.

2. The existence of an annular potential well is likely. Indeed

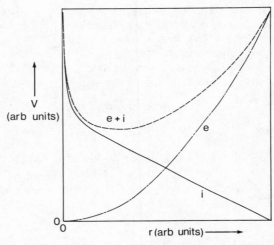

Fig. 1.   Calculated radial potential V(r) for cylindrical electron
          beam: e - electrons only, equation (8); i - trapped ions
          only, equation (13); e+i - arithmetical sum.

there is some evidence for it (Baker and Hasted 1966), deriving from
the unusual 'winged' form of the electron energy distribution unfolded
from a threshold linear ionization function. It is also well-known
that oxide cathode damage from ions within electron beams occurs
close to the axis rather than in a ring.

2.2   Inverse Cylindrical Triode

The inverse cylindrical triode is the most efficient form of
electrode impact ionizer for molecular beams, but its space charge
traps thermal ions and thereby destroys its efficiency, except for
fast molecular beams. However, the trapped ions could themselves be
used as a target for impacting beams passing parallel to the axis.
The problem of monitoring the collision product densities has not
been tackled experimentally, but experience obtained with other confinement systems could readily be applied. The Bayard-Alpert (Schulz
1957) ionization gauge is based on the collection of ion current
formed by the ionization of background gas. The use of the inverse
cylindrical triode for detection of fast neutral particles emerging
from pulsed hot plasma is of interest (Wynter 1974, Wynter and Hasted
1974). An application might be envisaged in neutral beam injection
into plasma machines.

The space-charge of the orbiting electrons produces much deeper
traps than those available from cylindrical beams, because the

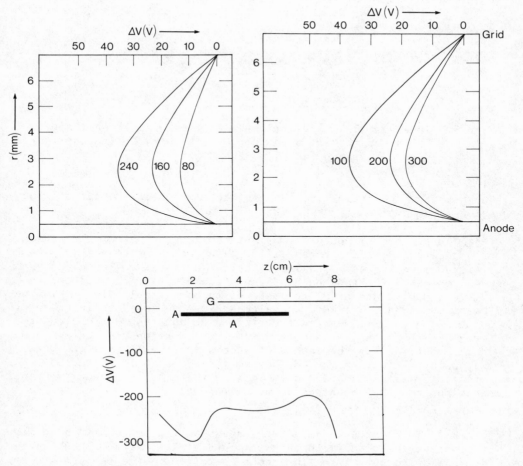

Fig. 2. Calculated potential functions for inverse cylindrical
triode. (a) $\Delta V(r)$ for specified values of I_{eff} (mA).
$V_a = V_g = 200$ V; (b) $\Delta V(r)$ for $I_{eff} = 160$ mA. Specified
values of $V_a(= V_g)$ (V); (c) longitudinal potential variation
at $r = 2.5$ mm, for $V_a = V_g = 200$ V, $I_{eff} = 160$ mA.

currents may readily be made much larger. There now exist relaxation
techniques (Wynter 1974, Wynter and Hasted 1974) of solving the
Poisson equation with charge density ρ

$$\frac{\partial^2 V}{\partial r^2} + \frac{1}{r} \frac{\partial V}{\partial r} + \frac{\partial^2 V}{\partial z^2} = -\rho/\varepsilon_0 \tag{17}$$

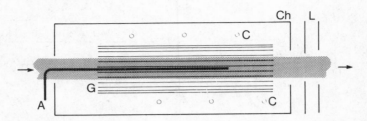

Fig. 3.   Inverse cylindrical triode with (ion)beam shown shaded.
          C, cathode helix; A, anode; Ch, chamber; L, lens apertures;
          G, grid wires parallel to the axis.

The boundary conditions are the electrode potentials and the axial
symmetry condition at r = 0, which is that dV/dr = 0.

     The solution is obtained by superimposing a net on the region
of interest and reducing the differential equation to approximate
difference equations by taking a Taylor series expansion about each
nodal point. Simultaneous difference equations are solved for the
entire net. Values of V are assigned to each nodal point including
the fixed boundary values, and a guess is made for each interior
point. The coefficients in the difference equations are calculated
for all mesh points and a solution of the equation connecting them
is then obtained by the method of successive over-relaxation (the
extrapolated Liebmann method), until the error is less than 1%. The
potential values so obtained are now substituted into the difference
equations and the method of solution repeated until a convergency
test is satisfied (Hornsby 1963).

     Results of such calculations, shown in figure 2, have been made
for an inverse triode with an axial anode, of construction shown in
figure 3. The purpose of the anode is to neutralize the space charge,
to improve stability, and to provide a situation in which the
calculation of space charge avoids divergence problems.

     It will be apparent from the data of figure 2 that the axial
trap depth is relatively weak in comparison with the radial potential
variation. It has been possible (Wynter 1974, Wynter and Hasted 1974)
to probe this potential variation by means of a fast beam of neutral
hydrogen atoms of known kinetic energy and approximately parallel
trajectories. The protons formed from this beam are energy-analyzed
before detection, and their deflection paths in the space-charge well
are computed. The resulting efficiencies of ionization and collection (figure 4) and energy losses (figure 5) correspond reasonably
to those calculated from the space-charge computations.

     In the computation of efficiencies the known cross-section for
electron impact ionization of atomic hydrogen is assumed, and the

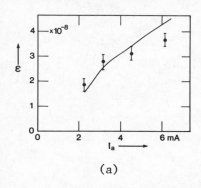
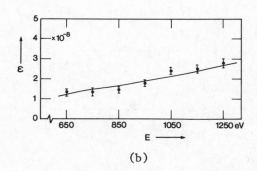

(a)   (b)

Fig. 4. Calculated (full line) and measured (points) efficiencies ε
of ionization of atomic hydrogen in inverse cylindrical
triode as functions of (a) atom beam energy E, at
I_{eff} = 117 mA, V_a = V_g = 200 V; I_a = 3.1 mA, I_g = 28 mA;
(b) anode current I_a, for E = 1250 eV, V_a = V_g = 200 eV.

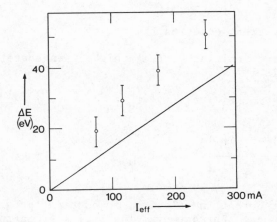

Fig. 5. Calculated (full line) and measured (points) space charge
energy losses incurred by fast neutral atomic hydrogen beam
on ionization in inverse cylindrical triode. V_a = V_g =
200 eV.

effective ionizing current calculated from the measured anode and
grid currents in the following way:

Electrons entering the anode-grid system may travel directly to
the anode or to the grid, or may traverse the region more than once.
A geometrical grid transparency t_g is assumed, and an analogous anode
transparency t_a is used in the analysis of the electron paths. The
grid and anode currents are respectively

$$I_g = I\{(1 - t_g) + t_g t_a (1 - t_g) + t_g{}^2 t_a (1 - t_g)$$
$$+ t_g{}^3 t_a^2 (1 - t_g) + \ldots\}$$

(18)

$$= I \left[\frac{(1 - t_g)(1 + t_g t_a)}{1 - t_g{}^2 t_a} \right]$$

and

$$I_a = I\{t_g (1 - t_a) + t_g{}^2 t_a (1 - t_a) + \ldots\}$$

(19)

$$= I \left[\frac{(1 - t_a) t_g}{1 - t_g{}^2 t_a} \right]$$

where I is the current entering the region.

The effective ionizing current which is encountered by the molecular beam is

$$I_{eff} = I\{t_g + t_g t_a + t_g{}^2 t_a + t_g{}^3 t_a{}^2 + \ldots\}$$

(20)

$$= I \left[\frac{(1 + t_a) t_g}{1 - t_g{}^2 t_a} \right]$$

Note that in each orbit the electron encounters the grid twice but the anode only once. Furthermore, the orbits are assumed always to pass through the molecular beam, which implies that the molecular beam diameter is as large as the grid diameter; the orbits are also assumed not to deviate appreciably in length from radial oscillations through the molecular beam.

The unknown t_a is eliminated from these equations, to give I_{eff} in terms of t_g, I_a, I_g and I.

The correspondence between calculated and observed ionization efficiencies (figure 4) and energy losses (figure 5) is sufficiently good to justify the claim that the above treatment is substantially correct. The values of I_{eff} are such that the electrons traverse the molecular beam some N=50 times. If the anode diameter is made smaller or the anode removed entirely, higher values of N can be achieved.

Thus the inverse cylindrical triode can be regarded as an electron confinement device, which also confines thermal ions; whilst it is unsuitable for electron-ion collision cross-section measurement, it can be applied in the ionizing of fluxes of charged or neutral beams.

2.3 Concentric Electron Beam Device

The most effective space charge confinement for beam trapped-ion collision cross-section measurement is within an annular 'hollow' cylindrical electron beam, such as can be obtained from a toroidal cathode. Within such a beam there is a cylindrical region free from beam electrons, but surrounded by potential gradients. In the central region the positive ions travel in oscillatory radial paths and may be impacted for collision studies by a second, axial, electron beam of energy different from the annular beam. Axial monitoring of positive ion densities is possible.

The toroidal cathode has been discussed in a series of papers by Hartnagel (1964a, b, 1965a, 1966). The annular cylindrical beam is treated in a more powerful way than that of simply bending a strip beam into a circle (Harris 1959). It is reduced to the problem of a toroidal concentric diode, in which the Poisson equation is solved in a toroidal coordinate system (figure 6), in which the variables R, R_0, r, α and ϕ are related as follows:

$$\begin{aligned}
z &= r \sin \alpha \\
x &= R \cos \phi \\
y &= R \sin \phi
\end{aligned} \tag{21}$$

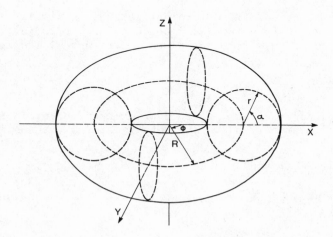

Fig. 6. Toroidal coordinate system.

with $R = R_0 + r \cos \alpha$.

The Laplacian operator is of the form

$$
\Delta\Phi = \frac{1}{r(R_0 + r \cos \alpha)}\left[\frac{\partial}{\partial r}\left(r(R_0 + r \cos \alpha)\frac{\partial\Phi}{\partial r} \right) \right.
$$

$$
+ \frac{\partial}{\partial\alpha}\left(\frac{R_0 + r \cos \alpha}{r} \cdot \frac{\partial\Phi}{\partial\alpha} \right)
$$

$$
\left. + \frac{\partial}{\partial\phi}\left(\frac{r}{R_0 + r \cos \alpha} \cdot \frac{\partial\Phi}{\partial\phi} \right) \right]
\tag{22}
$$

and since $\partial V/\partial\phi = 0$ the Poisson equation can be written in the form

$$
\frac{\partial}{\partial r}\left[r(1 + rH/r_c)\frac{V}{\partial r} \right] + \frac{\partial}{\partial\alpha}\left(1 + \frac{rH/r_c}{r} \cdot \frac{\partial V}{\partial\alpha} \right)
$$

$$
= -rR (1 + rH/r_c)\rho/\varepsilon_0
\tag{23}
$$

with $H = (r_c \cos \alpha)/R_0$. Using the approximation $\partial V/\partial\alpha = 0$, one eventually rearranges to

$$
\frac{\partial}{\partial r}\left[r(1 + rH/r_c)\frac{\partial V}{\partial r} \right] = \frac{J_c(1 + H)r_c}{\varepsilon_0\sqrt{2\eta V}}
\tag{24}
$$

where J_c is the radial current density.

This conception of the toroidal cathode is similar to the inverse cylindrical system described above. The radial current flow may be expressed in a form analogous to the normal and inverse diode:

$$
\frac{dI}{d\alpha} = \frac{\overline{K}V^{3/2}}{r\mu^2}
\tag{25}
$$

where $\overline{K}$ is a constant and μ a correcting coefficient similar to Langmuir's β. Since the current density at the cathode is J_c,

$$
\frac{dI}{d\alpha} = J_c 2\pi R_0 (1 + H)r_c.
\tag{26}
$$

Setting $\overline{K} = 8\varepsilon_0\sqrt{-2\eta}\ \pi R_0/9$ and substituting r/r_c by e^σ leads to the equation from which μ can be calculated:

$$3\mu \cdot \frac{d\mu}{d\sigma} + \left(\frac{d\mu}{d\sigma}\right)^2 + \mu \frac{d\mu}{d\sigma}\{4 + 3/(e^{-\sigma}/H + 1)\}$$
$$+ \mu^2\{1 + \frac{3}{2}\Big/(e^{-\sigma}/H + 1)\} - 1(1 + He^\sigma) = 0 \tag{27}$$

Solution of this equation allows the gun characteristics, in particular the perveance P, to be calculated, using

$$P = \frac{I_a}{V_a^{3/2}} = \frac{\overline{K}}{r_a} \int_{\alpha_2}^{\alpha_1} d\alpha/\mu^2(\alpha) \tag{28}$$

For potential V_a at the anode of radius r_a (figure 7), zero potential at the cathode, the current flowing between the aperture $\alpha_1 - \alpha_2 = \Delta\alpha$ is I_a. Figure 8 shows computed perveances as functions of $\Delta\alpha$, r_{ct}/R_0. Measurements of current have shown that perveances are in good agreement with these calculated values.

But there remains the problem of how the beam travels on its emergence from the toroid; the transition from radial flow to annular cylindrical flow. There are two problems involved: the lens action of the annular orifice, and the space charge expansion of the beam; the first affects the final inner radius r_{ib} and the second the final outer radius r_{ob} of the beam (figure 7). The ring aperture lens formula for the focal length is analogous to the Davisson-Calbick formula:

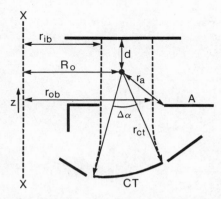

Fig. 7. Design parameters for toroidal electron gun.

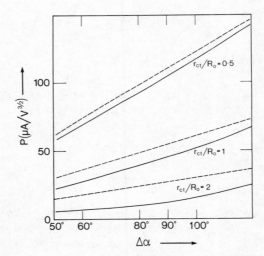

Fig. 8.   Computed perveances P of toroidal cathode, as functions of
          r_{ct}/R_0, $\Delta\alpha$ (full lines). Broken lines show equivalent
          cylindrical gun perveances.

$$\frac{1}{f} = \frac{V'_2 - V'_1}{2V_a} \tag{29}$$

where V'_2 and V'_1 are the electric fields respectively behind and in
front of the ring aperture anode, and V_a is its potential. It is
usual to aim for conditions under which the inner surface of the beam
is approximately parallel to the axis. The space charge forces must
be sufficiently strong to prevent the outer electrons crossing the
trajectories of the inner electrons.

To determine expected r_{ob} we use an integrated form of the
Poisson equation

$$\Delta V = \frac{1}{R} \cdot \frac{d(rE_r)}{dr} = \frac{I}{\pi(r_{ob}^2 - r_{ib}^2)v\varepsilon_0} \tag{30}$$

to obtain the radial electric field E_r

$$E_r = \frac{I\left[r - r_{ib}^2/r\right]}{2\pi(r_{ob}^2 - r_{ib}^2)v\varepsilon_0} , \quad \text{where } v = \sqrt{2\eta V} \tag{31}$$

On the assumption of no trajectory crossing, it is sufficient to consider only the paths of the outer boundary electrons. Their radial acceleration is calculated from E_r, and the final outer beam radius is calculated by integrating

$$\frac{dr_{ob}}{dz} = \left[2A \ln \frac{r_{ob}}{r_{ib}} \right]^2 \tag{32}$$

with $A = I/8\pi\varepsilon_0 \eta V_a^{3/2}$.

Experiments with pinhole collectors have shown that calculated current densities are achieved (figure 9).

We now discuss the space-charge configuration due to ions formed within the tubular beam. The radial field within the beam has been given as equation (31) above. We take the region within the hollow beam to be field-free. An ion formed with negligible velocity at a radius r_f will oscillate radially with amplitude $2r_f$. The velocity is described by

$$\dot{r} = \frac{-\eta J}{\varepsilon_0 v} \left[r^2 - r_f^2 - 2r_i^2 \ln r/r_f \right] \tag{33}$$

As before, (equation 11), the contribution to the ion distribution from these ions is proportional to the inverse of the velocity:

$$N(r) = \frac{Pr_0 dr_0}{\dot{r} r}$$

$$= \frac{P}{r} \int r_0 dr_0 \left[\frac{\eta J}{\varepsilon_0 v} (-r^2 + r_f^2 + 2r_i^2 \ln r/r_f) \right]^{\frac{1}{2}} \tag{34}$$

where P is the ion production rate.

The solution of this integral gives rise to infinities of space charge in the axial region $r = 0$; it is of the form

$$r_i > r, \quad N(r) \propto \mathrm{sech}^{-1} \left| \frac{r - r_i}{r_0 - r_i} \right| \tag{35}$$

As in the case of the cylindrical beam (equations 13-16) the potential arising from this distribution is sufficient to trap slow electrons produced by ionization (Hartnagel 1965b).

But it is to be noted that equation (31) is an inadequate solution of the Poisson equation for the entire radial system, since there is a discontinuity at $r = r_i$. When the relaxation methods

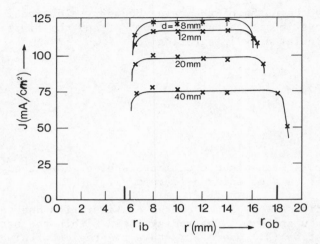

Fig. 9. Measured current densities obtained from toroidal cathode,
as functions of r, d.

described in section 2.2 are applied to an entire length of tubular
beam, including terminations, a potential minimum is found within
the beam and a maximum on the axis (Hasted and Awad 1972), as is
shown in figure 10(a). There is also a potential minimum at the
midpoint of the beam axis (figure 10(b)).

With this type of potential configuration a certain fraction of
ions formed and trapped within the beam will never reach the axis.
However, the space charge of the trapped ions modifies this potential
configuration. The minima fill up with ionic space charge, after
which full radial oscillations across the axis become possible.

Unfortunately the lifetime of ions in an electric confinement
device is nowhere near the length of lifetime of ions in the traps
described in later sections; it is normally rather less than a second, which is sufficiently long to enable collisional experiments to
be carried out, although not in the refined modes which minute or
hour lifetimes render possible. The cause of loss of ions from a
space-charge trap is the elastic collisions between electrons and
ions. These collisions, which have very large cross-sections, set
up diffusive losses, for which no general theory has been worked out,
but which might be treated by analogy with the diffusion across a
magnetic field of a hot plasma.

The concentric beam has been employed for the study of electron
impact ionization of positive ions (Hasted and Awad 1972; Hamdan et
al 1978) as well as for electron impact dissociation of molecular
ions and for electron-ion recombination (Mathur et al 1978). Under
steady-state conditions, when the diffusive losses balance the

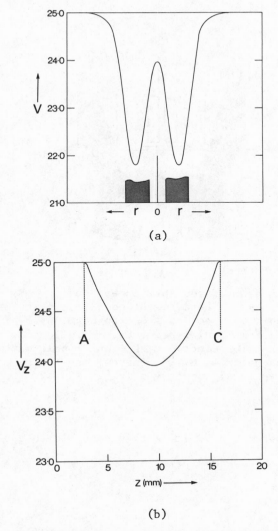

Fig. 10. Calculated potential functions for hollow electron beam.
(a) radial potential (volts); electron beam shown shaded;
(b) axial potential (volts) between anode A and collector
C (figure 8).

production of ions from gas or molecular beam by the tubular beam
electrons, a proportion of the ions is monitored axially by mass-
spectrometer (figure 11).

The assumption of perfect mass-spectrometric monitoring
efficiency is made. Detected current independence of the extracting

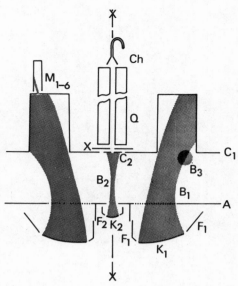

Fig. 11.   Scale diagram of concentric beam device. K_1 - toroidal
          cathode; K_2 - spherical cathode; $F_{1,2}$ - shaping electrodes;
          A - anode; B_1 - hollow electron beam; B_2 - axial electron
          beam; B_3 - molecular beam; $C_{1,2}$ - collecting electrodes;
          X - extraction electrode; Q - quadrupole mass filter;
          Ch - channeltron; M_{1-6} - current monitoring electrodes;
          ⁎ - cylindrical symmetry axis.

field must be demonstrated.   It is then possible to apply the
equation

$$\sigma_{12} = \frac{I_2 C}{I_1 I_e} \tag{36}$$

where

$$C = \frac{\ell e v_A v_e F}{N(v_+^2 + v_e^2)^{\frac{1}{2}}} \tag{37}$$

This equation is normally applied to crossed beam situations, in
which a current I_1 of singly charged ions is crossed by a current I_e
of electrons, being partly converted to a current I_2 of doubly
charged ions.   The beam velocities are v_+, v_e, the width of the
crossing area is ℓ, and the electronic charge is e.   The geometrical
form factor of the beam crossing (Dolder et al. 1961, 1963) is F.
In the particular geometry of the concentric beam device, the

trapped ions are taken to oscillate radially N times across the axial electron beam. Continual replenishment and leakage take place even in the absence of the axial beam I_e.

The quantities that make up the constant C are not all determinable; the constant can, however, be determined by measuring I_e, I_1, I_2 for an ionization process whose absolute cross-section $\sigma_{12}(v_e)$ is known accurately. A number of such determinations of C are summarized in Hamdan et al. (1978).

The presence of long lifetime excited states of positive ions in a crossed-beam experiment cannot always be avoided. In the trap, however, such states decay at rates which are not always known, but whose orders of magnitude can often be deduced.

We therefore require to have some indication of the mean trapping time of ions. Experiments have been carried out (Hamdan et al. 1978) by application of recurrent voltage pulses of length T ($0 \leqslant T \leqslant 700$ ms, recurrence time 800 ms) to the tubular beam anode.

In the initial filling stage the losses are small compared with the filling rate, which for singly charged ions is proportional to T, and for doubly charged ions produced in a two-step process, to T^2. Ultimately the losses and filling rate balance each other, and there is no further increase in dn_+/dT or dn_{2+}/dT, as shown in figure 12. The value of T for which this change takes place is a semiquantitative measure of mean trap lifetime of an ion, and is seen to be of the order of 300 ms under the conditions of figure 12. Such trap lifetimes, however, are quite adequate to ensure near-axial ion densities of the order of 10^6 cm^{-3} for the purposes of collision experiments.

However, it must be stressed that the experiments in which ions trapped in the hollow beam potential well are impacted by an axial electron beam yield only relative and not absolute cross-sections. These must be normalized against absolute measurements, and as we have seen, there is scatter in the normalisation constants C.

The previously published e Ar$^+$, e Ar^{2+} data (Hamdan et al. 1978) were normalized to the absolute measurements available at that time, but now superior data are available, some being given by Dr. Salzborn in the present volume. Renormalization must therefore be carried out, as in figure 13 below. The error bars have been omitted, but even so, the possible discrepancy is disturbing, and may be connected with the presence of different proportions of long-lived excited ions in the crossed beam and trap experiments.

A feature of the hollow beam trap is the collisional effects which arise from the large population of electrons which form the trap. When hydrogen molecular ions H$_2^+$ are confined, it is found (Mathur et al. 1979) that the cross-section function for their

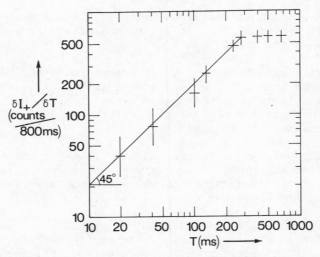

Fig. 12.   Linear filling rate and saturation of hollow beam.
            Variation of differential C^+ ion sample with pulse length
            T; hollow beam energy 35 eV, carbon monoxide pressure
            5×10^{-7} Torr.

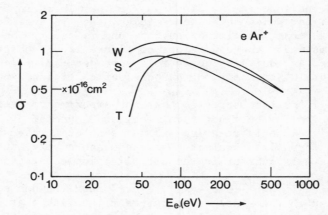

Fig. 13.   Comparison of crossed-beam and trap data for electron
            impact ionization of Ar^+. W - data from Woodruff et al.
            (1978); S - data reported by Salzborn in this volume;
            T - trap data (Hasted and Awad 1972).

dissociation into protons and hydrogen atoms rises sharply with
increasing energy and passes through a maximum, as shown in figure
14. This feature is not observed in crossed-beam experiments, but
is predicted (Peek 1967) only for ions in the ground vibrational

state, not for those in the range of states in which they are produced by vertical transitions from the molecule. The vibrationally
excited homonuclear molecular ions do not decay by radiative transitions, but it appears that many may be deactivated by electron
collisions in the trap. It is possible that the deactivation collision cross-sections are so large that any collisions which further
excite the ions, until they dissociate, play only a small role. It
is well-known that Franck-Condon overlap integrals between adjacent
levels decrease with increasing quantum number. However, the H_2^+
recombination cross-section measured in the trap (Mathur et al. 1978),
although a relative value, corresponds well in its energy variation
with the inclined beam measurements, as is shown in figure 15. Those
measurements are well-known to be more appropriate to a Franck-Condon
distribution of vibrational states. The lowest vibrational state of
H_2^+ is believed to be most unlikely to recombine with electrons. It
remains likely that at least for heteronuclear species the vibrational state distributions of molecular ions in traps will be fully
relaxed to v = 0 if the trapping time is sufficiently long. This
can be achieved in the Penning trap described in the section on
Radio-Frequency Traps.

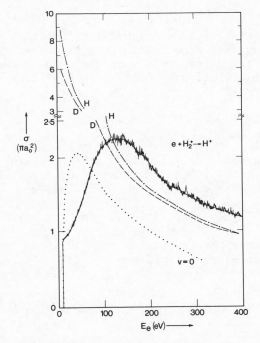

Fig. 14.   Concentric beam device cross-section functions for electron
           impact dissociation of H_2^+. Pen tracing – raw data; full
           line – numerically smoothed data; H, D – crossed-beam data
           (Dance et al 1967; Dunn and van Zyl 1967); dotted line –
           v = 0 calculations (Peek 1967).

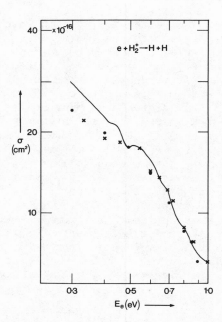

Fig. 15.   Recombination cross-section for H_2^+ with electrons. Full
line - raw data from hollow beam trap (Mathur et al 1978);
full circles - trap data unfolded with 300 meV electron
energy distribution; crosses - crossed beam data of Peart
and Dolder (1974).

3.   CONFINEMENT IN MAGNETIC FIELDS

Movement of charged particles in a uniform magnetic field is a
consequence of the force which is a cross product of charge velocity
and magnetic field.   The resulting acceleration is

$$m\dot{\underset{\sim}{v}} = e(\underset{\sim}{v} \times \underset{\sim}{B}) \tag{38}$$

neglecting any relativistic contribution.   It is convenient to solve
this equation by employing a cyclotron vector

$$\underset{\sim}{\omega} = e\underset{\sim}{B}/m \tag{39}$$

so that

$$\dot{\underset{\sim}{v}} = \underset{\sim}{v} \times \underset{\sim}{\omega} \tag{40}$$

A charged particle undergoes circular motion around the magnetic
lines, with velocity $v_\perp$ at right angles to them, and radius of
curvature

$$a = v_\perp/\omega = mv_\perp/eB \tag{41}$$

In addition the particle may possess, incidentally, velocity $v_{\parallel}$ parallel to the magnetic lines, so that the overall motion is helical, with pitch angle arctan $v_{\parallel}/v_\perp$. Both components of velocity, $v_{\parallel}$ and $v_\perp$, derive only from its initial velocity $\underset{\sim}{v}_0$.

Extensive mass-spectrometric studies have been made with the aid of ion cyclotron resonance, but these cannot truly be regarded as confinement experiments, so that discussion of them has been omitted from this section.

If an electric field $\underset{\sim}{E}$ were applied parallel to $\underset{\sim}{B}$, an acceleration eE/m would be added to the velocity $v_{\parallel}$, so that the pitch angle of the helix would vary with axial distance.

An electric field applied at right angles to the magnetic field would introduce a 'drift velocity', as follows. Suppose that $\underset{\sim}{B}$ is in the direction z, $\underset{\sim}{E}$ in the direction y. A transformation is made to a moving system of coordinates in which the electric field is zero. This system is moving with velocity v_D in the direction x:

$$|v_D| = \frac{|E|}{|B|} \,, \quad \text{or } v_D = \frac{\underset{\sim}{E} \times \underset{\sim}{B}}{B^2} \tag{42}$$

Vectors in this system, written as dashed, are $E' = 0$, and

$$\underset{\sim}{B}' = \underset{\sim}{B}\left(1 - \frac{E^2}{c^2B^2}\right) \tag{43}$$

In this coordinate system the motion is to be understood as in the purely magnetic case; but a drift velocity at right angles both to $\underset{\sim}{B}$ and $\underset{\sim}{E}$ is superposed. Note that this velocity is independent of sign or magnitude of charge, and of mass; in plasma or in systems containing various species, only a single drift velocity is found; no macroscopic current flows. Moreover, a similar treatment can be made of other force fields $\underset{\sim}{F}$ (such as the gravitational), in which $\underset{\sim}{E}$ would be replaced by F/e; here, drifts of electrons and positive ions in opposite directions are found, and macroscopic currents, such as earch centre currents, can flow.

Suppose now that, in the absence of electric or other force fields, the static magnetic field has spatial nonuniformity. Consider the case where B_0 is in direction z, but varies slowly in direction x: $\underset{\sim}{B} = B_0\{0, 0, h(x)\}$ with $h(x) \simeq 1 - \alpha x$ and $a\alpha \ll 1$. This is the condition

$$\nabla \underset{\sim}{B} \simeq \underset{\sim}{B}_0(-\alpha, 0, 0) \tag{44}$$

The drift velocity necessary for transformation to uniform magnetic field conditions is

$$\frac{v_D}{v_0} = \frac{a}{2B_0{}^2}(\underset{\sim}{B_0} \times \nabla \underset{\sim}{B}) \tag{45}$$

Note that this velocity is of opposite sign for opposite sign of charge, and can cause polarization of plasma confined in a torus. The charge spirals towards or away from a magnetic pole, on axes which are radial. For curved magnetic lines of radius $R(\gg a)$, with v_0 in direction y, R in the plane xy, B_0 in direction x, there is a drift velocity in direction z, given by

$$v_D = \frac{v_{\parallel}{}^2}{\omega R}\left(\frac{\underset{\sim}{R} \times \underset{\sim}{B}}{RB}\right) \tag{46}$$

which is also of opposite sign for opposite sign of charge, and will also cause plasma polarization in a torus.

For converging magnetic lines, in cylindrical symmetry (r,z) there is a deceleration of particle in direction z. Such a magnetic field can be expressed as

$$\underline{B} = \underset{\sim}{B_0}\left[\frac{\alpha}{2} r\underset{\sim}{e}_r + \underset{\sim}{e}_z(1 + \alpha z)\right] \tag{47}$$

where $\underset{\sim}{e}$ are unit vectors; for such deceleration, the condition $\alpha \ll 1$ must be assumed. It is readily shown that $v_{\parallel}$ decreases; and since v^2 is conserved, $v_{\perp}$ must increase; the cyclotron radii of the particles decreases as they are decelerated. Eventually there can be reflection, with reflection coefficient $R = 1 - B_0/B_{max}$. Thus a uniform magnetic field B_0 within a solenoid bounded by sufficiently strong convergence, to a value B_{max}, can form the simplest form of magnetic mirror, such as has been used in plasma magnetic mirror machines.

The possibility of using purely magnetic confinement in a magnetic mirror for collision studies is open; but the relatively large reflection coefficient of the linear configuration is a serious disadvantage. However, there is little reason why electric fields should not be added to magnetic fields for collisional confinement experiments.

4. CONFINEMENT IN STEADY MAGNETIC AND ELECTRIC FIELDS: THE PENNING
   CONFIGURATION

Radio-frequency electromagnetic traps, discussed in section 5, have been used for electron physics problems, particularly by Dehmelt

(Dehmelt 1969; Dehmelt and Walls 1968). They have the advantage of
mass specificity, and they also have a non-destructive ion density
diagnostic system. However, when the first feasibility study to
exploit such a trap for electron-ion collisions was made by Walls
and Dunn (1974) and their colleagues (Heppner et al. 1976; Dubois et
al. 1978) it was found that a Penning configuration of steady magnetic and electric field trap would offer greater advantages.
Although there is no mass specificity of confinement, the trapped
ions can be cooled to thermal energies, which is at present impossible
in a radio-frequency trap.

     In the Penning configuration positive ions are confined radially
by a magnetic field, and are confined axially within an electric
potential well. The latter is maintained by two positively biassed
hyperboloids of revolution with a negatively biassed hyperboloid of
revolution between them (figure 16).

     The dynamics of charged particle movement within this system
were worked out by Byrne and Farago (1972).

     The potential configuration inside the electrodes is

$$V = \frac{V_0(r^2 - 2z^2)}{r_0^2 + 2z_0^2} \tag{48}$$

with the dimensions (figure 16) typically $2r_0 = 1.25$ cm, $2z_0 = 0.76$
cm. Note that this trap can be flatter than the radio-frequency
trap, for which r_0^2 must equal $2z_0^2$. The dynamical equations of the
charged particles are

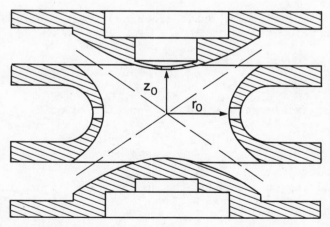

Fig. 16.   Cross-section through hyperboloid radio-frequency trap.

$$z = z_m \sin (\omega_z t + \delta) \tag{49}$$

$$\underset{\sim}{r} = \underset{\sim}{a} \exp \Omega_1 t + \underset{\sim}{b} \exp \Omega_2 t \tag{50}$$

with

$$\omega_z = \left(\frac{4eV_0/m}{r_0^2 + 2z_0^2} \right)^{\frac{1}{2}} \tag{51}$$

$$\Omega_{1,2} = i \left(\frac{\omega_c}{2} \pm \omega \right) \tag{52}$$

$$\omega_c = eB/m \tag{53}$$

$$\omega^2 = \frac{\omega_c^2}{4} - \frac{\omega_z^2}{4} \tag{54}$$

When $\omega^2 < 0$ (i.e. $\omega_z^2 > \omega_c^2/2$) the particle moves in an exponentially growing spiral, until it hits the walls; there is no trapping. But when $\omega^2 \geqslant 0$, the radial magnetic force dominates the radial electric force, and there is trapping, the ions following the radial equation

$$\underset{\sim}{r} = \underset{\sim}{a} \exp i \left(\frac{\omega_c}{z} + |\omega| \right) t + \underset{\sim}{b} \exp i \left(\frac{\omega_c}{z} - |\omega| \right) t \tag{55}$$

There is orbiting on a circle of radius $\underset{\sim}{a}$ with angular velocity

$$\omega_c' = \omega_c - \omega_m \tag{56}$$

The centre of the circular orbit precesses about a circle of radius $\underset{\sim}{b}$ at angular velocity

$$\omega_m = \frac{\omega_c}{z} - \omega \tag{57}$$

There is also axial oscillation with angular velocity ω_z, and the frequencies are related by the equation

$$2\omega_m^2 - 2\omega_m \omega_c + \omega_z^2 = 0 \tag{58}$$

For building such a trap, the accuracy of alignment of magnetic field with z is of importance. The cyclotron radius must be much smaller than radius r.

When the residual gas in the trap is at very low pressure, the ion trapping time can be of several minutes duration. An ion-molecule reaction rate coefficient of 10^{-9} cm^{3}sec^{-1} and a background pressure of 10^{-10} Torr would lead to a lifetime of $\simeq$300 sec.

The trap is filled by bombarding the background gas with electrons from a gun. Initially, the ion energy distribution extends right out to the limit of the trap depth, which is typically 0.4 V. However, Coulomb collisions eventually establish a Maxwellian distribution, in an equilibration time τ_e given by Spitzer (1962) as

$$\tau_e = \frac{11.4 A^{\frac{1}{2}} T^{3/2} \text{ sec}}{n_+ Z^4 \ln(\ell_D / b_c)} \tag{59}$$

where A is the atomic weight, n_+ is number density, ℓ_D is the Debye length and b_c is the impact parameter for Coulomb scattering through $\pi/2$. For $n_+ = 10^6$ cm^{-3}, $\ln(\ell_D/b_c) \simeq 12$, and the equilibration time is, typically, tens of milliseconds. Other cooling mechanisms also contribute.

Cooling also occurs by radiative coupling of the ionic motion to the external circuit in which the hyperboloid electrodes are placed (Dehmelt 1969). As the ions oscillate on the z axis, they induce image currents at pulsatance ω_z between the end caps, whose construction is such that across them is a stray resistance R_c. Joule heating of this resistance damps the motion of the ions, which thermalise at the temperature of the resistance R_c. Thus the control of the trap temperature serves as a control of the temperature of the trapped ions.

In addition to the control of ion temperature by means of the resistor R_c, heating of specific ions can be achieved by applying radio-frequency power at a frequency ω_c'. The mass resolution available to this technique is as high as one part in 10^4.

These image currents also provide an important method of monitoring the trapped ion densities non-destructively. The noise power spectrum in the circuit is measured, and shows a peak at the angular frequency ω_z (equation 51). The area under this peak is proportional to the product $n_+ T$. If either n_+ or T is constant during an experiment, then one can investigate a phenomenon which can be coupled to the other parameter. A mass resolution of a few tenths of a mass unit can be achieved in the detection technique. For $m_+ = 19$, $V_0 = 0.85$ V, $\nu_z \simeq 80$ kHz, $\nu_c \simeq 941$ kHz and $\nu_m \simeq 3$ kHz. The peak-to-background ratio is only 7:1 for total trapped ion populations limited by space charge to about 10^4, or number densities $n_+ \simeq 10^6$ cm^{-3}.

When an electron current I_e is directed at the trapped ions for a time t_e the population of ions is reduced from N_1 to N_2. The cross-section σ for destruction, which for molecular ions at low electron energies would normally be by dissociative recombination, is given by the equation

$$\sigma = \frac{eA}{I_e}\left\{\frac{\ln(N_1/N_2)}{t_e} - \frac{1}{\tau_n}\right\} \tag{60}$$

The collision overlap area A cannot be determined absolutely within a factor of two to three, so that relative cross-sections only are obtained. The time-constant τ_n is that for the decay of ion population in the absence of the electron beam.

The Penning trap has been used by Dunn and his collaborators for the study of electron-ion recombination of the molecular ions O_2^+, NO^+, H_3O^+, NH_4^+ over a range of electron energies approximately 0.1 - 5 eV, obtained by means of trochoidal monochromation. As an example of these data, the O_2^+ cross-section function is illustrated in figure 17. It will be seen that there is some variation in the cross-section when higher energy electrons are used to ionize the oxygen gas; the failure of all vibrationally excited states to relax in the trap is implied by this variation. It might be proposed that, ultimately, most ionic collisions will have to be studied as a function of age, in order to allow for decay of long lifetime excitation.

The possibility of employing the Penning trap for the study of

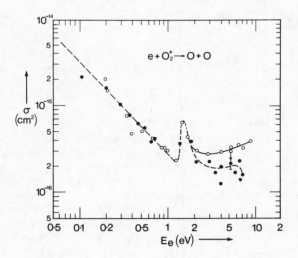

Fig. 17.   Recombination cross-section of O_2^+ measured in Penning trap (Walls and Dunn 1974). Open circles - ions produced by 30 eV electrons; full circles - ions produced by 14 eV electrons.

collisions at very low temperatures has now been exploited (Jeffries 1981). The trap is operated at liquid helium temperatures so that rate coefficients can be measured in the range 10-20 K. Although the data obtained have been for ion-molecule cross-sections, there is no reason why ion-ion collisions should not also be studied; both the mutual neutralization process and also inelastic collisions between two positive ions.

5. RADIO-FREQUENCY TRAPS

Radio-frequency confinement in quadrupole fields is based on the principles first established in cylindrical symmetry with radial but not axial trapping by Paul and his collaborators (Paul and Raether 1955; Paul and Steinwedel 1953).

This was extended to the full three-dimensional case (Paul et al. 1958; Wuerker et al. 1959) which has been extensively used (partly under the acronym QUISTOR, quadrupole ion store) for ionic collision studies; it has also been used by Dehmelt (1969) for studies in electron physics. The quadrupole mass-spectrometer, which util- izes mass-specific radial confinement, has itself been employed for many collision study purposes, including the monitoring of ion densities in afterglows and in electric traps.

We shall consider the dynamics of ions, of either sign, in radial confinement in a cylindrical quadrupole electric field. Close to the axis the electric potential can be approximated as propor- tional to the square of the radial variable; in Cartesian form the potential

$$V(x,y,z,t) = U(t)(\alpha_x x^2 + \alpha_y y^2 + \alpha_z z^2) \tag{61}$$

where $U(t)$ is the potential applied between the quadrupole electrodes. Since $\nabla^2 V = 0$ and $\alpha_z = 0$, $\alpha_y = -\alpha_x$ for hyperbolic cylindrical rods. For the application between the rods of a composite potential $U + V_0 \cos \omega t$,

$$V(r,z,t) = (U + V_0 \cos \omega t)(y^2 - z^2)/r_0^2 \tag{61a}$$

where r_0 is the inner hyperbola radius; in practice circular rods of radius $r_r = 1.2 r_0$ are a sufficiently good approximation to the hyper- bolic cylinders; however, hyperbolic rods are now commercially available.

In this field the radial and azimuthal motion of the ions is governed by the equations:

$$md^2x/dt^2 = 0 \tag{62}$$

$$\frac{d^2y}{dt^2} - \frac{2\eta_+}{r_0^2}(U + V_0 \cos \omega t)y = 0 \tag{63}$$

$$\frac{d^2z}{dt^2} - \frac{2\eta_+}{r_0^2}(U + V_0 \cos \omega t)z = 0 \tag{64}$$

The solutions of these equations, which are of the Mathieu type, fall into two classes, corresponding to stable orbits and diverging orbits: for stable orbits

$$a = \frac{8\eta_+ U}{r_0^2 \omega^2} \quad \text{and} \quad q = \frac{4\eta_+ V_0}{r_0^2 \omega} \tag{65}$$

must lie within the limits set by the equations

$$a = -q^2/2 + 7q^4/128 - \dots \tag{66}$$

and

$$a = 1 - q - q^2/8 + q^3/64 - \dots \tag{67}$$

At constant frequency, a peak-shaped region of stability is found on a graph of a versus q (figure 18). When U and V_0 are chosen to be within this region, ions are able to pass along the space in between the rods in stable orbits; otherwise they diverge and hit the rods. By adjustment of U (always keeping U/V_0 constant), a choice can be made of the mass of ion which will remain in a stable orbit, passing through the mass region between the quadrupole rods, rather than striking one of them. An additional feature is that one can choose by adjustment of the ratio U/V whether to make the choice of U near to the tip of the stability peak, or whether to work further down the peak where it is wider and the region of stability is greater. Under these conditions the mass resolution is less good, but there is less mass discrimination.

A feeling for the frequencies f necessary for maintaining the tip of the stability peak may be obtained from the equation

$$\frac{V}{r_0^2 f^2} = 7.15\left(\frac{m}{e}\right) \tag{68}$$

with $m/e = \eta_+^{-1}$ in amu, f in MHz, r_0 in cm, V in volts (peak to peak).

Thus the quadrupole mass spectrometer is tuned for mass by adjusting U, and the resolution is adjusted by varying the ratio $U/V = a/2q = u/2$. The principles, including the important choice of mass-specificity, can be extended to a three dimensional trap, which is illustrated in figure 16.

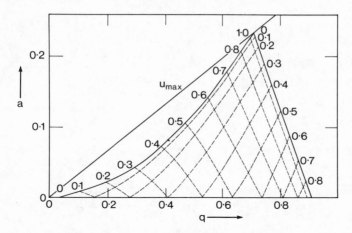

Fig. 18.   Calculated a(q) diagram for quadrupole mass filter, with
           iso-β lines.

The quadrupole ion storage trap (Dawson 1976a) (Quistor) is a
three electrode structure comprising a combination of a hyperboloid
of one sheet with a hyperboloid of two sheets; it is a solid of
revolution generated by rotating a section through the electrode
array of an ideal two-dimensional quadrupole mass filter about an
axis perpendicular to the axis of the filter.

The configuration of potential in the three-dimensional trap is
as follows:

$$V(x,y,z,t) = (U + V \cos \omega t)\alpha_z(z^2 + y^2 - 2x^2) \tag{69}$$

or

$$V(r,z,t) = (U + V_0 \cos \omega t)(r^2 - 2z^2)/2r_0^2; \tag{70}$$

note that

$$r_0^2 = 2z_0^2 \tag{71}$$

The conditions for stable orbiting in three dimensions are similar
to those in the cylindrical system, and the stable region is
illustrated in figure 19, with $a_z = -2a_r$ and $q_z = -2q_r$.

It is current practice with quadrupole mass filters to operate
with positive a, but with ion traps to operate with negative a. For
the simultaneous trapping of both positive and negative ions the
smaller region, with positive a, must be used. Although the stability

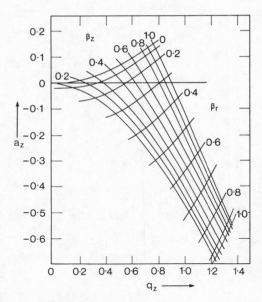

Fig. 19. Calculated a(q) stability diagram for quadrupole ion
storage trap, with iso-β lines.

regions in a, q space are obtained mathematically, they are found to
hold with high precision for quadrupole mass filters, and may be
investigated experimentally for an ion trap, using an accurate mass
filter. It is found (Todd 1981) that distortion of the stable region
can occur, an example being shown in figure 20; space charge repulsion, thermal velocities of ions and geometrical imperfections in
the electrodes all contribute to such distortion.

Current practice with quadrupole mass filters is to tune for
mass number at constant a/q (i.e. U/V). However with traps, other
modes of operation are possible, and operation has been achieved with
$a = 0$, that is with unbiassed radio-frequency, and $0 \leqslant q_z \leqslant 0.91$.

The dynamics of ion trapping have in recent years been studied
by matrix methods and phase space dynamics (Baril and Septier 1974;
Waldren and Todd 1978). These studies provide information about the
pulse extraction techniques for monitoring trap number densities.
The maximum possible excursion of an ion, u_{max}, occurs at a phase
angle γ such that $\dot{u}(\gamma) = 0$, and all components of the motion reach
their maxima in phase. This condition is applied to the general
solution of the Mathieu equation in the form

$$u(\gamma) = \alpha' \sum_{n=-\infty}^{\infty} C_{2n} \cos(2n + \beta) + \alpha'' \sum_{n=-\infty}^{\infty} C_{2n} \sin(2n + \beta)\gamma \qquad (72)$$

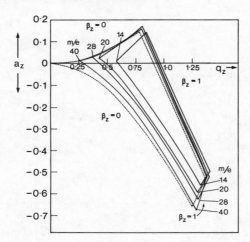

Fig. 20. Experimental stabilitv diagrams for $^{14}N^+$, $^{20}Ne^+$, $^{28}N_+^+$ and $^{40}Ar^+$ stored for 1 ms at 0.83 MHz. Broken line - calculated stability diagram.

where α' and α'' are parameters which depend on the conditions at ion formation, and which are determinable by matrix algebra methods, or from the Wronskian determinant W.

The application yields

$$u_{max} = (\alpha'^2 + \alpha''^2)^{\frac{1}{2}} \sum_{n=-\infty}^{\infty} |C_{2n}| \tag{73}$$

where

$$\alpha' = \{u(\gamma_0)\dot{u}_2(\gamma_0) - \dot{u}(\gamma_0)u_2(\gamma_0)\}/W \tag{74}$$

$$\alpha'' = \{u(\gamma_0)\dot{u}_1(\gamma_0) - \dot{u}(\gamma_0)u_1(\gamma_0)\}/W$$

so that

$$\frac{W^2 u_{max}^2}{|\Sigma C_{2n}|^2} = u^2(\gamma_0)\left[\dot{u}_1^2(\gamma_0) + \dot{u}_2^2(\gamma_0)\right] + \dot{u}^2(\gamma_0)\left[u_1^2(\gamma_0) + u_2^2(\gamma_0)\right]$$
$$- 2u(\gamma_0)\dot{u}(\gamma_0)\left[u(\gamma_0)\dot{u}_1(\gamma_0) + u_2(\gamma_0)\dot{u}_2(\gamma_0)\right] \tag{75}$$

This equation represents an elliptical relationship between $u(\gamma_0)$ and $\dot{u}(\gamma_0)$. For a given phase angle γ_0 of entry of the ion into the radio-frequency field there are pairs of initial coordinates $(u,\dot{u})$ which lie on an ellipse characteristic of the phase angle, and which

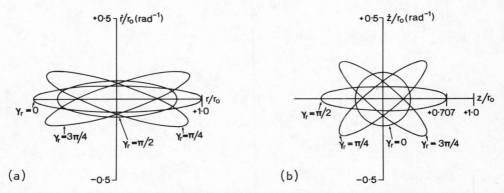

Fig. 21. Phase-space ellipses for quadrupole ion trap operating at
$a_z = 0$, $q_z = 0.64$. (a) with $\beta_r = 0.23$; (b) with $\beta_z = 0.50$;
ellipses scaled according to $z_0 = 0.707r_0$. All phases are
quoted in terms of γ_r.

give rise to stable trajectories having at some phase angle γ a
displacement equal to u_{max}.

The motion of ions may be represented as a family of ellipses
with a common centre (figure 21), each ellipse having the same area
but different orientation. The two families in figure 21 represent
respectively the radial and axial motions, and the phase angles γ_r
refer to the canonical time variable in the Mathieu equation for the
r direction, which is related to the phase angle of the radio-
frequency drive potential by

$$\gamma_r = \frac{\omega_r t}{2} - \frac{\pi t}{4} \tag{76}$$

When the maximum displacement is equal to r_0 or to z_0 $(0.707r_0)$,
the emittance ε (area of the ellipse) is equal to the acceptance, and
ions will only be trapped if the $(r, \dot{r})$ and $(z, \dot{z})$ coordinates lie on
or within the respective ellipse at every phase angle. The results
of experiments with phase-synchronized ejection pulses can be
explained in terms of these concepts. The z motion determines the
arrival time of the ions and the r motion determines whether, on
application of the ejection pulse, the ions are directed towards or
away from the exit hole in the end-cap. At $\gamma_r = 3\pi/4$ most ions lie
within the quadrants $(+r, +\dot{r})$ and $(-r, -\dot{r})$ so that a z impulse will
cause loss of ions, whereas at $\gamma_r = \pi/4$ there is a focussing effect,
so that ions can pass through the hole in the end-cap.

The phase-space technique can also be used to calculate mean
kinetic energies of the trapped ions (Dawson 1976b); the technique has

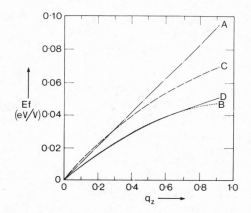

Fig. 22.   Energy factors Ef as functions of q_z. A - ion oscillating
          in pseudo-potential well; B - smoothed general solution of
          Mathieu equation; C - average r.m.s. velocity at each phase
          angle, on phase-space model; D - energy averaged over all
          q_z, on phase space model.

been compared (Todd et al., 1980a) with previous calculations, in which
pseudo-potentials were used, and also with smoothed versions of the
general solution of the Mathieu equation (Todd et al. 1976; Lawson et
al. 1975).   It is seen from the graphical comparison of 'energy
factors' in figure 22 that superior agreement is obtained between
the phase-space and the smoothed general solution techniques.   The
former results have been curve-fitted to give

$$\left\langle E_{max} \right\rangle = V_0 \times \text{energy factor} \tag{77}$$

$$= 0.3059\ V_0 \ln(1.662q_z + 1)$$

$$\left\langle E_{rms} \right\rangle = 0.0765\ V_0 \ln(1.662q_q + 1) \tag{78}$$

$$\left\langle E_{av} \right\rangle = 0.0551\ V_0 \ln(1.662q_z + 1) \tag{79}$$

Some Doppler shift investigations of trapped ion energies have been
reported (Ifflander and Werth 1977).

    The phase-space approach has also been applied (Todd et al.
1980b) to the calculation of spatial ion distributions in the trap;
contour diagrams have been obtained of the form shown in figure 23.
The validity of this approach is demonstrated by the good correspon-
ence between calculated and measured ejected ion currents (figure 24)
taken as a function of q_z.

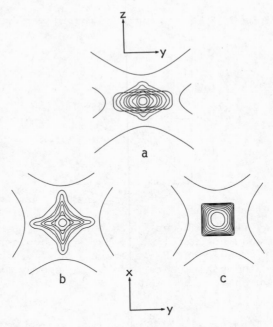

Fig. 23.   Contour diagrams of ion distribution in quadrupole ion
            store.   a – time-averaged, r-z plane, $a_z = -0.667$,
            $q_z = 1.236$; b – time-averaged distribution, $a_z = 0$,
            $q_z = 0.9$; c – quadrupole mass filter, $a = 0$, $q = 0.9$.

        The radio-frequency trap can be operated in a mass-specific mode
by working sufficiently close to the peak of the a(q) graph.   The
trapped ion densities may be monitored either by pulsed extraction
or in the resonance absorption mode; as with the Penning trap, the
non-destructive monitoring of ion densities within the radio-
frequency trap can be carried out by measuring the peak in the noise
spectrum; alternatively, the damping of applied radiofrequency
potential at an appropriate orbiting frequency is measured.   These
angular frequencies

$$\omega_d = \beta_z \omega/2 = \beta_r \omega/2 \tag{80}$$

may be calculated for the appropriate point on the a(q) graph given
in figure 19.

        One technique (Fischer 1959) has been to operate on a $\beta = 0.6$
line and measure the power absorbed from an auxiliary oscillator
connected between the end caps.   Another (Rettinghaus 1967) has been
to operate at $\beta = 0.5$ and use a sensitive frequency-tuned detector
in place of the auxiliary oscillator.

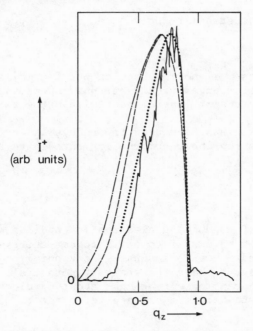

Fig. 24.   Measured ejected $^{84}Kr^+$ ion current as function of q_z,
           compared with various calculations (Dawson 1976a).

    One of the difficulties associated with the quadrupole ion store
is that of its construction. It is necessary to machine and align
the hyperboloids with precision; even when they are built, the moun-
ting can still pose problems of pumping speed within the trap. In
order to surmount such problems, quadrupole ion stores made of metal
mesh have been developed.

    An original approach has been the development of the radio-
frequency cylindrical trap (Fulford et al. 1980), consisting of two
plates at either end of a cylinder. Hemispherical end plates have
also been used.

    The possibility of injection into a quadrupole ion storage trap
was the subject of a claim in a patent submitted by Dawson and Whetten
(1970), but it is only recently that computational analysis has been
made of phase-synchronized pulsed injection (Nand Kishore and Ghosh
1976). Phase-space dynamics have been used to demonstrate the
optimum phase and geometry for injection. For the conditions investigated, radial injection appears to be the most advantageous mode.
The limiting factor for axial injection is the small range of displacement u_z for a given value of the velocity $\dot{u}_z$. In the case of
radial injection, values of u_z are small and $\dot{u}_z$ is close to zero;
the orientations and shapes of the phase-space ellipses for the

z-component limit the stability only at an initial phase angle of
$\gamma = \frac{3\pi}{8}$.

Detailed studies must now proceed on the feasibility of producing populations of trapped highly stripped species, whether by
injection or by electron or laser impact within the trap.

Photodissociation has been used to produce $T\ell^+$ and I^- ions from
thallium iodide; the ion-ion recombination rates have been measured
(Major and Schermann 1978).

6. CONCLUSIONS

The necessity of conducting electron-ion and ion-ion collision
studies with the ions relaxed into their lowest energy levels ensures
a future for confined ion experiments. The Penning and radio-
frequency traps appear to have the greatest potential. However, an
understanding of ion confinement brought about by electron space
charge would appear to be of importance for other types of
experimentation.

7. REFERENCES

Baker, F. A. and Hasted, J. B., 1966, Phil. Trans. Roy. Soc. A,
        261:3
Baril, M. and Septier, A., 1974, Rev. Phys. Appl., 9:525.
Byrne, J. and Farago, P. S., 1972, Proc. Phys. Soc. B, 86:2320.
Cuthbert, J., Farren, J., Prahallada Rao, B. S. and Preece, E. R.,
        1966, Proc. Phys. Soc., 88:91.
Daly, N. R. and Powell, R. E., 1967, Phys. Rev. Lett., 19:1165.
Dance, D. F., Harrison, M. F. A., Rundel, R. D. and Smith, A. C. H.,
        1967, Proc. Phys. Soc., 92:577.
Dawson, P. H., in: "Quadrupole Mass Spectrometry and Its Applica-
        tions," P. H. Dawson, ed., Elsevier, Amsterdam (1976a).
Dawson, P. H., 1976b, Int. J. Mass Spect. Ion Phys., 20:237.
Dawson, P. H. and Whetten, N. R., 1970, US Pat. 3521939.
Dehmelt, H. G., 1967, Adv. Atom. Mol. Phys., 3:53.
Dehmelt, H. G., 1969, Adv. Atom. Mol. Phys., 5:109.
Dehmelt, H. G. and Walls, F. L., 1968, Phys. Rev. Lett., 21:127.
Dolder, K. T., Harrison, M. F. A. and Thonemann, P., 1961, Proc.
        Roy. Soc. A, 264:367.
Dolder, K. T., Harrison, M. F. A. and Thonemann, P., 1963, Proc.
        Roy. Soc. A, 274:546.
Dubois, R. D., Jeffries, J. B. and Dunn, G. H., 1978, Phys. Rev. A,
        17:1314.
Dunn, G. H. and van Zyl, B., 1967, Phys. Rev., 154:40.
Fischer, E., 1959, Z. Physik, 156:26.

Fulford, J. E., March, R. E., Mather, R. E., Todd, J. F. J. and
    Waldren, R. M., 1980, Can. J. Spect., 25:85.
Hamdan, M., Birkinshaw, K. and Hasted, J. B., 1978, J. Phys. B,
    11:331.
Harris, L. A., 1959, J. Appl. Phys., 30:826.
Hartnagel, H., 1964a, Proc. Inst. Elec. Eng., 111:57.
Hartnagel, H., 1964b, Proc. Inst. Elec. Eng., 111:1821.
Hartnagel, H., 1965a, Int. J. Electronics, 18:431.
Hartnagel, H., 1965b, Proc. Inst. Elec. Eng., 112:1015.
Hartnagel, H., 1966, Int. J. Electronics, 21:277.
Hasted, J. B. and Awad, G. L., 1972, J. Phys. B, 5:1719.
Heppner, R. A., Walls, F. L., Armstrong, W. T. and Dunn, G. H., 1976,
    Phys. Rev. A, 13:1000.
Hornsby, J., 1963, CERN Report, 63-7.
Iffländer, R. and Werth, G., 1977, Metrologia, 13:167.
Jeffries, J. B., 1981, Ph.D. Thesis, University of Colorado,
    University Microfilms, Ann Arbor, No. 8113973.
Lawson, G., Todd, J. F. J. and Bonner, R. F., "Dynamic Mass
    Spectrometry," vol. 4, p. 39, D. Price and J. F. J. Todd,
    eds., Heyden, London (1975).
Lloyd, O., 1965, Private Communication, UKAERE Harwell, quoted in
    Baker and Hasted, 1966.
McGuire, M. D., Petsch, R. and Werth, G., 1978, Phys. Rev. A, 17:1999.
Major, F. G. and Dehmelt, H. G., 1968, Phys. Rev., 170:91.
Major, F. G. and Schermann, J. P., 1978, Appl. Phys., 16:225.
Mathur, D., Hasted, J. B. and Khan, S. U., 1979, J. Phys. B, 12:2043.
Mathur, D., Khan, S. U. and Hasted, J. B., 1978, J. Phys. B, 11:3615.
Nand Kishore, M. and Ghosh, P. K., 1976, Int. J. Mass Spect. Ion
    Phys., 29:345.
Paul, W. and Raether, M., 1955, Zeit. Phys., 140:2621.
Paul, W., Reinhardt, H. P. and von Zahn, U., 1958, Zeit. Phys., 152:
    143.
Paul, W. and Steinwedel, H., 1953, Zeit. Naturf., 8a:448.
Peart, B. and Dolder, K. T., 1974, J. Phys. B, 7:236.
Peek, J. M., 1967, Phys. Rev., 154:52.
Pierce, J. R., "Theory and Design of Electron Beams," Van Nostrand,
    Princeton (1954).
Redhead, P. A., 1967, Can. J. Phys., 45:1791.
Redhead, P. A., 1969, Can. J. Phys., 47:2449.
Rettinghaus, G., 1967, Z. angew. Phys., 22:321.
Schuessler, H. A., in "Progress in Atomic Spectroscopy," Part B:799,
    Manle and Kleinpoppen, eds., Plenum, New York (1979).
Schulz, G. J., 1957, Rev. Sci. Inst., 28:105.
Spitzer, L., "Physics of Fully Ionized Gases," 2nd edn., Ch. 5,
    Interscience, New York (1962).
Todd, J. F. J. in "Dynamic Mass Spectrometry," vol. 6, D. Price and
    J. F. J. Todd, eds., Heyden, London (1981).
Todd, J. F. J., Lawson, G. and Bonner, R. F., in "Quadrupole Mass
    Spectrometry and Its Applications," p. 181, P. H. Dawson, ed.,
    Elsevier, Amsterdam (1976).

Todd, J. F. J., Waldren, R. M. and Bonner, R. F., 1980a, Int. J. Mass
        Spect. Ion Phys., 34:17.
Todd, J. F. J., Waldren, R. M., Freer, D. A. and Turner, R. B., 1980b,
        Int. J. Mass Spect. Ion Phys., 35:107.
Waldren, R. M. and Todd, J. F. J., "Dynamic Mass Spectrometry," vol.
        5, p. 14, D. Price and J. F. J. Todd, eds., Heyden, London
        (1978).
Walls, F. L. and Dunn, G. H., 1974, J. Geophys. Res., 79:1911.
Woodruff, P. R., Hublet, M.-C. and Harrison, M. F. A., 1978, J. Phys.
        B, 11:L305.
Wuerker, R. F., Shelton, H. and Langmuir, R. V., 1959, J. Appl. Phys.,
        30:342.
Wynter, A. P., 1974, Ph.D. Thesis, University of London.
Wynter, A. P. and Hasted, J. B., J. Phys. E, 7:627.

STUDIES OF ION-ION RECOMBINATION USING FLOWING AFTERGLOW PLASMAS

David Smith and Nigel G. Adams

Department of Space Research
University of Birmingham
Birmingham B15 2TT, England

1. INTRODUCTION

Binary positive-ion negative-ion mutual neutralization viz:

$$A^+ + B^- \longrightarrow C + D \tag{1}$$

can be an important loss process for ionization in low pressure
ionized gases containing electron attaching species, for example
oxygen, oxides of nitrogen, halogens, etc. It is therefore potentially
important in many familiar plasma media such as flames, gas lasers
and the Earth's ionosphere. Because of the Coulombic nature of these
interactions the cross-sections, σ_i, for the process at low energies
can be very large ($\gtrsim 10^{-12} \text{cm}^2$) but as we shall see σ_i varies over a
few orders of magnitude depending on the nature of the ionic species
involved in the reactions and on the interaction energy. At the onset
it should be stated that in this paper for the sake of simplicity we
use the term binary ionic recombination to describe the process
although this is clearly a misleading description.

At gas pressures which are sufficiently high such that three-body
collisions occur, the process of ternary ionic recombination viz:

$$A^+ + B^- + M \longrightarrow C + D + M \tag{2}$$

occurs and becomes increasingly dominant over the binary process as
the gas pressure (or gas number density, M) increases. In the Earth's
atmosphere, both the binary and ternary processes occur; between the
altitudes 30-60 kms (in the lower mesosphere and upper stratosphere)
the binary process dominates and below 30 kms (in the lower

stratosphere and the troposphere) the ternary process dominates. It was to fulfil the need for data for atmospheric deionization rate calculations that the ionic recombination work in our laboratory (which is summarised in this paper) was initiated. The relevance of our data to atmospheric research will be referred to from time-to-time. To date only the binary process has been studied in detail in our laboratories, largely due to the limitations of the present flowing afterglow technique used, but some preliminary data have been obtained on the ternary process and these are briefly referred to at the end of this paper.

The major objective of the paper is to describe the advances that have been made in the study of thermal energy ionic recombination using the flowing afterglow/Langmuir probe (FALP) technique which has been developed and exploited during the last few years to study ionic recombination and other low temperature plasma processes. However, to set this work in context, it is pertinent first to briefly refer to the important contributions to ionic recombination studies which were made prior to the FALP work.

The earliest studies of ionic recombination were made in weakly ionized gases at high pressures (~ 1 atmosphere) and as such were concerned with the ternary process (2) (Mächler, 1936; Sayers, 1938). No mass identifications of the ions involved in the reactions were made although it seems clear now that both the positive and negative ions were probably "cluster ions", that is consisting of a strongly bonded core ion (e.g. H_3O^+, NO_3^-, etc.) to which molecules (usually polar molecules such as H_2O) are bonded via weaker polarization forces. These cluster ions are expected to be the dominant species in ionized gases containing polar molecules and are known to be the dominant species in the lower atmosphere. We shall be concerned with the binary reactions of cluster ions such as $H_3O^+.(H_2O)_n$ and $NO_3^-.(HNO_3)_m$ in a later section of this paper. Reviews of the early work have been given by Sayers (1962) and by Flannery (1976), the latter including the development of the theoretical aspects of the subject.

Experiments carried out in collision-dominated media are concerned with the measurement of rate coefficients rather than cross-sections (these were measured in later beam experiments). Thus by observing the rate of loss of the ionization in the media the binary ionic recombination coefficient, α_2, or the ternary ionic recombination coefficient, α_3, can be obtained, since for process (1):

$$\frac{dn_{+,-}}{dt} = - \alpha_2 \, n_+ \, n_-$$

(3)

and for process (2):
$$\frac{dn_{+,-}}{dt} = - \alpha_3 \, n_+ \, n_- \, [M]$$

(4)

where n_+ and n_- are the number densities of the positive and negative
ions $[A^+]$ and $^-[B^-]$ respectively and $[M]$ is the number density of the
"inert" third body support gas. Additionally in these experiments
it is assumed that charge neutrality exists in the medium, i.e.
$n_+ = n_-$. An 'effective binary coefficient' can be determined for
process (2) as $\alpha_3 [M]$, and the earlier experiments established that
the magnitude of this varied with pressure. At about 0.1 atmosphere
it was $\sim 1 \times 10^{-6} cm^3 s^{-1}$ passing through a maximum value of $\sim 3 \times 10^{-6} cm^3 s^{-1}$
at about 1 atmosphere and then decreasing with further increase in
pressure. In later similar but more sophisticated experiments by
Mahan and Person (1964), studies were again made over a wide pressure
range (10 – 700 Torr) but extended towards lower pressures and values
of α_2 were obtained by extrapolation of the $\alpha_3 [M]$ data to the limit
of zero pressure. Values for α_2 of typically $2 \times 10^{-7} cm^3 s^{-1}$ were
obtained in several different media but as before no mass analysis was
carried out to determine the nature of the reactant ions. However,
as Mahan (1971) recognized in his excellent review of ionic recombination, it is probable that the ions were cluster ions in several cases.
This emphasised the necessity for mass identification in these and
indeed all recombination experiments, and such was incorporated in the
later experiments by Mahan and co-workers (Fisk et al, 1967), and also
in pulsed afterglow plasma experiments (Eisner and Hirsh, 1971; Hirsh
and Eisner, 1972).

With the development of the merged beam technique for ionic
recombination studies (Aberth et al, 1968; Aberth and Peterson, 1970)
a determined effort was made to study reactions involving various
combinations of atomic and molecular positive and negative ions.
Cross-sections were determined as a function of the centre-of-mass
energy of two ion beams (with high laboratory energies), and in order
to obtain thermal energy rate coefficients the cross-section data were
extrapolated to lower energies and then averaged over a Maxwellian
velocity distribution. A summary of the data obtained using this
technique is given in the review by Moseley et al (1975) in which
comparisons are made with theoretical calculations of the cross-
sections and rate coefficients for particular reactions. Good
agreement was obtained between theory and experiment for reactions
involving only atomic ions whereas discrepancies were apparent for
reactions involving molecular ions, the experimental values being
consistently larger than the theoretical predictions. Some of the
results of the merged beam and pulsed afterglow experiments will be
discussed further in relation to the FALP data.

The basic ideas in the theoretical description of binary ionic
recombination are also included in the review by Moseley et al (1975).
These ideas are based on the premise that electron transfer occurs
at avoided crossings of the reactant ionic and product covalent
potential curves for the system, and Landau-Zener theory is used to
calculate the cross-sections. Varying degrees of sophistication are
possible depending on the nature of the reactant ions and the number

of possible product neutral states. For the simplest systems, that is for some atomic systems in which the number of product states is few then detailed close coupling calculations are possible (Olson 1977). However, for molecular systems in which a large number of product states are possible then this rigorous procedure is not practicable and a less rigorous procedure has been developed (Olson, 1972) - the so-called absorbing sphere model. This approach avoids the protracted procedure of calculating transition probabilities for all curve crossings and instead defines an 'absorbing sphere' inside which reaction occurs with unit probability. The predictions of this model will be related to the FALP data below. The theory relating to the ternary process is summarised in the reviews by Mahan (1973) and Flannery (1976).

2. THE FLOWING AFTERGLOW/LANGMUIR PROBE (FALP) TECHNIQUE

The thermalised afterglow plasma is a medium in which a wide range of ionic processes can be investigated under well defined conditions. The time-resolved or pulsed (stationary) afterglow was first exploited by S.C. Brown and co-workers using microwave cavity and mass spectrometric diagnostics and this technique has since been exploited to great effect by M.A. Biondi and co-workers to study electron-ion dissociative recombination. Pulsed afterglows have also been used by H.J. Oskam and co-workers, D. Smith and co-workers and by W.C. Lineberger and co-workers to study a variety of plasma processes. Brief summaries of this early work are given in the books by McDaniel (1964) and by McDaniel and Mason (1973).

The flowing afterglow technique was first developed and exploited by Ferguson, Fehsenfeld and Schmeltekopf (1969) to study ion-molecule reactions at thermal energies. This technique and other flow tube techniques have been discussed in a recent review (Smith and Adams, 1979a). In essence, a flowing afterglow consists of a flow tube along which a carrier gas is constrained to flow by the action of a large Roots-type pump, and in which ionization is created in the upstream region by a gas discharge or some other type of ion source. An afterglow plasma is thus distributed along the flow tube and under favourable conditions the charged particle energies will be relaxed in the afterglow to those appropriate to the carrier gas temperature. Distance along the flow tube and afterglow time are coupled via the flow velocity of the plasma which can readily be determined (see Section 2.2). The flowing afterglow has the precious advantage over the stationary afterglow that reactant gases can be introduced into the thermalized plasma and thus are never exposed to the high-energy electrons in the upstream discharge/ion source. Thus unwanted molecular excitation and dissociation can be avoided and greater control can be exercised in the creation of suitable plasma media.

A standard diagnostic technique used in most flowing afterglow experiments is quantitative mass spectrometry which is achieved by positioning a differentially pumped mass spectrometer at the downstream end of the flow tube. This is the only diagnostic required in the determination of ion-molecule reaction rate coefficients but for the determination of the rate coefficients for ionic and electronic recombination, absolute charged particle number densities are required and it is for this purpose that we developed the Langmuir probe technique (described in Section 2.2) for use in afterglow plasmas. The successful application of the FALP technique to the study of ionic recombination is the culmination of several years of development and exploitation of the probe technique first in stationary afterglows and then in flowing afterglows. This work has led to an understanding of afterglow plasma processes which has proved invaluable in our efforts to create appropriate plasma conditions for ionic recombination studies and in the interpretation of FALP data. These aspects of the work will now be discussed briefly.

2.1 The Flowing Afterglow Plasma

In the FALP experiments, the carrier gas most often used is pure helium at about 1 torr pressure, and a microwave cavity discharge is created in it to establish a sufficiently high ionization density in the afterglow to permit the study of recombination loss processes above other loss processes. In these experiments, small quantities of reactant gases can be added either upstream of the discharge or downstream into the afterglow (see Fig.1). Often two different reactant gases are used simultaneously, one being introduced upstream and one downstream. For the ionic recombination studies one of the gases is invariably an electron attaching gas from which negative ions are created. Other processes which can occur in these plasmas are discussed below and it is an appreciation of their relative rôles which allows the creation of suitable plasma media to study the various individual processes. A major objective is to establish afterglow plasmas in which the loss of the particular charged particle species occurs almost solely by the process which it is desired to study.

2.1.1 <u>Ambipolar diffusion</u>. This is always a finite loss process for ions and electrons in bounded plasmas. It has been studied for its own sake (e.g. see McDaniel and Mason, 1973) and we have studied it in both stationary afterglows (e.g. Smith and Cromey, 1968; Smith and Copsey, 1968; Smith et al, 1972a), and in the FALP apparatus (Smith et al, 1975) by observing the electron density gradient along a pure helium electron-ion afterglow plasma. The time decay of the electron density, n_e, and the positive ion density, n_+, is described by:

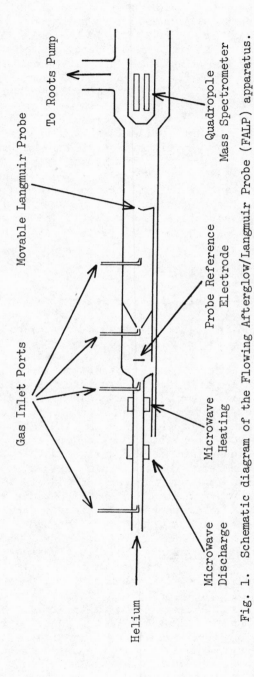

Fig. 1. Schematic diagram of the Flowing Afterglow/Langmuir Probe (FALP) apparatus.

$$\frac{dn_{+,e}}{dt} = D_a \nabla^2 n_{+,e} \tag{5}$$

The solution of equation (5) is well understood for the cylindrical geometry of the flowing afterglow (Bolden et al, 1970; Adams et al, 1975) and thus the ambipolar diffusion coefficient, D_a, can be determined. D_a is inversely proportional to the pressure, p, ($D_a p \sim$ const.) and so in order to study other charged particle loss processes, ambipolar diffusion is inhibited by operating at a suitably high pressure. Additionally $D_a p$ is directly dependent on the electron temperature and thus low electron temperatures diminish ambipolar diffusion losses.

2.1.2 <u>Electron temperature relaxation</u>. The 'hot' electrons which exist in the microwave discharge are cooled in collisions with positive ions and carrier and reactant gas atoms or molecules during their passage down the flow tube. It is important to be able to estimate the rate of cooling of the electrons so that the position in the afterglow at which they are thermalized can be determined. Under favourable conditions, the Langmuir probe can be used to measure the electron temperature (Dean et al, 1972; Smith, 1972) and this has been achieved in both stationary and flowing afterglows (Smith et al, 1972a; Dean et al, 1974). The rate of electron temperature relaxation is described by:

$$\frac{dT_e}{dt} = -\frac{1}{\tau} (T_e - T_{+,n}) \tag{6}$$

T_e, T_+ and T_n are the temperatures of the electron gas, positive ion gas and carrier gas respectively and $T_+ \approx T_n$ in these plasmas. The time-constant τ describes the net effect due to electron-ion (time-constant τ_{e+}) and electron-neutral (time-constant τ_{en}) collisions such that $\tau^{-1} = \tau_{e+}^{-1} + \tau_{en}^{-1}$. The small fractional ionization density in these afterglow plasmas ($\sim 10^{-6}$) ensures that electron-neutral collisions are the most effective. This is especially so when the neutral gas is molecular since then τ_{en} is very short (Smith and Dean, 1975; Dean and Smith, 1975). It should be noted that T_e relaxation will be inhibited in afterglows in which metastable excited species are present due to superelastic collisions of electrons with these species. These are particularly effective in pure helium afterglows in which an electron colliding with a helium metastable can gain about 20 eV of energy. Such, however, is usually unimportant when other gases are present since Penning reactions rapidly destroy the metastable atoms.

2.1.3 <u>Electron attachment</u>. Three different processes can be distinguished, two binary and one ternary (see e.g. Massey, 1976, 1979). Thermal dissociative attachment, e.g.:

$$Cl_2 + e \longrightarrow Cl^- + Cl \tag{7}$$

can occur rapidly when the electron affinity E_a of the electron
acceptor (Cl) exceeds the dissociation energy E_d of the molecule.
Non-thermal dissociative attachment, e.g.

$$O_2 + e \longrightarrow O^- + O \tag{8}$$

can occur only for electron energies in excess of $(E_d - E_a)$, which
for reaction (8) is 3.65 eV, and so this cannot occur in a thermalised afterglow. However, it can occur when oxygen is passed through
the gas discharge which generates the afterglow.

The ternary attachment process, e.g.

$$O_2 + e + He \longrightarrow O_2^- + He \tag{9}$$

is obviously promoted by high pressures of the catalysing third
body (in this case He). The rate of loss of electrons via these
processes and hence the rate of formation of negative ions is
described by the relation:

$$\frac{dn_e}{dt} = -\beta_2 \, n_e \, n_n \quad \text{or} \quad -\beta_3 \, n_e \, n_n \, [M] \tag{10}$$

for the binary and ternary processes respectively. The attachment
coefficient β_2 amd β_3 have been measured for a large number of
reactions (Christophorou, 1971) and are valuable in estimating the
concentrations of attaching gases needed to generate negative ions
in plasmas.

2.1.4 <u>Electronic and ionic recombination</u>. Dissociative
electronic recombination e.g.

$$O_2^+ + e \longrightarrow O + O \tag{11}$$

is the only effective electronic recombination process in laboratory
thermalised afterglow plasmas. It is especially rapid when cluster
ions are involved (Biondi, 1973; Leu et al, 1973; Bates, 1979) and
can rapidly remove ionization from the plasma. The recombination
coefficient, α_e, is defined by:

$$\frac{dn_{e,+}}{dt} = -\alpha_e \, n_e \, n_+ \tag{12}$$

and is analogous to the ionic recombination coefficient α_2
defined by equation (3). Electrons and negative ions often co-exist
in plasmas and so electronic and ionic recombination will be

occurring simultaneously. However α_e is invariably much greater than α_2. Ionic recombination will be discussed in detail later.

2.1.5 <u>Ion-molecule reactions</u>. Positive ions and negative ions usually react rapidly with neutral molecules at thermal energies generating new ionic species and such reactions are exploited in the ionic recombination studies described in this paper to generate appropriate ionic species in the plasmas. Ion-molecule reactions can result from either binary or ternary collisions, e.g.:

$$O^+ + N_2 \longrightarrow NO^+ + N \tag{13}$$

$$O_3^- + CO_2 \longrightarrow CO_3^- + O_2 \tag{14}$$

$$NO^+ + H_2O + N_2 \longrightarrow NO^+.H_2O + N_2 \tag{15}$$

$$NO_3^- + H_2O + N_2 \longrightarrow NO_3^-.H_2O + N_2 \tag{16}$$

Reactions (15) and (16) are examples of association or clustering reactions and these and many other reactions are important in the Earth's atmosphere, interstellar clouds and many other plasma media (see e.g. Ferguson et al, 1979; Smith and Adams, 1980a; Dalgarno and Black, 1976; Smith and Adams, 1979b). A very large literature exists relating to reaction types, rate coefficients etc. (see for example the data compilation by Albritton, 1978) and a knowledge of this is essential when considering how to generate afterglow plasmas containing specific positive and negative ion types for recombination studies. The loss rate of a given ionic species due to, for example, ternary ion molecule reactions is described by

$$\frac{dn_{+,-}}{dt} = -kn_{+,-} n_g [M] \tag{17}$$

where k is the rate coefficient, n_g is the reactant gas concentration and [M], as before, is the concentration of the third body.

2.1.6 <u>The creation of ion-ion afterglow plasmas</u>. In order to determine ionic recombination coefficients in plasmas it is very desirable to ensure that no plasma electrons are present since dissociative electronic recombination seriously influences the loss of ionization and confuses the interpretation of the probe data These complications are eliminated by ensuring that sufficient electronegative gas is present in the plasma medium to promote rapid electron attachment. In the FALP experiments this is achieved by introducing a suitable concentration of the electronegative gas (e.g. NO_2, Cl_2, SF_6) either upstream or downstream of the gas

discharge when negative ions are formed as described in Section
2.1.3. Additionally, as the afterglow plasma passes down the flow
tube, loss of electrons occurs from the plasma via ambipolar
diffusion whereas the negative ions remain trapped in the plasma
by the ambipolar space charge field. Eventually at some position
down the flow tube the electron density decreases to a value such
that the electron Debye length in the plasma is comparable to the
flow tube radius. The ambipolar field then collapses and the
remaining electrons freely diffuse to the flow tube walls. Thus
a positive ion/negative ion plasma remains which is devoid of
electrons. We have discussed this phenomenon in detail previously
in relation to both stationary and flowing afterglow plasmas (Smith
et al, 1974; Smith and Church, 1976). The essential requirement in
the FALP measurements of ionic recombination coefficients is to
generate ion-ion plasmas consisting of appropriate ion types in
sufficient densities such that loss by ionic recombination dominates
diffusive loss. Clearly knowledge of the ionic reactions which
generate the desired ions is required. In this respect careful use
of the mass spectrometer (see Fig. 1) is essential to monitor the
complex ion chemistry which occurs and to establish the nature of
both the positive and the negative ions. Also both a sufficiently
low carrier gas pressure and the addition of sufficient electronegative gas are required so that a rapid transition to an ion-ion
plasma occurs, yet at a carrier gas pressure sufficiently high that
ionic recombination loss dominates over diffusive loss. Sometimes
these requirements are counteractive and a compromise has to be
adopted such as accepting the presence of more than one species of
positive and/or negative ion and tolerating a reduced range of n_+ and
n_- values from which to obtain α_2 (see Section 3 1).

2.2 The Langmuir Probe Diagnostic

Since their inception Langmuir (electrostatic) probes have been
used to great effect in the study of steady state discharges
(Mott-Smith and Langmuir, 1926). The theory and practice of their
operation is discussed in the book by Swift and Schwar (1970).
Prior to our work however, probes had not been used satisfactorily
to study decaying (afterglow) plasmas which are easily disturbed
by the presence of a current-collecting electrode. In essence, the
Langmuir probe is simply a small regular-shaped electrode immersed
in a plasma, the potential of the probe being swept positively or
negatively relative to the plasma (or space) potential. In
practice, plasma potential is referenced to another electrode
(see Fig. 1) of surface area much greater than that of the probe.
Thus from the electron current (i_e)vs voltage (V_p) characteristics
of an appropriately shaped probe immersed in an electron/positive-
ion plasma the electron temperature, T_e, and the electron density,
n_e, can be determined. Conventionally, n_e is determined from i_e
at plasma potential. However plasma potential cannot be

unambiguously determined accurately (Goodall and Smith, 1968) and so we have adopted the so-called orbital-limited-current mode of current collection which has the advantage that, for a cylindrical probe, the n_e values are determined from the linear slopes of i_e^2 vs V_p curves and not from a point on the i_e vs V_p curve. Thus in the electron-collecting orbital-limited-current regime;

$$i_e^2 = \frac{2A^2 n_e^2 e^2}{\pi^2 m_e} (eV_p + kT_e) \tag{18}$$

and so n_e can be determined if the probe area, A, is known. A similar expression (Smith and Plumb, 1972) can be used to deduce the positive ion density from the ion current to the probe when it is negatively biased with respect to plasma potential, although in this case the probe current is much smaller. This asymmetry in the i vs V characteristic is due to the smaller mobility of positive ions relative to electrons. The successful application of this probe technique to stationary and flowing afterglow plasmas results largely from the recognition that very small probes must be used in these decaying plasmas in order to avoid disturbing the plasma by drawing excessive current. A typical cylindrical probe used in our afterglow work is a short tungsten wire of length 0.5 cm and diameter 2.5×10^{-3} cm. These probes have been used successfully to study a variety of afterglow plasma problems and the results have been reported in the literature. For example, the probes have been used to study the phenomenon of diffusion cooling of electrons and have measured electron temperatures as low as 90 K (Smith et al, 1972a), to determine α_e for reaction (11) which is in excellent agreement with the α_e obtained using microwave diagnostics (Smith and Goodall, 1968; Plumb et al, 1972), and to follow the transition from electron-ion to ion-ion afterglow plasma (as discussed in Section 2.1.6) by monitoring the ambipolar fields within afterglow plasmas (Smith et al, 1974). A detailed appraisal of the application of the technique to the study of electron/ion afterglow plasmas has also been reported (Smith and Plumb, 1972).

This probe technique has also been adopted for use in ion-ion afterglow plasmas. In such plasmas, equation (18) is modified to:

$$i_{+,-} = \frac{2 A^2 n_{+,-}^2 e^2}{\pi^2 m_{+,-}} (eV_p + kT_{+,-}) \tag{19}$$

Here the electron mass, m_e, has been replaced by the positive ion or negative ion mass, m_+ or m_- and to determine n_+ and/or n_- then obviously m_+ and/or m_- must be known. In practice, the probe is swept both positively and negatively relative to plasma potential and so both negative and positive ion currents are collected in accordance with equation (19). The slopes of the i_+^2 vs V_p and the i_-^2 vs V_p provide a value for the mass ratio m_+/m_- of the ions in the plasma. When more than one species of positive or negative ion is present then a mean ionic mass ratio is obtained (Smith et al, 1978) and this can be a useful indicator of the ion content of the flowing afterglows in positions upstream of the mass spectrometer sampling position (Fig. 1).

All of the data obtained to date using the FALP technique have been acquired using a glass flow tube about 100 cm long and 10 cm diameter. The Langmuir probe can be located anywhere along the flow tube axis from a position some 10 cm downstream of the micro-wave discharge to about 10 cm upstream of the mass spectrometer sampling orfice, a movement of about 70 cms. Thus the axial ion and/or electron density gradients can be determined over this appreciable length, z, of the afterglow column. To relate z to afterglow time in order to determine α_2, the plasma flow velocity, v_p, is measured by pulse-modulating the microwave discharge and monitoring the passage of the plasma density disturbance along the flow tube using the probe. v_p is typically 10^4 cm s^{-1}. An experimental and theoretical appraisal of the dynamics of these afterglow plasmas has been carried out and this has been reported in detail in the literature (Adams et al, 1975). Further reference to the essential details of the FALP technique will be given as appropriate in the following sections.

3. DETERMINATION OF BINARY IONIC RECOMBINATION COEFFICIENTS: SIMPLE MOLECULAR IONS

We define here "simple" molecular ions as those which are organised by normal covalent bonds (e.g. NO^+, NH_4^+, NO_2^-, SF_6^-) in contrast to "cluster" ions (e.g. $H_3O^+ \cdot (H_2O)_n$, $NO_3^- \cdot (HNO_3)_m$). The reactions of cluster ions are considered later.

3.1 Room Temperature Measurements

Several reactions involving molecular ions have been studied using the FALP but none more thoroughly than the reaction:

$$NO^+ + NO_2^- \longrightarrow products \tag{20}$$

We will therefore briefly discuss the experimental details which were involved in the study of reaction (20) to exemplify the experimental approach (discussed in detail by Smith and Church, 1976). Pure helium was used as the carrier gas at a pressure, p_o, of typically 0.6 Torr, although p_o was varied over a limited range (0.4 to 1.0 Torr) to check for any three-body contribution to the ionic recombination loss. Nitrogen dioxide, NO_2, at a partial pressure $\sim$1 mTorr, was added to the carrier gas upstream of the microwave discharge, this being a sufficient concentration to establish an ion-ion plasma some 20 cm downstream of the discharge. The mass spectrometer indicated that NO^+ and NO_2^- were essentially the only ion species present in the downstream afterglow region. The NO^+ was formed via direct electron collisional ionization of NO_2, and via the $He^+ + NO_2$ and $He^m + NO_2$ reactions. The NO_2^- was largely formed in the reaction sequence:

$$NO_2 + \text{hot discharge electrons} \longrightarrow O^- \xrightarrow{\quad NO_2 \quad} NO_2^- + O \qquad (21)$$

The Langmuir probe was used to measure both n_+ and n_- along the ion-ion afterglow plasma column and also the plasma flow velocity, v_p. Ambipolar diffusion and binary ionic recombination are the competing loss processes for ions in the plasma. Therefore:

$$v_p \frac{dn_{+,-}}{dz} = \frac{dn_{+,-}}{dt} = D_{+,-} \nabla^2 n_{+,-} - \alpha_2 n_{+,-}^2 \qquad (22)$$

At low pressures and low n_+, n_- the diffusive term dominates the loss of ionization whereas at higher pressures and higher n_+, n_- then the recombination term dominates as is required for these studies. The requirement is that $D_{+,-}/\Lambda^2 \ll \alpha_2 n_{+,-}$ (where Λ is the characteristic diffusion length, Adams et al, 1975) and if $D_{+,-}$ and α_2 can be estimated then the pressure and ionization density required to ensure that ionic recombination is the dominant process may be estimated. For a recombination controlled plasma:

$$\left(\frac{1}{n_{+,-}}\right)_t - \left(\frac{1}{n_{+,-}}\right)_o = \alpha_2 t \qquad (23)$$

and a plot of reciprocal density versus time will be linear and α_2 can be obtained from the slope. Example plots are given in Fig. 2. Note that for the $NO^+ + NO_2^-$ reaction the initial ionization density (n_+) is about 3×10^{10} cm^{-3}, which is typical of FALP experiments, and that for large z (late afterglow times) then characteristic upcurving of the plots occurs due to the increasing importance of diffusive loss at the lower ionization densities. For the $SF_3^+ + SF_5^-$ reaction, $(n_{+,-})_o$ is significantly larger this being a manifestation of the more rapid electron attachment in plasmas containing SF_6. Note also the close equivalence of the derived n_+ and n_- indicating the consistency of the probe technique in the positive ion and negative ion collecting modes and that the correct ionic masses were used to calculate n_+ and n_-. Since the probe data provides a value of the mass ratio, m_+/m_-, then in circumstances where the mass of only one ionic species can be ascertained then that of the other can be deduced. This approach

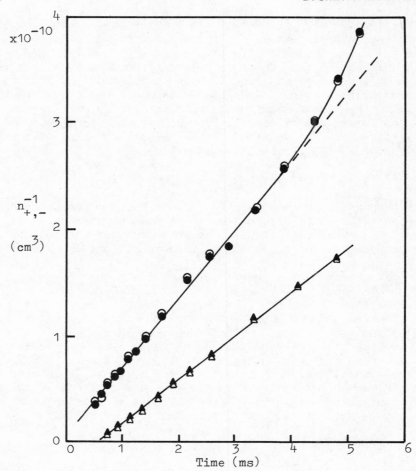

Fig. 2.   Reciprocal positive ion densities (filled symbols)
          and reciprocal negative ion densities (open symbols)
          as a function of afterglow time for the reaction
          $NO^+ + NO_2^-$ (circles) at 0.5 Torr and the reaction
          $SF_3^+ + SF_5^-$ (triangles) at 0.7 Torr at room temper-
          ature.   The binary ion-ion recombination coefficients
          are deduced from the slopes of the linear portions
          of the curves and the upcurving of the $NO^+ + NO_2^-$
          data at late afterglow times is a manifestation of
          the increasing importance of diffusion losses at the
          lower ion densities.   The time scale has an arbitrary
          zero.   Note the higher initial density in the $SF_3^+ +$
          SF_5^- study due to rapid electron attachment in SF_6.

is especially valuable in the determination of the mean ionic mass
when more than one species of positive or negative ion is present
in the plasma (Smith et al, 1978), and also can act as a confirmation
that the ion content of the plasma as determined by the downstream
mass spectrometer also applies to the upstream recombination
controlled reaction region.

The value of α_2 for reaction (20), i.e. $\alpha_2(20)$, at 300 K
was measured to be $(6.4 \pm 0.7) \times 10^{-8}$ $cm^3 s^{-1}$ (Smith and Church,
1976). The uncertainty in α_2 is largely due to the uncertainty in
the probe area ($\sim 10\%$) and to smaller errors in v_p and in the best
fit slope to the reciprocal density plots. The first measurement of
α_2 was perplexing in that it was very much smaller than both the
value of $(51 \pm 15) \times 10^{-8}$ $cm^3 s^{-1}$ deduced from merged beam data
(Peterson et al, 1971) and the value of $(17.5 \pm 6) \times 10^{-8}$ $cm^3 s^{-1}$
obtained from stationary afterglow data (Eisner and Hirsh, 1971) yet
within a factor of two of the absorbing sphere theoretical upper-
limit estimate of $(12 \pm 3) \times 10^{-8}$ $cm^3 s^{-1}$ (Olson 1972). Clearly
further work was necessary to resolve such serious discrepancies.

The reaction:

$$NO^+ + NO_3^- \longrightarrow products \qquad\qquad (24)$$

was also studied in the FALP (Smith and Church, 1976). The
conversion of NO_2^- to NO_3^- in the afterglow was achieved simply by
increasing the NO_2^- partial pressure to about 10 mTorr which
promoted the reaction of NO_2^- with the HNO_3 impurity in the NO_2
(Fehsenfeld et al, 1975):

$$NO_2^- + HNO_3 \longrightarrow NO_3^- + HNO_2 \qquad\qquad (25)$$

without changing the identity of the positive ion, NO^+ (further
increase in the NO_2 partial pressure resulted in the production of
cluster ions, see below). Thus α_2 (24) was determined under very
similar conditions (pressure, flow velocity etc.) to $\alpha_2(20)$ and
was found to be $(5.7 \pm 0.6) \times 10^{-8}$ $cm^3 s^{-1}$ at 300 K again much
smaller than the merged beam value $(81 \pm 23) \times 10^{-8}$ $cm^3 s^{-1}$,
Aberth et al, 1971) and within a factor of two of the absorbing
sphere theoretical estimate $((11 \pm 3) \times 10^{-8}$ $cm^3 s^{-1})$. It is
however somewhat larger than the stationary afterglow value
$((3.4 \pm 1.2) \times 10^{-8}$ $cm^3 s^{-1})$.

What are the causes of these large discrepancies? Little
comment can be made on the stationary afterglow results since a
detailed critique of them has not been published. However it can
be said that since these measurements were carried out at
pressures of $\gtrsim 10$ Torr then a three-body contribution to α_2 is to
be expected and such was not accounted for. Our preliminary data

TABLE 1

Compilation of the binary ion-ion recombination coefficients, α_2, for simple molecular ions as determined in the FALP apparatus at room temperature. Also included are the average cross-sections, $\overline{\sigma}_2$, calculated by dividing α_2 by the mean relative velocity $\overline{v}_r \ (=(8kT/\pi\mu)^{\frac{1}{2}})$ of the reactant ions. In cases where there is more than one reactant positive ion their percentage concentrations are given in parentheses.

Positive Ions	Negative Ions	$\alpha_2 \times 10^8 \mathrm{cm}^3 \mathrm{s}^{-1}$	$\overline{\sigma}_2 \times 10^{12} \mathrm{cm}^2$	References
NO^+	NO_2^-	6.4	1.1	Smith and Church (1976)
NO^+	NO_3^-	5.7	1.0	
$CCl_3^+(80), CCl_2^+(20)$	Cl^-	4.5	0.9	
$CClF_2^+(90), CCl_2F^+(10)$	Cl^-	4.1	0.8	
SF_3^+	SF_5^-	4.0	1.2	Church and Smith (1977)
SF_5^+	SF_6^-	3.9	1.3	
NH_4^+	Cl^-	6.7	0.9	Smith et al (1978)
$NO^+(70), NH_4^+(30)$	NO_2^-	6.3	1.0	
Cl_2^+	Cl^-	5.0	1.0	Church and Smith (1978)
O_2^+	CO_3^-	9.5	1.7	
CF_3^+	F^-	5.8	0.9	Unpublished data
NF_2^+	F^-	7.5	1.1	
N_2F^+	F^-	8.5	1.2	

on collision-enhanced recombination which is referred to later
(reaction (33), Section 7) shows that at about 10 Torr in helium,
the effective binary recombination coefficient is about a factor
of two above the low pressure value, $\alpha_2(20)$, and that it could be
even greater for a nitrogen third body (as was used in the
stationary afterglow experiments). Thus the large stationary
afterglow value for $\alpha_2(20)$ can be accounted for, but the low value
for $\alpha_2(24)$ is difficult to understand (indeed it would be further
reduced by accounting for the collision-enhanced contribution).
We can only tentatively suggest that not all of the NO_3^- production
processes had been accounted for in the analysis and that perhaps
a further reaction such as reaction (25) was also occurring in
the afterglow.

The discrepancy between the FALP and merged beam data almost
certainly lies in the differing nature of the techniques. Whilst
the flowing afterglow is a collision dominated medium in which the
internal states of the reactant ions are most probably thermalised,
this may not be the case for the ions in the merged beams.
Moseley et al (1975) have discussed the possibility of excitation
in the NO^-, NO_2^- and NO_3^- ions and also have shown by experiment
that no appreciable fraction of the NO^+ is excited to the metastable
state ($a^3\Sigma$). They concluded that a possible reason for their
surprisingly large values of $\alpha_2(20)$ and $\alpha_2(24)$ may be the presence
of internal excitation of the NO_2^- and NO_3^- which, in effect, lowers
the electron affinities of the NO_2 and NO_3 or, equivalently, reduces
the electron detachment energy, D_E^-, of the negative ions (for
example NO_3^- may be in the peroxy form viz. $O^-.NO_2$). According
to theory (Olson, 1972; Hickman, 1979) this would lead to an
enhancement of α_2 for these reactions. Unfortunately there are
no other reactions which have been studied in both the FALP and the
merged beam experiments from which to search for further insight
into the reasons for these discrepancies.

α_2 for several other reactions involving simple molecular ions
have been determined with the FALP and these are listed in Table 1.
Note the small variation in α_2. Note also that data have been
obtained from plasmas in which no single positive ion is dominant.
In such cases an average α_2 can be obtained provided that the
n_- vs time plots are linear, n_- being calculated using m_-, the
mass of the single negative ion species. The mean positive ion mass
can be obtained from the ion mass ratio measured by the probe and
compared with that obtained from the mass spectrometer data.

It is in the nature of the present FALP technique that it is
only practicable to study reactions involving negative ions with
relatively large values of D_E such as those in Table 1. The
absorbing sphere theory predicts that α_2 will be largest for
reactions involving negative ions with small D_E, and Hickman (1979)

has shown from such theoretical considerations that α_2 is expected to vary as $\mu^{-\frac{1}{2}}D_E^\lambda$, where μ is the reduced mass of the ion pair. From a consideration of the merged beam and FALP data he has suggested that the best fit to the data is when $\lambda \approx -0.4$ although when all of the experimental data are considered, a dependence of α_2 on D_E is not obvious. It remains an important objective to develop the FALP technique in order to study reactions of ions with small D_E.

The small spread in α_2 for the reactions listed in Table 1 is expected from the absorbing sphere model because of the multiplicity of possible product states in all of these reactions. Thus α_2 at 300 K is seen to lie within the narrow range $(4 - 10) \times 10^{-8} cm^3 s^{-1}$ for these simple molecular ion reactions and also, somewhat surprisingly, for the cluster ion reactions discussed in Section 4 (see also Table 2). Also included in Table 1 are the equivalent mean thermal cross-sections $\overline{\sigma_2}$ $(= \alpha_2/\overline{v})$ for the reactions for direct comparison with beam measurements. Note the very large values of $\overline{\sigma_2}$ $(\sim 10^{-12} cm^2)$. Note also that the few reactions which have been studied which evolve only atomic ions have much smaller α_2 values consistent with the availability of fewer product states (Section 5).

3.2 Temperature Dependence Studies

Only a limited amount of FALP data have been obtained to date on the temperature dependence of α_2 (Smith and Church, 1977). These relate to only two reactions i.e. reaction (20) and also the reaction:

$$NH_4^+ + Cl^- \longrightarrow \text{products} \tag{26}$$

Data below room temperature were obtained by surrounding the flow tube with solid carbon dioxide and above room temperature by surrounding the flow tube with an oven. The maximum temperature range covered was 180 K to 530 K. During the low temperature measurements great care had to be taken to ensure that as little of the reactant gases as possible were used to prevent appreciable clustering of the positive and negative ions. This was especially the case when studying reaction (26) since $NH_4^+.(NH_3)_n$ clusters can form rapidly at low temperatures. The data obtained for reactions (20) and (26) can be described by a simple power law relationship of the form $\alpha_2 \sim A_1 T^{-n}$ or alternatively as $\overline{\sigma_2} \sim A_2 T^{-(n+0.5)}$. For reaction (20):

$$\alpha_2 = 6.8\ T^{-0.4} \times 10^{-7} cm^3 s^{-1} \tag{27}$$

$$\overline{\sigma_2} = 2.0\ T^{-0.9} \times 10^{-10} cm^2 \tag{28}$$

The absorbing sphere model (Olson, 1972) predicts that $\alpha_2 \sim T^{-\frac{1}{2}}$ and $\overline{\sigma_2} \sim T^{-1}$ in the low temperature regime. Thus the FALP data for reaction (20) are quite consistent with these predictions. Also the data for reaction (26), whilst being more limited, are not inconsistent with the theoretical predictions.

The satisfactory agreement between the FALP data and theory, both in the magnitude and the temperature dependence of α_2, lends credence to both theory and experiment. Since for many reactions α_2 can be determined at 300 K but not at lower temperatures it is extremely useful to be able to assume that α_2 varies as $T^{-\frac{1}{2}}$ and thus to obtain estimates of α_2 at low temperatures. We have adopted the $T^{-\frac{1}{2}}$ dependence to estimate appropriate values of α_2 for both simple and cluster ion reactions which occur in the Earth's atmosphere (Smith and Church, 1977). A variable-temperature FALP recently constructed in our laboratories which can operate readily over a temperature range of 80 K to 600 K will be used to provide further information on the variation of α_2 with temperature for other reactions.

4. DETERMINATION OF BINARY IONIC RECOMBINATION COEFFICIENTS: CLUSTER IONS

Cluster ions are important components of ionized gases at pressures sufficiently high to promote ternary association reactions and especially important when polar molecules are present. Cluster positive and negative ions are the only ion types present in the lower atmosphere and this was a prime motivation for determining the ionic recombination coefficients for cluster ion reactions in the FALP apparatus. Additional motivation followed from the suggestion by F.T. Smith and his co-workers (Smith et al, 1973; Bennett et al, 1974) that recombination of cluster ions containing an appreciable number of clustered molecules, n, could be much more efficient than that of the simple 'core' ions alone. Bennett et al (1974) have pointed out that if n is large enough then the clustered ions would be more stable than the product neutrals and consequently that recombination via electron transitions near crossings of potential curves could not occur. Instead contact or hard-impact collisions would occur followed by coalescence of the ion pair and 'boiling-off' of neutral molecules. Based on this model, an expression was derived for the lower limit to α_2 which contained a parameter R_C, the sum of the effective radii of the ions, analogous to the radius of the absorbing sphere, R_T, in Olson's model. We will discuss this model briefly in relation to the FALP data below. F.T. Smith et al (1973) also introduced the concept of 'tidal trapping' of the ions in which they envisaged the excitation of internal modes of vibration and rotation of the ions as they orbit each other in the Coulombic field. Thus kinetic energy of motion is converted into internal energy and the ions can become bound this

TABLE 2

Compilation of binary ion-ion recombination coefficients, α_2, for cluster ions as determined in the FALP apparatus at room temperature. For other information concerning nomenclature refer to the caption to Table 1. The range of $\overline{\sigma_2}$ values for reactions involving more than one ionic species of either sign is a reflection of the different masses of these ions.

Positive Ions	Negative Ions	$\alpha_2 \times 10^8 cm^3 s^{-1}$	$\overline{\sigma_2} \times 10^{12} cm^2$	Reference
$H_3O^+(H_2O)_3$	NO_3^-	5.5	1.3	Smith et al (1976)
$H_3O^+(H_2O)_3$	$NO_3^-\cdot HNO_3$	5.7	1.5	
$NO^+(NO_2)_2$	$NO_3^-(HNO_3)_3$	4.5⁺	2.1⁺	Smith and Church (1977)
$NH_4^+(NH_3)_2$	Cl^-	7.9	1.4	
$NH_4^+(NH_3)_2$	NO_2^-	4.9	0.9	
$NH_4^+(NH_3)_2(67)$, $NH_4^+(NH_3)_3(33)$	NO_2^-	5.5	1.1-1.3	
$NH_4^+(31),NO^+(25)$, $NH_4^+NH_3(25)$, $NH_4^+(NH_3)_2(19)$	$NO_2^-(47),NO_3^-(53)$	9.6	1.3-2.0	Smith et al (1978)
$NH_4^+(16),NO^+(16),NO_2^+(19)$, $NH_4^+NH_3(23)$, $NH_4^+(NH_3)_2(26)$	$NO_3^-(19)$, $NO_3^-\cdot HNO_3(38)$, $NO_3^-(HNO_3)_2(43)$	5.8	0.9-1.5	
$NH_4^+NH_3(27)$, $NH_4^+(NH_3)_2(33)$, $NH_4^+(NH_3)_3(40)$	$NO_3^-(30)$, $NO_3^-\cdot HNO_3(33)$, $NO_3^-(HNO_3)_2(37)$	6.1	1.2-1.7	
$H_3O^+(H_2O)_3$	$HSO_4^-(30),NO_3^-\cdot H_2O(30)$, $NO_3^-\cdot HNO_3(15)>140amu(25)$	5.9	1.4-1.6	
$H_3O^+(H_2O)_3$	$HSO_4^-(35),HSO_3^-\cdot H_2O(25)$, $HSO_2^-\cdot H_2O(20),SO_3^-(20)$	6.6	1.6-1.7	
$H_3O^+(H_2O)_3$	$Cl^-(>95)$	6.8	1.3	Smith et al (1981)
$H^+(H_2O)\cdot(CH_3CN)_3$	$NO_3^-(65),NO_2^-(35)$	6.3	1.5-1.7	
$H^+(H_2O)\cdot(CH_3CN)_3$	$NO_3^-\cdot HNO_3(60),NO_3^-(40)$	5.9	1.5-1.9	
$H^+(H_2O)\cdot(CH_3CN)_3$	$NO_3^-\cdot HNO_3(40),HSO_4^-(30)$, $>140amu(30)$	5.9	1.8-2.0	
$H^+(H_2O)\cdot(CH_3CN)_3$	$HSO_4^-(30),SO_3^-(30)$, $HSO_2^-\cdot H_2O(25)$, $HSO_3^-\cdot H_2O(15)$	5.8	1.7-1.8	
$NO^+(>90)$	$NO^-\cdot SO_2(40),NO_3^-\cdot SO_2(35)$, $>140amu(25)$	6.5	1.2-1.3	

⁺ the α_2 and $\overline{\sigma_2}$ values quoted are the values at 182 K. When calculated for a temperature of 300 K, assuming that $\alpha_2 \sim T^{-\frac{1}{2}}$ and $\overline{\sigma_2} \sim T^{-1}$ (See section 3.2), they become 3.5 x $10^{-8}cm^3s^{-1}$ and 1.2 x $10^{-12}cm^2$ respectively.

ultimately resulting in recombination. This process can take place irrespective of whether or not the ion pair is more stable than the neutrals, but the magnitude of α_2 depends on the probability that trapping occurs which in turn depends in an uncertain way on n but is expected to be appreciable only for $n \gtrsim 6$. An upper limit to α_2 has been formulated from this model for water cluster ions as $5 \times 10^{-7} (300/T) (6/n)^2 cm^3 s^{-1}$.

The experimental determination of α_2 for cluster ion reactions is more challenging experimentally since it is difficult to create ion-ion plasmas of sufficient ionization density containing a single species of either cluster positive or negative ions. However this has been achieved in several cases; notably ion-ion plasmas have been created containing $H_3O^+.(H_2O)_3$ as the only positive ion species as well as plasmas containing only the ammonia cluster ions $NH_4^+.(NH_3)_{1,2,3}$ together with single negative ion species both simple and clustered (Smith et al, 1976, 1978). It is fortunate that the $H_3O^+.(H_2O)_3$ ion is a stable terminating positive ion under flowing afterglow conditions at 300 K and that even in the presence of large concentrations of water vapour the next highest order cluster $H_3O^+.(H_2O)_4$ does not exist in significant concentrations. Thus at 300 K the equilibrium in the reaction sequence :

$$H_3O^+.(H_2O)_2 \underset{\longleftarrow}{\overset{H_2O}{\longrightarrow}} H_3O^+.(H_2O)_3 \underset{\longleftarrow}{\overset{H_2O}{\longrightarrow}} H_3O^+.(H_2O)_n \qquad (29)$$

is overwhelmingly in favour of the $H_3O^+.(H_2O)_3$. Similarly plasmas can be created which favour the production of the $NH_4^+.(NH_3)_2$ ion. Chlorine is an especially useful gas in these studies since it rapidly produces Cl^- ions yet it is not very prone to cluster to ions. Similarly NO_2 is a valuable source of NO_2^- and NO_3^- ions. Thus it has been possible to determine α_2 for the reactions of several cluster positive ion species with these simple negative ion species and these data are included in Table 2 together with our other data on cluster ion reactions, including some reactions in which both the positive and negative ion species are clustered. For those plasmas in which more than one species of positive and/or negative ion clusters co-exist, the Langmuir probe determination of the mean mass ratio m_+/m_- is a vital addition to the mass spectrometry data, since it can indicate where any change in the ion types occurs along the afterglow plasma column. Only those systems were studied seriously for which there was no such change in the ion types and for which reciprocal density plots were linear over at least about a factor of 4. Under conditions for which more than one ionic species of either charge were present then an average α_2 was obtained. The most recent data included in Table 2 relates to the reactions of some positive and negative ions which have recently been detected in the stratosphere, specifically the reactions of $H^+(H_2O)(CH_3CN)_3$ ions (Smith et al, 1981).

A glance at the α_2 values given in Table 2 shows that they are not significantly different than those for the simple molecular ion reactions (Table 1). In view of the theoretical postulates referred to above these were somewhat surprising results. The reaction involving the most heavily-clustered ions :

$$NO^+ \cdot (NO_2)_2 + NO_3^- \cdot (HNO_3)_3 \longrightarrow products \qquad (30)$$

which was studied at 182 K in order to enhance the degree of clustering, has the smallest α_2 but the largest $\overline{\sigma_2}$ (a consequence of the large reduced mass and the lower temperature). When $\overline{\sigma_2}$ for this reaction is calculated for a temperature of 300 K (assuming $\overline{\sigma_2} \sim T^{-1}$), for comparison with the other data, then it is not appreciably larger than the $\overline{\sigma_2}$ for the other cluster ion reactions. For the reactions of these moderately-sized cluster ions there is no indication of significant enhancement of α_2 above those for the simple molecular ions, and the $\overline{\sigma_2}$ are only just a little larger on average. Therefore as expected, tidal trapping is not occurring to an appreciable extent in these cluster ion reactions.

It remains to ask to what extent contact collisions followed by coalescence are involved in these reactions. Bates (1979) has considered this problem and has calculated values for R_T (the effective absorbing radii for the interactions referred to above) using the experimentally determined α_2 for several of the reactions given in Table 2. He showed that the values obtained (typically 8Å) are greater than would be expected for R_C ($\sim$ 4Å) but within the range of R_T that Olson (1972) calculated for simple ion reactions It seems probable therefore that ionic recombination via electron transitions near curve crossings occurs in most of the reactions studied, although for those reactions involving the largest cluster ions (such as reaction (30)) then this may be energetically unfavourable (as mentioned above) since the ion pair can become more stable than their neutral counterparts. This results when the recombination energy of the positive ion (which is effectively reduced by the presence of cluster molecules) becomes less than the electron detachment energy of the negative ion (which is effectively increased by the presence of cluster molecules). Under these circumstances coalescence of the ion pair can occur within which the charges remain   separated forming a kind of "zwitterion".

What are now needed are data relating to reactions at much larger cluster ions (i.e. for $n > 6$) for which the tidal trapping model predicts significantly larger recombination coefficients. This would assist theorists in this most difficult area. However the experimental difficulties involved in such measurements are great, although further experiments at lower temperatures in our laboratory offer some hope of determining α_2 for larger clusters. Unfortunately, it seems unlikely that we will be able to directly obtain the variation of α_2 with temperature for a given cluster ion reaction because the rapid change with temperature of ternary ion-molecule association reactions results in changes in the degree of clustering to both positive and negative ions.

5. IONIC RECOMBINATION REACTIONS: ATOMIC IONS

The large rate coefficients for the ionic recombination reactions involving both simple and cluster ions are largely due to the large number of possible product channels in these reactions. However, when both the positive and negative ions are atomic then from such considerations the α_2 may be much smaller since there are many fewer possible product states. For such systems detailed curve crossing calculations may be made. Olson (1977) has calculated the α_2 and their temperature dependencies for the reactions $Na^+ + Cl^-$ and $K^+ + Cl^-$. The values at 300 K are $9.3 \times 10^{-9} cm^3 s^{-1}$ and $1.8 \times 10^{-10} cm^3 s^{-1}$ respectively, more than two orders-of-magnitude smaller than those typical of molecular ion recombination reactions. Experimental support for Olson's calculations has been obtained by Burdett and Hayhurst (1977) from their flame plasma studies. More data on atomic ion systems are required so that more such comparisons can be made with theory.

Determination of α_2 for atomic ion reactions in the FALP necessarily requires the generation of afterglow plasmas comprised of sufficiently high densities of atomic ions only, an interesting challenge in itself. The addition to an helium afterglow of any gas which generates stable negative ions also generates molecular positive ions. For example, Cl_2 addition to the helium afterglow initiates the following reaction sequence :

$$He^+, He^m, e, \xrightarrow{Cl_2} \begin{array}{c} Cl^+ \xrightarrow{Cl_2} Cl_2^+ \\ \\ Cl^- \end{array} \qquad (31)$$

Thus, the Cl^- ion forms via thermal dissociative attachment (Section 2.1.3) but unfortunately Cl_2^+ ions are formed also. This problem has been avoided however by the following procedure (see Church and Smith, 1978). The usual microwave cavity discharge was used to generate the primary ionization and a short distance downstream the afterglow plasma was exposed to a second microwave cavity in which the field was insufficiently intense to discharge the gas but intense enough to heat the plasma electrons to greatly enhance the ambipolar diffusive loss of both electrons and positive ions. Thus all of the electrons and positive ions were excluded from the carrier gas and only helium metastable atoms, He^m (mainly 2^3S with some 2^1S), remained to be convected down the flow tube (Smith et al, 1975). A rare gas (Ar, Kr or Xe) was then added in a sufficient quantity

to destroy all of the He^m thus generating the rare gas atomic ions
(Ar^+, Kr^+ or Xe^+) and electrons but insufficient to result in the
conversion of the atomic rare gas ions to molecular rare gas ions
(Smith et al, 1972b). A suitable halogen gas (F_2 or Cl_2) was then
added further downstream which generated the negative ions (F^- or
Cl^-) via thermal dissociative attachment but which could not be
ionized by the rare gas positive ions. By this technique
$Ar^+ + F^-$, $Kr^+ + F^-$, $Xe^+ + F^-$ and $Xe^+ + Cl^-$ afterglow plasmas were
generated and the decay of ionization in them studied. $Kr^+ + Cl^-$
and $Ar^+ + Cl^-$ plasmas could not be created since both Kr^+ and Ar^+
charge transfer with Cl_2 producing unwanted Cl_2^+ ions.

It was immediately apparent that the α_2 for all of the reactions
studied were much smaller than those for molecular ion reactions
because the gradients of ionization density along the afterglow
columns were very much smaller than in the molecular ion plasmas.
In fact the axial gradient of ionization density was closely
exponential (see Church and Smith, 1978) indicating that the only
significant loss process for the ions was ambipolar diffusion.
Values of the ambipolar diffusion coefficients $D_{+,-}$ were obtained
which are in good agreement with those calculated from mobility
data. Thus only upper limits to α_2 could be obtained these being
$\sim 5 \times 10^{-9} cm^3 s^{-1}$ for the $Xe^+ + Cl^-$, $Xe^+ + F^-$ and $Kr^+ + F^-$ reactions
and $\sim 1 \times 10^{-8} cm^3 s^{-1}$ for the $Ar^+ + F^-$ reaction.

In order to place a closer limit on α_2 for the $Ar^+ + F^-$
reaction it was further investigated using argon as the carrier gas
rather than helium, since diffusion coefficients of ions in argon
are much smaller than in helium (see McDaniel and Mason, 1973).
Thus a small amount of F_2 was introduced into a pure argon afterglow
(argon pressure $\sim$ 0.3 Torr) whence F^- rapidly formed via thermal
dissociative attachment. As before the Ar^+ ions were unaffected
since they cannot charge transfer with F_2. Again the measurement
of the axial ion density gradient indicated only diffusive loss of
ions but at a much slower rate than in the helium carrier gas thus
allowing a reduced upper limit to be placed on α_2 for the $Ar^+ + F^-$
reaction of $\sim 2 \times 10^{-9} cm^3 s^{-1}$. Unfortunately the other rare gas
positive ion/halide negative ion reactions cannot be studied using
an argon carrier gas since Ar^+ charge transfers with Cl_2 and the
Penning ionization technique is not possible because Ar^m atoms cannot
ionize Kr or Xe. The use of higher pressures of the argon carrier
gas can in principle assist in obtaining closer limits to α_2 for
slow reactions, although at much higher pressures in argon the probe
technique becomes suspect due to collisions of ions in the space
charge sheath which forms around the probe.

The α_2 for these atomic ions reactions are at least one order-
of-magnitude smaller than those typical of molecular ion reactions.
Although to our knowledge no calculations of α_2 have been made for
these reactions the small values are consistent with expectations

because in these systems very few possible product channels are
available. Other atomic ion reactions studied using the merged
beam technique (Moseley et al, 1975) have much larger ionic
recombination cross-sections (and hence larger α_2 values). So
whilst the α_2 values for reactions involving molecular ions do not
show great variations, those for atomic ions clearly vary widely
and each reaction needs individual consideration.

6. PRODUCTS OF BINARY IONIC RECOMBINATION REACTIONS

In addition to the determination of α_2 for ionic recombination
reactions it is clearly desirable to determine the nature of the
neutral products and the energy partition in the reactions. This
very difficult problem has received little attention from experimentalists and to our knowledge only two reactions have been studied.
Weiner et al (1970, 1971), using a merged beam apparatus, have
studied as a function of interaction energy the optical emissions
from the atomic energy levels in the Na atoms which are populated
following the neutralization reaction $Na^+ + O^-$ and deduced reaction
cross-sections for the population of individual atomic levels.
Moseley et al (1972) have measured the total cross-section for this
reaction as a function of energy and obtained values irreconcilably
lower than those of Weiner et al for the single $Na(3^2P)$ product.
The determination of absolute cross-sections from optical emission
studies is difficult and the inconsistency may be due to calibration
errors in these experiments. However this cannot detract from the
considerable importance of this pioneering work. Further useful
work on atomic ion systems is clearly possible with this kind of
technique and would be of great value.

Optical emission studies of molecular ion recombination
reactions are expected to be considerably more complicated than
for the atomic ion reactions but a start has been made in a study
of the emission from a NO^+/NO_2^- thermalized afterglow (Smith et al,
1978) in an attempt to identify the products of reaction (20).
The emission spectrum of the afterglow plasma was investigated in
the wavelength range 180 - 600 nm and the only significant radiations
identified were the γ-bands of NO. Proof that this radiation was
being emitted from the NO formed in the reaction was obtained from
the close correlation between the emitted intensity in the γ-bands
and the square of the ion density in the afterglow volume from which
the radiation was emitted, as is expected for a recombination
controlled plasma (see equation (17)). Therefore it was argued
that the reaction proceeded thus :

$$NO^+(X^1\Sigma^+) + NO_2^-(^1A_1) \longrightarrow NO(A^2\Sigma^+(\nu' = 0 \text{ to } 5)) + NO_2(X^2A_1)$$

$$(32)$$

Some population of NO $(C^2\Pi)$ and NO $(D^2\Sigma^+)$ or some vibrational excitation of the NO_2 $(X\ ^2A_1)$ could not be ruled out entirely (although from energy considerations both cannot occur in a single interaction). A discussion of the details of the measurements is given by Smith et al (1978). Inspection of the potential curves shows that an electron transfer from the NO_2^- into the lowest vibrational state of NO $(A^2\Sigma^+)$ will occur at an internuclear separation of about 1 nm and therefore that the product molecules will possess a net kinetic energy of about 1.4 eV. If the product NO and/or NO_2 are vibrationally excited then, of course, the net kinetic energy in the products will be correspondingly smaller. Sufficiently high resolution spectroscopy could perhaps detect the Doppler broadening on the rotational structure of the NO γ-bands.

This is just the beginning of what could be a very profitable approach to the study of the products of ionic recombination reactions. Further work along these lines is planned for simple ion reactions. Laser resonance fluorescence studies on the product neutrals should also be a viable proposition in flowing afterglow plasmas considering the relatively high density of product neutrals involved ($\sim 10^{10} cm^{-3}$).

7. PRELIMINARY STUDIES OF COLLISION-ENHANCED IONIC RECOMBINATION

It was mentioned in Section 1 that the ternary ionic recombination process (2) was studied extensively during the first half of this century, most effort then being directed towards the so-called Thomson regime (the intermediate pressure region $\sim 100 - 1000$ Torr) and the so-called Langevin-Harper regime (the high pressure region above about 1000 Torr). More recent work has extended the investigations to pressures as low as a few tens of Torr (Mahan and Person, 1964; McGowan, 1967; Sennhauser and Armstrong, 1978a,b). There is a need to study ionic recombination at the lower pressures at which the pure binary process just begins to be enhanced by collisional effects. Such a situation obtains in the middle stratosphere, a regime currently being probed using balloon-borne instruments and in which ionic recombination rate data are required for de-ionization rate calculations (Arnold, 1980). Previous estimates of "effective binary recombination coefficients" in the low pressure regime were obtained by the doubtful procedure of extrapolating higher pressure data.

In order to provide data in this low pressure regime we have studied the collision-enhanced ionic recombination coefficient, α'_2, for two reactions up to a pressure of 8 Torr in helium (the maximum available with the present FALP technique, Smith and Adams, 1980b).

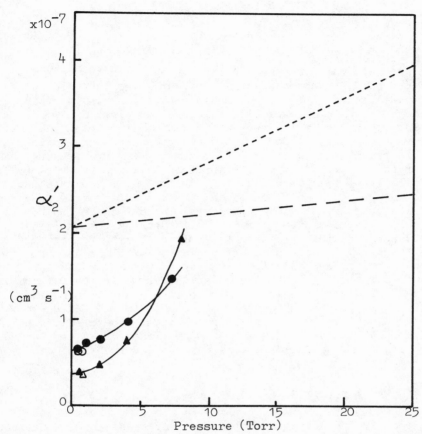

Fig.3   Collision-enhanced ionic recombination coefficients,
          α_2', as a function of pressure at room temperature
          for the reactions $NO^+ + NO_2^- + He$ (filled circles)
          and $SF_3^+ + SF_5^- + He$ (filled triangles). Also
          included are the binary ionic recombination
          coefficients, α_2, of Smith and Church (1976) for
          the $NO^+ + NO_2^-$ reaction (open circles) and Church
          and Smith (1977) for the $SF_3^+ + SF_5^-$ reaction (open
          triangle). The dashed lines represent a linear
          extrapolation to lower pressures of the data of
          Mahan and Person (1964) ostensibly for the reactions
          $NO^+ + NO_2^- + He$ (long dashes) and $NO^+ + NO_2^- + N_2$
          short dashes); see Section 7.

They are :

$$NO^+ + NO_2^- + He \longrightarrow products \tag{33}$$

$$SF_3^+ + SF_5^- + He \longrightarrow products \tag{34}$$

The ions were generated in the flowing afterglow in the usual way
and the α_2' were obtained at several pressures from the slopes of
the reciprocal density plots as described previously. The results
obtained are presented graphically in Fig. 3 as plots of α_2' against
helium pressure where it is apparent that α_2' is sensibly independent
of pressure below about 1 Torr. Note the good agreement between the
present and previously published values for α_2, corroborating the
previous result that $\alpha_2(33)$ is significantly larger than $\alpha_2(34)$.
The increase in $\alpha_2'(34)$ is about a factor of 5 over the pressure
range whereas $\alpha_2'(33)$ only increases by about a factor of 2 and
neither increases linearly with pressure. Also included in Fig. 3
is a linear extrapolation of the data of Mahan and Person (1964)
ostensibly for reaction (33) from their lower limit of pressure
of about 50 Torr. Clearly this extrapolation of their data to
pressures below 8 Torr would indicate very different values of
$\alpha_2'(33)$ than those obtained in the FALP experiment. This forcibly
illustrates that extrapolation of high pressure data to the limit
of zero pressure to obtain α_2 is unacceptable (Mahan and Person
obtained a value for $\alpha_2(20)$ of $2.1 \times 10^{-7} cm^3 s^{-1}$ compared to the
low pressure FALP value of $6.4 \times 10^{-8} cm^3 s^{-1}$ Table 1). Also included
in Fig. 3 is an extrapolation of the Mahan and Person data ostensibly
for the reaction :

$$NO^+ + NO_2^- + N_2 \longrightarrow products \tag{35}$$

and this clearly demonstrates that N_2 is a more effective third body
catalyst for the recombination reaction (N_2 is also found to be more
effective than He as a stabilizing third body in ion-molecule
association reactions - see the review by Good, 1975). It must be
stated again however that no mass analysis was performed in the Mahan
and Person experiments and so the ions may not be the simple ions
NO^+ and NO_2^- but perhaps clusters (e.g. $NO^+ \cdot NO$). There is an
obvious need for further study of more ionic recombination reactions
in this low pressure regime, and for investigations of the dependence
of α_2' on the nature of the third body. Also, theory predicts a much
more rapid change of the ternary recombination coefficients with
temperature ($\sim T^{-3}$, Flannery, 1976) than that for the binary
coefficients ($T^{-\frac{1}{2}}$). It is clearly desirable to study the temperature
dependence of α_2' to assist in the understanding of this interesting
recombination regime.

8. SUMMARY AND CONCLUSIONS

Using the FALP apparatus the first reliable data have been obtained on the pure binary ionic recombination coefficients (or mutual neutralization coefficients), α_2, for the reactions of mass-identified ground-state reactant ions at thermal energies. Thus values of α_2 have been obtained for several reactions involving simple molecular ions and for several reactions involving cluster ions. The values of α_2 at 300 K for all of these reactions are within the limited range $(4 - 10) \times 10^{-8} cm^3 s^{-1}$ and are consistent with theoretical expectations for such reactions. The study of the temperature dependence of α_2 for two reactions has shown that they also vary in accordance with theoretical predictions. Upper limits to α_2 have been obtained for a few reactions involving atomic ions only; the very small values for α_2 for these particular reactions are consistent with expectations although this must not be taken as a generalisation for all atomic ion reactions since others are expected to have larger values of α_2 (this has been demonstrated in merged beam experiments). The neutral products and the energy partition in these reactions have not been seriously studied to date; the products of only one reaction ($NO^+ + NO_2^-$) have been studied spectroscopically in the FALP apparatus. A start has also been made in the study of collision enhanced ionic recombination in the FALP. This very versatile apparatus can be exploited much more in the study of ionic and electronic recombination reactions at thermal energies.

ACKNOWLEDGEMENTS

We are indebted to our co-authors of the papers referenced here for their invaluable contributions to this work. We are also grateful to the Science and Engineering Research Council and to the United States Air Force for their financial support of the FALP research programme.

REFERENCES

Aberth, W. and Peterson, J.R. 1970, Phys.Rev., A1, 158.
Aberth, W., Peterson, J.R., Lorents, D.C. and Cook, C.T. 1968, Phys.Rev.Letts., 20, 979.
Aberth, W., Moseley, J.T. and Peterson, J.R. 1971, AFGRL Rept.No. 71-0481, Air Force Cambridge Research Laboratories, Bedford, Mass.
Adams, N.G., Church, M.J. and Smith, D. 1975, J.Phys.D., 8, 1409.
Albritton, D.L. 1978, At.Data. Nucl.Data Tables, 22, 1.
Arnold, F. 1980, Proceedings of the ESA Symposium on Rocket and Balloon Programmes, Bournemouth, England.
Bates, D.R. 1979, Adv.At.Mol.Phys., 15, 235.

Bennett, R.A., Huestis, D.L., Moseley, J.T., Mukherjee, D., Olson,
        R.E., Benson, S.W., Peterson, J.R. and Smith, F.T. 1974,
        Rept.No. TR-74-0417, Air Force Research Laboratories,
        Cambridge, Mass.
Biondi, M.A. 1973, Comments At.Mol.Phys., 4, 85.
Bolden, R.C., Hemsworth, R.S., Shaw, M.J. and Twiddy, N.D. 1970,
        J.Phys.B, 3, 45.
Burdett, N.A. and Hayhurst, A.N. 1977, Chem.Phys.Letts, 48, 95.
Church, M.J. and Smith, D. 1977, Int. J. Mass Spectrom. Ion.Phys.
        23, 137.
Church, M.J. and Smith, D. 1978, J.Phys.D, 11, 2199.
Christophorou, L.G. 1971, Atomic and Molecular Radiation Physics,
        Wiley - Interscience, London, p.465.
Dalgarno, A. and Black, J.H. 1976, Rept.Prog.Phys., 39, 573.
Dean, A.G. and Smith, D. 1975, J.Atmos.Terres.Phys., 37, 1419.
Dean, A.G., Smith, D. and Plumb, I.C. 1972, J.Phys. E., 5, 776.
Dean, A.G., Smith, D. and Adams, N.G. 1974, J.Phys. B., 7, 644.
Eisner, P.N. and Hirsh, M.N. 1971, Phys.Rev.Letts., 26, 874.
Fehsenfeld, F.C., Howard, C.J. and Schmeltekopf, A.L. 1975,
        J.Chem.Phys., 63, 2835.
Ferguson, E.E., Fehsenfeld, F.C. and Albritton, D.L. 1975, Gas
        Phase Ion Chemistry, Vol. 1, ed. M.T. Bowers, Academic Press,
        New York, p. 45.
Fisk, G.A., Mahan, B.H., and Parkes, E.K. 1967, J.Chem.Phys., 47,
        2649.
Flannery, M.R. 1976, Atomic Processes and Applications, eds. P.G.
        Burke, and B.L. Moiseiwitsch, North Holland, Amsterdam,
        p. 408.
Good, A. 1975, Chem.Rev., 75, 561.
Goodall, C.V., and Smith, D. 1968, Plasma Phys., 10, 249.
Hickman, A.P. 1979, J.Chem.Phys.,70, 4872.
Hirsch, M.N. and Eisner, P.N. 1972, Radio Sci., 7, 125.
Leu, M.T., Biondi, M.A. and Johnsen, R. 1973, Phys.Rev., A7, 292.
McDaniel, E.W., 1964, Collision Phenomena in Ionized Gases, Wiley:
        New York.
McDaniel, E.W. and Mason, E.A. 1973, The Mobility and Diffusion of
        Ions in Gases, Wiley : New York.
Mächler, W. 1936, Z. Physik, 104, 1.
McGowan, S. 1967, Can J.Phys., 45, 439.
Mahan, B.H. 1971, Adv. Chem. Phys. 23, 1.
Mahan, B.H. and Person, J.C. 1964, J.Chem.Phys., 40, 392.
Massey, H.S.W. 1976, Negative Ions. 3rd Edn., Cambridge Univ. Press.
Massey, H.S.W. 1979, Adv. Atom. Molec. Phys., 15, 1.
Moseley, J.T., Aberth, W. and Peterson, J.R. 1972, J. Geophys. Res.
        77, 255.
Moseley, J.T., Olson, R.E. and Peterson, J.R. 1975, Case Studies
        in Atomic Phys., 5, 1.
Mott-Smith, H.M. and Langmuir, I. 1926, Phys.Rev., 28, 727.
Olson, R.E. 1972, J.Chem.Phys., 56, 2979.
Olson, R.E. 1977, Combustion and Flames, 30, 243.

Peterson, J.R., Aberth, W., Moseley, J.T. and Sheridan, J.R. 1971,
        Phys. Rev., A3, 1651.
Plumb, I.C., Smith, D. and Adams, N.G. 1972, J.Phys. B, 5, 1762.
Sayers, J. 1938, Proc. Roy. Soc. (Lond), A169, 83.
Sayers, J. 1962, Atomic and Molecular Processes, Vol. 13, D.R. Bates,
        ed., Academic Press : New York, p. 272.
Sennhauser, E.S. and Armstrong, D.A. 1978a, Radiat.Phys.Chem., 11,
        17.
Sennhauser, E.S. and Armstrong, D.A. 1978b, Radiat.Phys.Chem., 12,
        115.
Smith, D. 1972, Planet Space Sci., 20, 1717.
Smith, D. and Adams, N.G. 1979a, Gas Phase Ion Chemistry, Vol. 1,
        ed. M.T. Bowers, Academic Press : New York, p. 1.
Smith, D. and Adams, N.G. 1979b, Kinetics of Ion-Molecule Reactions,
        ed. P. Ausloos, Plenum Press : New York, p. 345.
Smith, D. and Adams, N.G. 1980a, Topics in Current Chemistry Vol. 89,
        ed. S. Veprek and M. Venugopalan, Springer-Verlag : Berlin,
        p. 1.
Smith, D. and Adams, N.G. 1980b, AFGL, Rept. No. TR-81-0035, Air Force
        Geophysics Laboratory, Hanscom Air Force Base, Mass.
Smith, D. and Church, M.J. 1976, Int.J.Mass Spectrom Ion. Phys., 19,
        185.
Smith, D. and Church, M.J. 1977, Planet Space Sci., 25, 433.
Smith, D. and Copsey, M.J. 1968, J.Phys. B, Ser 2, 1, 650.
Smith, D. and Cromey, P.R. 1968, J.Phys. B, Ser 2, 1, 638.
Smith, D. and Dean, A.G. 1975, J.Phys. B, 8, 997.
Smith, D. and Goodall, C.V. 1968, Planet Space Sci., 16, 1177.
Smith, D. and Plumb, I.C. 1972, J.Phys. D, 5, 1226.
Smith, D., Dean, A.G. and Adams, N.G. 1972a, Z. Physik, 253, 191.
Smith, D., Dean, A.G. and Plumb, I.C. 1972b, J.Phys. B, 5, 2134.
Smith, D., Dean, A.G. and Adams, N.G. 1974, J.Phys. D., 7, 1944.
Smith, D., Adams, N.G., Dean, A.G. and Church, M.J 1975. J.Phys.
        D, 8, 141.
Smith, D., Adams, N.G. and Church, M.J. 1976, Planet Space Sci.,
        24, 697.
Smith, D., Church, M.J. and Miller, T.M. 1978, J.Chem.Phys., 68,
        1224.
Smith, D., Adams, N.G. and Alge, E. 1981, Planet Space Sci., 29, 449.
Smith, F.T., Huestis, D.L. and Benson, S.W. 1973, Electronic and
        Atomic Collisions, eds. B.C. Cobić and M.V. Kurepa, Inst.
        Phys. Belgrade, Yugoslavia, p. 895.
Swift, J.D. and Schwar, M.J.R. 1970, Electrical Probes for Plasma
        Diagnostics, Iliffe : London.
Weiner, J., Peatman, W.B. and Berry, R.S. 1970, Phys.Rev.Letts.,
        25, 79.
Weiner, J., Peatman, W.B. and Berry, R.S. 1971, Phys.Rev., 4, 1825.

THE OLD BARN

I've been a WILD ROVER for many a year
And I've spent all my money on whiskey and beer
But now I'm returning to CAPE BRETON'S shore
I never will be a WILD ROVER no more,
And its NO, NAY, NEVER,
* * * *
No, nay, never no more,
Will I be a WILD ROVER
No never no more.

— Johannes Hasted composuit

Adiabatic states, 355
  a priori determination, 359
Afterglow, 282 (see also Plasma
      and Flowing afterglow)
Ambipolar diffusion, 505-507
  losses, 282
Angular distribution of
      deexcitation light,
      452-453
Atmosphere
  gas ions in dissociative
      recombination, 306
  of Jupiter, 28
  of Mars, 28
  of Saturn, 28-29
  of Titan, 29
  of Venus, 28
Autoionization
  of Al^{+2}, 229
  of alkali-like ions, 225
  of Ga^+, 225
  of N^{+4}, 222
  of Na-like ions, 227
  inverse, 280

Background
  in coincidence measurements,
      443
  determination, 436-442
  detection and elimination, 384
  methods to reduce or eliminate,
      443-445
  modulation, 381
Beam experiments (see also
      Intersecting beams,
      Crossed beams, Merged
      beams)

Beam experiments (continued)
  choice of angle of intersection,
      377-380
  electron-ion, 211-217, 247-251
  initial excitation of ions, 388
  intersecting charged particles,
      286-288, 376-380
  intensity modulation, 380
  kinematics, 286-288, 378-379,
      415-428
  overlap integral, see Form
      factor
  superimposed beams, see
      Merged beams
  position modulation, 433-436,
      438-439
Bethe
  approximation, 117, 241
  cross section for ionization,
      148, 241
  parameters, 118
      for the excitation of
      Li-like ions, 121
Born approximation, 103
  Coulomb-Born, 107-109, 207
  distorted wave, 109-110, 207,
      326-335
  plane wave, 104-107
  second Born, 156-157
Bremsstrahlung, 49, 93

Capture
  into excited states, 451-454
  for excited states, 454-458
Cathode
  toroidal, 474, 478

Charge exchange, 12, 79-80
  between positive ions, 404-406
  double, 396
  leading to excited states, 448
  theory
    H+O^{8+} electron capture, 346
    for He+He^{++}, 351-360
    for non symmetrical systems,
      343-346
    quantum distorted wave, 320-
      335
    relation between quantal and
      semi classical, 319
Charge state distribution, 80
  for Ar, 246
  for Fe, 80
  for Li, 74
  for tungsten, 50
Charge transfer see Charge
      exchange
Chemical reactions, 15
Close-coupling approximation,
      158-159, 205
Clouds
  interstellar, 16-22
Collision energy
  in inclined beams, 378, 421
  in merged beams, 421
Collisions between protons and
      positive ions, 396-402
Collision strengths, 208
  of e$^-$ impact excitation of
      H$^+$(1s-2s), 218
Comets, 22-24
Configuration interaction, 178-
      181, 351-353
Confinement, 461
  electrical, 462-482
  inertial, 82-93
  in magnetic fields, 42-49,
      482-484
  in steady fields, 484-489
Cooper minimum, 135
Coronal equilibrium, 12
  ionic species abundances, 56
  emissivities of various
      elements, 54
Correlation
  effects in electron-atom

Correlation (continued)
      collision theory, 153-156
  rules for a one e$^-$ diatomic
      molecule, 243, 339-340
  rules for He^{++}+H, 340
Coulomb Born approximation,
      107-109, 207, 243
Coupled state theory, 205-206
Coupling
  electronic, 336
  dynamical, 336
  rotational, 337
Crossed beam method, 247-251,
      286-287, 318 (see also
      Beam experiments)
  electron-ion, 211-213, 247-251
Cross section
  differential, 417
  general definition, 415-416
  total, 417

Debye-Huckel transition
      strengths, 87
Detachment
  photodetachment, 2
  from negative ions, 407-410
Diabatic states, 353
  determination
    a priori, 359
    from adiabatic states, 355
  double perturbation problems,
      357
  in H$^+$-H$^-$, 361
  pseudo-crossings, 354
Dielectronic recombination (see
      also Recombination)
  general process, 315
  intersecting beam method,
      318-320
  plasma techniques, 316-317
Differential cross section
  triple, 126-127
Dissociating states, 171
Dissociation
  in flight, kinematic effects,
      427-428
  of H$_2^+$ by electron impact, 481-
      482
  photodissociation, 4, 15

Dissociative recombination (see also Recombination)
  of atmospheric gas ions, 306–310
  of cluster ions, 312–314
  cross sections, 295–298
  of H_2^+, 192–194, 294–303
  mechanism, 169
  of O_2^+, 181–192
  of polyatomics, 310–312
  potential energy curves, 167–200
  rate coefficients, 295–298
  wavefunctions used for O_2 and H_2, 172–181
Distorted wave
  Born approximation, 109–110, 207
  classical limit, 329–332
  theory, 326–335
Divertor, 61
  poloidal, 64
Dynamical coupling, 336

Electron
  attachment, 507–508
  beam
    hollow, 477
    concentric device, 471–482
    cylindrical, as ion trap, 462–466
    cylindrical, potential distribution, 466
  Capture
    dielectronic, 13
    into excited states, 451–454
    for excited states, 454–458
  temperature relaxation, 507
Electronic coupling, 336
Emissivities of various elements, 54
Energy levels
  of $e^- + Fe^{+15}$, 232
  of $e^- + He$ like ions, 204
Energy resolution
  distribution of secondary e^-, 132
    from He, 139, 141
  in merged beams experiments, 385

Excitation
  autoionization, 222–227
    of Fe^{+4}, 227
    of Na-like ions, 227
  initial ion ion beams, 388
  electron impact, 10, 89
    of Ba^+, 223
    of $Be^+(1s-2s)$, 220
    of C^{3+}, 124
    of $C^{+3}(2s-2p)$, 221
    of $He^+(1s-2s)$, 218
    of Li^+, 218
    of Li-like ions, 222
    of N^{+4}, 215, 222
    theory, 202–209
  proton impact, 14
  rate measurements, 209
Excited states
  capture for, 454–458
  capture into, 446, 451–454
  doubly, 14
  formed in charge exchange, 448
Experimental design
  pitfalls, 380–385

Fano plot, 120–125
  for angular distribution of secondary electrons, 133
  for excitation
    of He, 120, 122
    of S^{3+}, 155
  for total ionization
    of C^{3+}, 149
    of H^-, 145
    of N^{4+}, 150
    of O^{5+}, 151
Field ionization, 457–458
Flame sampling, 285
Flowing afterglow, 504–510
  ion densities, 514
  plasma, 505–510
Form factor, 212, 249, 421
  determination, 429–431
  circumventing the determination, 433–436
Fusion
  ignition, 39, 41
  inertial confinement, 38–42
  magnetic confinement, 38–42

Fusion (continued)
  mirror machine, 43, 45
  reactors reaction rates, 40
  relevant experiments in ion-
     ion collisions, 402
  schemes, 38-39
  toroidal systems, 43-46

Gaunt factors, 208
Gaussian basis set, 174
Glauber approximation, 157-158

Heating of plasmas, 55
  negative ions, 73
  neutral injection, 67-70
  positive ions, 73
Hollow electron beam trap, 294

Impurities
  in fusion experiments, 49-60
  maximum allowed concentration,
    58
  molecular, 66
  tungsten density, 52
Inclined beams, see Intersecting
    beams
Independent electron model, 349-
    351
Inelastic collisions
  energies of product particles,
    425-428
Injection of neutrals, 68
Interaction geometry in beam-
    beam experiments, 428-
    436
Intersecting beams (see also
    Beam experiments)
  choice of angle, 377-380
  energies of product particles,
    425-428
  kinematics, 286-288, 378-379
  types of reactions investi-
    gated, 376-377
Ion densities in flowing after-
    glow, 514
Ion-ion
  afterglow plasmas, 509-510
  collisions, measurements,
    373-413

Ion-ion (continued)
  neutralization
    measurements, 389-396
    rate, 394
    theory, 360-362
Ionization
  associative, 5
  charge transfer, 12, 68
  by electric field, 457-458
  by electron impact, 145
    of Al^{+2}, 229
    of alkali-like ions, 225
    of atomic hydrogen, 6
    of Ar^{n+}, 247
    of C^{3+}, 149
    of Fe^{+4}, 230
    of H^-, 145
    of He, 146
    of He^+, 250-252
    of Li^+, 147
    of multiply charged ions,
      257-269
    of molecular ions, 270-271
    of O^{5+}, 151
    of singly charged ions, 252-
      256
    theory, 241-244
    of Ti^{+3}, 226
    total cross-section, 141-
      151
    of W^{4+}, 150
  penning, 5
  photoionization, 7, 9, 11
Ion pair formation, 303, 311
Ions, doubly negative, 407-410
Ion trapping in electron
    beams, 439-442

Kinematics of beam-beam inter-
    actions, 421-428

Langmuir probe, 507
Laser
  excitation, 455-457
  induced photofluorescence, 284
Line strength, 120
Lippman-Schwinger equation, 320
Lotz formula, 226, 244

Mass filter, quadrupole, 491
Merged-beams, 380 (see also Beam experiments)
  collision energy, 421
  electron-ion, 288-291
  energy resolution, 385-388
Modulated background, 382 (see also Background)
  detection and elimination 384, 437
Molecular theory of atomic collisions, 335-372
Mott cross section, 135
Multichannel scattering, 327-329
Multicharged ions on atoms, theory 362-364
Multiconfiguration self consistent field, 177

Nebulae, gaseous, 7-11
Neutralization
  efficiency for D^-, D^+, D_2^+, D_3^+, 72
  ion-ion, 360-362
  mutual, of ions, 2,5,6, 360-362, 389-396
  oscillator strength, 208
    generalized, 110
Overlap integral in beam-beam experiments, see Form factor

Penning trap, 484-489
Plane wave Born approximation, 104-107
  integrated cross section, 116-120
Plasma
  decay measurements, 282-285
  diagnostics, 1, 75-82
  edge, 60-67
  flowing afterglow, 504-512
  heating, 55, 67-75
  high temperature, 11-14
  inertial confinement, 82-93
  Langmuir probe technique, 504-512
  magnetic confinement, 42-48
  technique for dielectronic recombination, 316

Plasma (continued)
  spectroscopy method, 244
Platzman plot, 135-138
Potential energy curves
  for adiabatic bound states, 170
  curve crossing and pseudo-crossing, 343-346
  diabatic states, 356
  of a diatomic ion, 170
  for dissociative recombination of H_2^+, 195
  for H_2 and H_2^+, 193
  purely electronic, 341-345

Quadrupole ion trap, 291-293

Radio-frequency traps, 489-498
Radiative association, 245
Radiative attachment, 2
Radiative lifetime, 49
Radiative power loss density, 52
Rate coefficients
  for dielectronic recombination, 90
  for dissociative recombination, 295-299
  for excitation by electron impact, 209
  for ionic recombinations, 523-528
Reaction rates of fusion reactors, 40
Recombination
  of atomic ions, 523-525
  charge transfer, 10
  coefficients, 512-523
  dielectronic, 5, 10, 11, 51, 88, 280, 314-320
    to excited states, 90
  dissociative
    of atmospheric gas ions, 306-310
    of H_2^+, 192-194
    of He_2^+, 194
    of O_2^+, 181-192
    of rare gas ions, 304-306
    of polyatomics, 310
  electronic, 508
  of H_2^+ with e^-, 482

Recombination (continued)
  ionic, 508,523–528
Relativistic electron–atom
        collisions theory, 151–
        153
R–Matrix method, 159–160
Rotational coupling, 337

Scaling laws
  in e⁻ impact detachment for
        negative ions, 409
  in a one electron diatomic
        molecule, 339
  Thomson's theory, 241
Scattering
  from two potentials, 326–329
  zero angle of fast primary
        electrons, 127
Secondary electrons
  angular distribution, 128–132
  energy distribution, 132–141
Semi–classical theories, 325–326
Shock tubes, 285
Sum–rule method, 143–145
Superimposed beams, see Merged
        beams

Theory
  of atomic collisions
      molecular, 335–365
      semi–classical, 325–335
  of electron–atom collisions,
        101–165
      fast collisions, 103–115
      slow collisions, 156–160
Transition
  amplitude in semi–classical
        theory, 332–335
  strengths of Debye–Huckel, 87
Translational spectroscopy
  for capture into excited
        states, 451
  kinematics, 426–427
Trap
  radio–frequency, 489–498
  penning configuration, 484
Trapped–ion method, 245
Trapping cross section for
        neutral hydrogen, 70
Tokamak, 44–48, 50

Date Due

			UML 735